the sixth AUDIO anthology

Compiled from **AUDIO**, *the original magazine about high fidelity,* from January 1960 to December 1961, by **C. G. McProud**, Editor. **RADIO MAGAZINES, INC.**, Mineola, New York • 1962.

Audio Amateur Press
Publishers
Peterborough, New Hampshire

ALPHABETICAL LISTING OF AUTHORS

Author	Page
Abilock	80
Boegli	27, 44
Bernard	11, 17
Bosselaers	14
Burstein	85, 90, 97
Buscher	137
Canby	64
Chen	51
Cooper	24
Crowhurst	72
Csicsatka	66
Cunnningham	106
Dalzell	120
Dilley	114
Eilers	69
Ferres	125
Goeller	7
Greenleaf	127
Greiner	132
Herlocker	58
Jackson	49
Kauder	41
Klipsch	143
Lerner	104
Linz	66
Malan	119
Malmstedt	129
Mangels	48
McProud	6
Moss	17
Pruitt	111
Saslaw	61
Shirer	33
Shottenfeld	80
Snape	109
Snow	102
von Recklinghausen	77

CONTENTS

Article & Author	Issue	Page
FOREWORD, C. G. McProud		6
A Feel for Transistors, L. Goeller	10/61	7
A Transistorized Stereo Phono Preamplifier, W. B. Bernard	06/60	11
Designing a Transistorized Preamp, R. J. Bosselaers	01/61	14
A Transistorized Stereophonic Control Unit, Richard Y. Moss	10/60	17
A Transistor Protector, George Fletcher Cooper	12/61	24
Amplifiers with Positive and Negative Feedback, Charles Boegli	04/61	27
Feedback Techniques in Low-Level Amplifiers, Donald Shirer	05/61	33
Universal Feedback Amplifier Circuit, Arnold Kauder	01/60	41
The Anode Follower, Charles Boegli	12/60	44
Hanging Hi-Fi System, Harold Mangels and Wife	12/61	48
A Case for the Custom Console, F. H. Jackson	02/61	49
Common-Bass Stereo Speaker System, Francis F. Chen	11/61	51
New Design Chart for Bass-Reflex Enclosures, R. D. Herlocker	04/60	58
What Hath the FCC Wrought?, David Saslaw	06/61	61
The Good News--Stereo Broadcast, Edward Tatnall Canby	06/61	64
FM Stereo--The General Electric System, Antal Csicsatka & Robert Linz	06/61	66
FM Stereo: Time-Division Approach, Carl Eilers	08/61	69
Filters for FM-Stereo, Norman Crowhurst	08/61	72
An FM Multiplex Stereo Adaptor, Daniel R. von Recklinghausen	06/61	77
Signal Sampling for FM Stereo, R. Shottenfeld & S. Abilock	12/61	80
Distortion in Tape Recording, Herman Burstein	05 & 06/60	85
Maintaining Frequency Response in Recorders, Herman Burstein	03 & 04/60	90
Improving the Signal-to-Noise Ratio, Herman Burstein	01 & 02/60	97
Characteristics of Tape Noise, William Snow	02/61	102
Tape Indexing Nomograph, Jerry Lerner	01/61	104
Stereo Misconceptions, James Cunningham	09/61	106
To Phase or Not to Phase?, E. A. Snape, III	07/61	109
Those Crazy Mixed-up Currents, Almus Pruitt	04/60	111
Regulate That Voltage!, Willam Dilley	01/61	114
Graphical Solution to the Tracking Problem, W. B. Bernard	09/60	117
Determination of Tracking Angle in Pickup Arm Design, Niel Malan	02/60	119
The Silicon Diode in Audio Equipment, L. B. Dalzell	07/60	120
Equipment Failure Alarms, Allan Ferres	12/61	125
FM Sweep Alignment Unit--Austerity Model, Allen Greenleaf	10/61	127
"Ersatz Stereo" Unlimited, C. H. Malmstedt	02/61	129
Another Power Amplifier, R. A. Greiner	09/61	132
Computers in Audio Design, R. G. Buscher	02 & 03/61	137
Speaker Power, Paul W. Klipsch	10/61	143

Foreword In this, the SIXTH AUDIO ANTHOLOGY, we are proud to include articles on two most significant milestones in the field of high fidelity. FM STEREO and TRANSISTORS IN AUDIO EQUIPMENT. The FM STEREO articles which appeared in AUDIO — *the original magazine about high fidelity* — were written by the men who actually worked on the system approved by the FCC. The articles pertaining to TRANSISTORS IN AUDIO APPLICATIONS cover interesting aspects of designing with the semi-conductor.

•

As in previous editions of the AUDIO ANTHOLOGY, the SIXTH is a compilation of important articles which appeared in AUDIO over a period of about two years. And, all of the articles were written by knowledgeable and experienced authorities in the field.

•

We know the SIXTH AUDIO ANTHOLOGY will serve as a meaningful reference for everyone in the diverse fields of audio engineering, recording, broadcasting, manufacturing and servicing of components and equipment, and for the audio fans who made this business of high fidelity what it is today.

C. G. McProud, Publisher,

AUDIO *magazine*

December, 1962.

A Feel for Transistors

L. GOELLER

Reduced prices, off-the-shelf availability, and special properties make transistors desirable in many audio circuits. All that remains is for the designer to develop the "feel." Here's how . . .

ALMOST EVERYONE SEEMS to approach transistor circuits with fear and trembling. This attitude is doubtless induced by the relationship between transistors and solid-state physics, a subject of considerable complexity. It is apparently not generally realized that solid-state physics—and indeed all physics beyond Ohm's Law—can be avoided entirely while designing quite satisfactory audio-frequency transistor circuits. Only three basic rules are needed:

1. If the transistor is going to work at all, its base is at almost the same voltage as its emitter. This applies to both d.c. (bias) voltages and a.c. (signal) voltages.

2. The currents in the base, emitter, and collector leads are related by the three symbols α, β, and γ as summarized in *Fig.* 1. It is convenient to remember that α is nearly unity and β is nearly equal to γ and numerically equal to 20 or more.

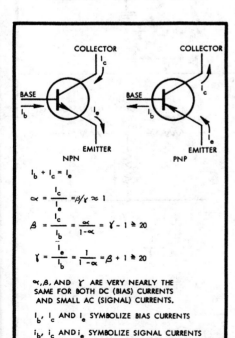

Fig. 1. Relationships between α, β, and γ.

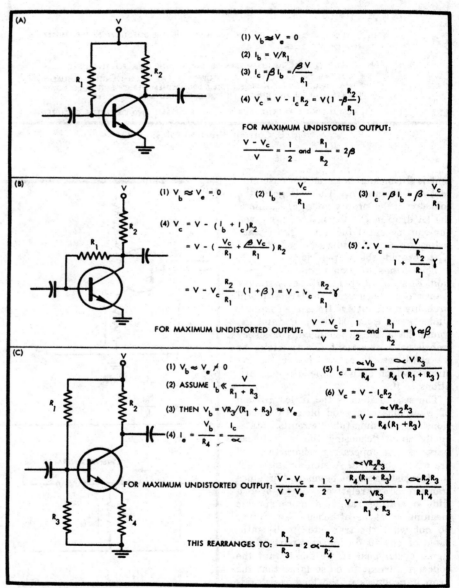

Fig. 2. Bias circuits: (A) simple, (B) better, (C) much better.

3. The resistance between the base and emitter leads is very low and corresponds to a forward-biased diode. The resistance between the base and collector is very high and corresponds to a back-biased diode.

Biasing

With these three rules and a firm grasp of Ohm's Law, a surprisingly complete knowledge of transistor audio amplifiers can be deduced. As an example, consider the typical bias circuit shown in (A) of *Fig.* 2. If the voltage at the base is equal to the voltage at the emitter, and the emitter is grounded, it follows that the current in R_1 is obtained by dividing R_1 into V. The collector cur-

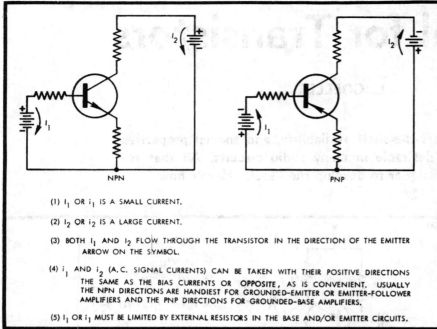

Fig. 3. Current direction rules.

peak-to-peak swing of nearly 5.7 volts. If the supply voltage happened to be 6 volts, a popular value in transistor work, nonlinear operation could be avoided only if the collector d.c. voltage in the absence of signal were adjusted to one half the supply voltage in the circuit shown in (A) of *Fig.* 2. Then the positive and negative swings would both remain just short of trouble. The general conclusion we can draw is this: the collector should be biased to a value halfway between the emitter and supply voltages in resistance-coupled amplifiers.

It should now be a simple matter to follow a similar analysis through for (B) of *Fig.* 2. This circuit is interesting because, if we adjust R_1 and R_2 for the maximum undistorted output as discussed above, the collector voltage will be only half as sensitive to variations in β as the circuit of (A). Proof of this is left as an exercise for mathematics professors.

The ratio R_1/R_2 has been related to β in these two examples. Successful biasing, however, requires particular values for these resistors and the proper polar-

rent is β times as large as the base current and flows in R_2. The voltage at the collector is the supply voltage, V, minus the IR drop in R_2. Thus, if we are given (or can measure) β for the given transistor, we can choose values of R_1 and R_2 to provide the proper bias.

This brings up two points of digression. First, if β in ten supposedly similar transistors is measured, the largest β will probably by about five times the smallest. Under such circumstances, R_1 and R_2 would have to be changed for every transistor replacement. Fortunately, as we shall see, there are better biasing circuits which do not depend upon variations in β.

The second item is the little problem of what bias is a good bias. In discussions of vacuum-tube circuits, statements about "choosing the most linear part of the operating characteristic" are often observed. A similar statement for transistors might be made, but, since transistors are relatively linear over a wide operating range, a more specific meaning must be attached. We simply do not want the transistor to be saturated or cut off during any part of the signal cycle. That is, we don't want the collector current to be so large that the drop it produces in the load resistor is larger than the supply voltage, and we don't want it so small that it fails to flow at all.

This is a situation seldom encountered in vacuum tube work. With a 200 volt supply and a signal of 2 volts rms at the plate, there is little fear that the signal voltage will overpower the bias. The same 2-volt signal at the collector of a transistor would, however, mean a

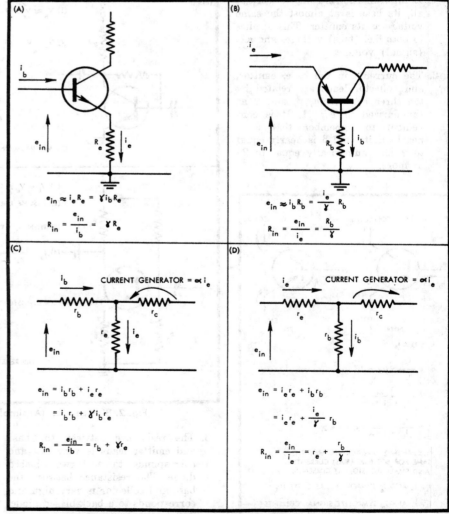

Fig. 4. Input impedances: (A) and (B) first approximations, (C) and (D) better approximations.

ity for the bias supply. Considering the latter first, *Fig.* 3 summarizes the current direction rules. Base and collector current combine to form emitter current and emitter current flows in the direction of the arrow on the emitter lead in the transistor symbol. In a PNP transistor the arrow points in; it points out for a NPN transistor. The base current is small; the collector current is large. Base current controls collector current. To rephrase this for emphasis, the collector current is very nearly as large as the emitter current; the difference between them is the base or control current which is usually less than 5 per cent of the emitter current.

The actual sizes of R_1 and R_2 in *Fig.* 2 are determined by the signal currents involved and also by leakage currents which are often an unpleasant transistor by-product. For the time being a rule-of-thumb can be offered: for supply voltages between about 6 and 20 volts, R_2 should be somewhere between 1000 and 20,000 ohms and R_1 should be on the order of β times as large.

The circuit of (C) in *Fig.* 2 can now be attacked. If we make the assumption, which will be justified later, that $V/(R_1+R_2)$ is much greater than I_b, or about the same size as I_c, the analysis is easy. V_e is just about equal to V_b and V_b is fixed by the voltage divider. Therefore, the current in the emitter is V_e/R_4. This same current flows through R_2, or, to be more exact, αI_e gets there. It follows that V_c is fixed by α and the four resistors. Since α varies from about 0.95 to 0.999, its variation is of negligible importance. The factor β is derived from α as indicated in Fig. 1. The two values of α given above correspond to β's of approximately 20 and 1000. This illustrates the importance of depending on α rather than β for bias stabilization.

Input Impedance

Now that the transistor is biased properly, new data must be obtained if an amplifier is to be designed successfully. In particular, input and output impedances are required. Here we come to the main difference in transistor and vacuum-tube theory. A vacuum tube has very nearly infinite input impedance. At audio frequencies, it presents no loading to the previous stages, and is not loaded by the stages which follow. In contrast, transistor input impedances are quite low. The output impedance of the driving stage and the input impedance of the following stage affect the behavior of the stage under consideration. It is this basic situation which causes 99 per cent of the troubles encountered by individuals oriented to vacuum tubes.

Since the low input impedance is the factor which causes the trouble, let's look at it in terms of Ohm's Law. Con-

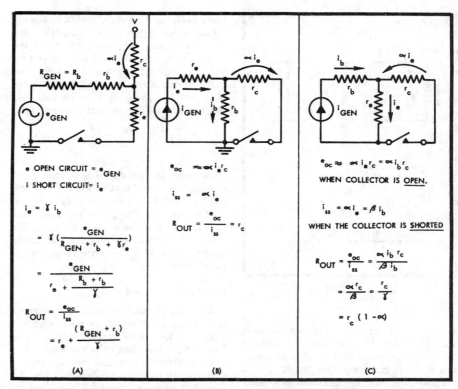

Fig. 5. Output impedances by the open-circuit short-circuit method: (A) emitter follower, (B) grounded base, (C) grounded emitter.

sider the circuit of (A) in *Fig.* 4. Suppose i_b flows into the transistor. e_{in}/i_b is the input impedance. But e_{in} is given by the voltage drop across R_e. This, we know, is $i_e R_e$, or $\gamma i_b R_e$. To a first approximation, therefore, the input impedance looking into the base of a transistor is γ times the impedance in the emitter circuit.

If we look into the transistor between the emitter and ground in the grounded base configuration of (B) in *Fig.* 4, we can apply the same technique. $e_{in}=i_b/R_b$ and $i_b=i_e/\gamma$. Thus, the impedance looking in at the emitter is the base resistance divided by γ while the impedance looking in at the base is the emitter resistance multiplied by γ.

Unfortunately, things are not quite this simple. Suppose, for instance, the circuit of (A) in *Fig.* 2 is encountered, and no resistance is present in the emitter lead. The resistances internal to the transistor must now be examined. They are small, but sometimes they are dominant. The circuit of (C) in *Fig.* 4 shows the transistor internal resistance and how it is divided between the base and the emitter. If we proceed as in (A) of *Fig.* 4, $r_{in}=e_{in}/i_b=r_b+\gamma r_e$. Similarly, the procedure in (B) gives $r_{in}=e_{in}/i_e=r_e+r_b/\gamma$. Usually, r_b is of the order of 100 ohms or so. r_e is about 25 ohms, and can be approximated fairly accurately for small a.c. signals by $26/I_e$ where I_e is the d.c. bias current in *ma* when the transistor is operated at room temperature.

If both internal and external resistors are present, as is usually the case, there is no reason why they can't be lumped together in calculations (following the usual rules for series or parallel circuits as the case may be).

Note that for the first approximation the collector circuit has not even been considered. This is usually an excellent approximation, and can be justified by the fact that the collector resistance is on the order of 1 megohm or so. Thus, the collector circuit is effectively decoupled from the input circuits.

Output Impedance

So far, Ohm's Law has given the input impedance of the transistor in a simple and straightforward manner. But how about the output impedance? It must be obtained in a slightly different manner. The technique is called "the open-circuit short-circuit method," and is often better applied on paper than on actual circuit elements.

Consider a black box with two wires coming out of it. We know that inside the box a generator exists within a network of resistors. First, measure the open-circuit voltage. This gives an equivalent voltage generator. Then, measure the current which flows when the output leads are shorted. This current is obviously limited by the internal resistance and the generator voltage. Define a new resistance which is given by dividing the open-circuit voltage by the short-circuit current. If this resistance is

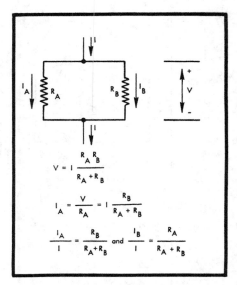

Fig. 6. Current divider.

placed in series with the open-circuit voltage generator, we have a simple functional representation of the whole circuit. Alternatively, we could use a current-generator approximation (based on the short-circuit current) in parallel with the same resistance.

If we try this idea on the emitter-follower circuit, we can get a check on its validity. We already know the impedance looking into the emitter; the output impedance should be the same. Consider the circuit of (A) in *Fig.* 5. If we open the emitter, the voltage at the base is e_{gen}. If we short the emmitter, the current is $i_e = \gamma i_b$. Further, $i_b = e_{gen}$ divided by $(R_b + r_b + \gamma r_e)$. R_{out} must be, therefore, $\frac{R_b + r_b}{\gamma} + r_e$. This checks with our previous result. The conclusion we must draw is this: The output impedance of the emitter follower is roughly the generator impedance divided by the current gain, γ.

Now we can attack the collector impedance. This is the most difficult of the lot, but the open-circuit short-circuit method provides an excellent approximation. There are two configurations: grounded base, with the signal inserted at the emitter, and grounded emitter with the signal inserted at the base. The first is the easier of the two. From the circuit of (B) in *Fig.* 5, the open circuit voltage is $\alpha i_e r_c$. There is also a very small voltage due to the base current in r_b, but it can be neglected. The short-circuit current is simply αi_e. Thus, the output impedance is nothing but r_c. This is usually a very high value, on the order of 1 to 10 megohms. It can be seen that the grounded-base transistor acts very much like the ideal current source described in the textbooks.

For the grounded-emitter stage, the circuit of (C) in *Fig.* 5 is required. The open circuit voltage is $\alpha i_e r_c$ as before. But there is an important difference If the external collector circuit is open, the emitter current and the base current are the same. Therefore, we must remember that the open-circuit voltage is $\alpha i_b r_c$. The short-circuit current is $\alpha i_e = \beta i_b$, since i_b is *not* the same as i_e when collector current flows. The output impedance is

$$R_{out} = \frac{e_{oc}}{i_{ss}} = \frac{\alpha i_b r_c}{\beta i_b} = \frac{r_c}{\gamma} = r_c(1-\alpha)$$

This is much lower than the output resistance of the grounded-base stage, but much higher than that of the emitter follower. This approximation is not as good as the one for the grounded-base configuration since the voltage drops in the base and emitter resistors are effectively larger. However, the approximation is good enough for most purposes.

The Amplifier Stage

The input and output impedances of the transistor alone are not enough to permit satisfactory circuit design. The transistor must always be embedded in a biasing circuit, and the biasing circuit will always have some effect on gain. To see what happens, consider the current-divider circuit of *Fig.* 6. A current I flows in the parallel combination R_A and R_B. As the calculations show, the ratio of current in one resistor to the total current is as the *other* resistor to the sum of the resistance. By dividing the numerator and denominator of the left-hand side of the equation by the numerator, a slightly more useful form is obtained:

$$\frac{I_B}{I} = \frac{1}{1 + (R_B/R_A)}$$

This points out clearly that most of the current will flow in R_B only when R_B is much smaller than R_A. If R_B is the load and R_A is a bias resistor, the moral of the story should be clear.

As a specific example, the circuit of (A) in *Fig.* 2 is reproduced in *Fig.* 7, along with the equivalent circuit for a.c. signals as derived before. If we consider the output circuit first and assume the impedance of the coupling capacitor is small compared with the sum of the load and source impedance, the circuit reduces to three resistors in parallel driven by a current generator. We are interested in the part of the current available, βi_b, which we can get into the load.

At this point, there is usually a strong temptation to call R_2 the load resistor, by analogy with vacuum tube circuits. Nothing but trouble will result unless the temptation is strongly resisted. R_2 is a *bias* resistor. The *load* is the next stage or a loudspeaker or some such.

Thus, the output impedance of the grounded-emitter *stage*, as opposed to that of the grounded-emitter transistor, is R_2 in parallel with $r_c(1-\alpha)$. This is the impedance the load sees looking back

(Continued on page 16)

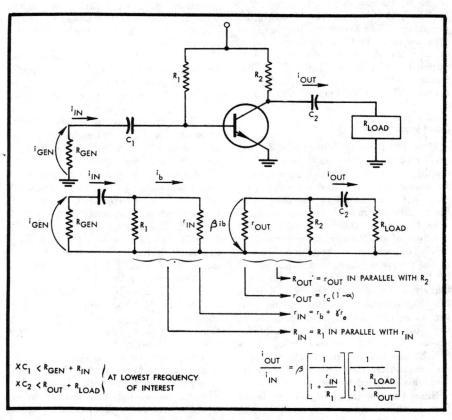

Fig. 7. Calculations for simple amplifier.

A Transistorized Stereo Phono Preamplifier

W. B. BERNARD, Capt. USN

You can avoid both hum and the effect of long leads from cartridge to amplifier by building this simple preamplifier.

THE PLACEMENT OF the phono preamplifier at the turntable carries with it some very desirable advantages. The principal one being the elimination of long low-level cables which ordinarily connect the phono cartridge to the preamplifier at the control position. These long cables are a probable source of hum and they result in a capacitive load which may put a resonant hump in the response characteristic of the magnetic cartridge. When tubes were required in the preamplifier, the problem of supplying power to the preamplifier remotely located from the control unit had to be balanced against the undesirability of the long connecting cables. Objections resulting from the complications of supplying power to the preamplifier were practically eliminated by the advent of transistors.

In the case of the preamplifier described here the a.c. input to the power supply is connected to the phono motor power leads so that anytime the motor is turned on, power is applied to the transistors. Further economies could be achieved by winding a secondary of about 6 volts on one of the legs of the motor stator which would permit the elimination of the power-supply transformer.

Circuit Description

Figure 1 is the schematic of one of the two identical preamplifier modules and the power supply unit used in the stereo system. Each preamplifier consists of three cascaded stages connected in the common-emitter mode. Each stage is stabilized by the use of resistors in the emitter circuit and d.c. feedback from the collector terminal to the base.

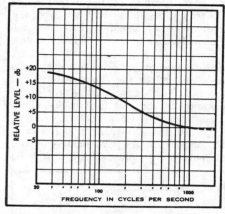

Fig. 2. Low frequency compensation for RIAA characteristic.

The first transistor, Q_1, is a 2N104 which is operated with the emitter resistor unbypassed. The current feedback developed in this resistor increases the input impedance of Q_1 to the point that the resistors of the bias network offer most of the loading to the phono cartridge. This is desirable since it permits us to know what the loading on the cartridge is, and this knowledge is essential if we wish to use resistance loading on the cartridge to achieve the high-frequency attenuation required to equalize the RIAA recording characteristic.

The second transistor, Q_2, is a 2N405. In this stage the emitter resistor is bypassed to secure the maximum signal gain. The third stage is also a 2N405. Like the first stage, the emitter resistor of the third stage is unbypassed. A feedback network from the emitter of the third transistor to the emitter of the first transistor is used to produce the low-frequency boost needed to compensate for the chararistics of the magnetic pick-up cartridge.

The output impedance of the third stage is approximately 18,000 ohms, the resistance of the collector resistor of this stage. This impedance is low enough to allow a reasonable length of cable to be connected to the output of the unit if it is to be used as a voltage source, that is, used to feed a conventional tube-type amplifier. The impedance is high enough to consider the unit as a constant-current source if it is to be used to feed a

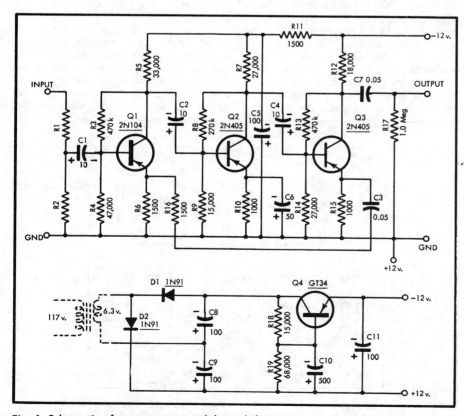

Fig. 1. Schematic of one preamp module and the power supply that furnishes power for two such modules.

transistor amplifier. If it is to be used to feed a transistor amplifier the output capacitor, C_7, should be increased to at least 3 μf.

The high-frequency attenuation needed to compensate for the RIAA characteristic may be accomplished either by loading the cartridge or by loading the output of the preamplifier. Space is left on the module for the components needed to use either of these methods. However, loading of the cartridge is the more desirable since this method cuts down on the level of the signal which must be handled by the preamplifier, while the loading of the output offers the remote possibility that a very strong high-frequency signal might exceed the dynamic range of the output transistor. The necessary components for RIAA high-frequency compensation may also be connected in the cables external to the unit. This eliminates the problem of opening up the unit if it is desired to change the type of cartridge feeding it.

The unit described here has a mid-frequency gain of approximately 50 and an output capability of about 3 volts

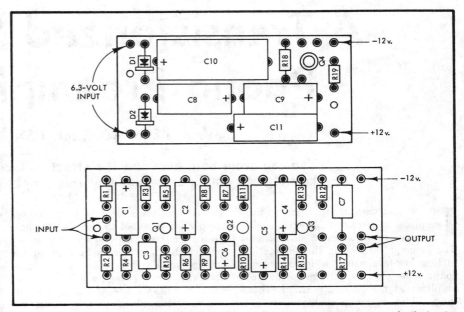

Fig. 4. Parts placement on preamplifier "chassis" (above) and power supply (below).

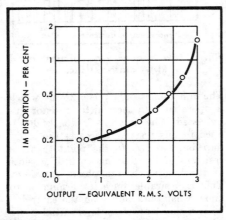

Fig. 3. Curve showing intermodulation distortion vs. output voltage.

r.m.s. when operated from a 12-volt d.c. power supply. The low-frequency response of the amplifier is shown in *Fig. 2* and the distortion characteristic in *Fig. 3*. There is little change in the characteristics of the unit until a temperature of over 120° F. is reached at which time the dynamic range of the unit begins to decrease. By keeping the heat producing units of the stereo installation, such as the power output amplifiers, away from the phono turntable the temperature should stay well below the 120° point.

Output Capability

It should be appreciated that a transistor amplifier working with a 12-volt supply will have a much lower output voltage capability than will an electron tube amplifier operating from a 100- or 200-volt supply. If both units have about the same voltage gain the transistor unit will overload at a lower input voltage than will the electron tube unit. With an output capability of 3 volts and a gain of 50 the unit described will accept an input of about 60 millivolts r.m.s. at 1000 cps and above without overloading. Such a capability is compatible with most of the magnetic cartridges available today. Should a cartridge having a nominal output of more than 25 or 30 millivolts be employed with this preamplifier it is recommended that the output voltage of the cartridge be reduced by a resistance network at the input of the preamplifier. The same resistance may, in many cases, be used to provide the high-frequency roll-off for RIAA compensation. The method of determining the values of the resistors needed is given in the appendix.

Because of the stabilizing effect of the feedback circuitry, the performance of the unit is not greatly affected by a change in the type of transistors used. A number of types have been plugged into the sockets for the last two stages with very little change in performance. 2N107's, 2N109's, 2N270's, and GT34's all seemed to work equally as well as the 2N405's shown in the diagram. Because of the low signal conditions under which the first transistor operates, the principal requirement placed on it is that it be a low-noise type. Although no other type was substituted for the 2N104 it is considered that any low noise PNP transistor would work satisfactorily in this position.

The power transformer is a 6.3-volt, 1-ampere filament transformer. This is considerably underloaded since the total current drawn by the two preamplifier modules is about 1 milliampere. However, the transformer is small and inexpensive so no great effort was expended

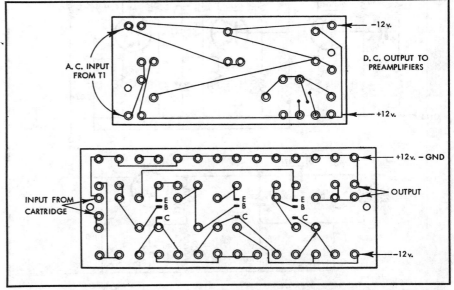

Fig. 5. Underside of preamp and power supply modules showing strapping between eyelets.

Fig. 6. Modules mounted inside enclosures.

to find a transformer which was theoretically better suited to the application. As stated previously, in many cases it should be possible to wind an additional winding on the phono motor to furnish current to the rectifiers. The 6.3 volts from the secondary of the transformer is run into a full-wave voltage doubler consisting of two 1N91 rectifiers and then through a GT-34 transistor which is used as a filter element. A transistor has one of the desirable characteristics as a filter choke—low d.c. resistance combined with a high impedance to a.c. In a capacitance-input filter system where the energy storage characteristic of a choke is not necessary, a transistor may be substituted for the choke with a considerable saving in space, weight, and cost. In such an application the designer must be careful not to exceed any of the limiting ratings of the transistor. In the power supply described here this did not operate as a restriction.

The output current from the emitter of the transistor will be just about as free of ripple as is the base-bias current so a fairly large capacitor is used in conjunction with the bias resistor network to ensure a smooth current supply to the base. The output of the supply is 12.5 volts with a ripple that is less than 1 millivolt.

Construction

Figure 4 shows the layout of the preamplifier and power supply modules. The module cards are constructed by drilling at the proper locations, inserting eyelets in the holes, and flaring the ends of the eyelets protruding from the bottom of the board. This flaring holds the eyelet in the card and permits #22 tinned bare wire to be wrapped around the flared portions of the eyelets to permit connections between the parts which have their leads passed through the eyelets from the top of the boards— a sort of "poor man's printed circuit." *Figure* 5 shows the connections on the bottom of the board. This construction method is much faster than the printed circuits where only a few units are desired. It has the same topological problems as a printed circuit and therefore can easily be converted to a printed circuit if a large quantity of the units is desired.

The two preamplifier modules are mounted inside a $5 \times 7 \times 2$ in. aluminum chassis. They are held away from the chassis by $\frac{1}{2}$-in. spacers. The power supply module is mounted inside a $2\frac{1}{2} \times 2\frac{1}{2} \times 6$ in. Minibox with a thin sheet of phenolic material between the bottom of the card and the inside of the box. The power transformer is mounted outside the box to keep the heat from the transformer away from the other power-supply parts.

The chassis and the box, *Fig.* 7, are mounted under the motorboard of the turntable, taking care that the input jacks of the preamplifiers are near the pivot of the pickup arm and that the power transformer is as remote as possible from the position of the cartridge during operation. The primary leads of the power transformer are connected in parallel with the turntable motor leads, and cables are connected from the preamplifier outputs to high-level input jacks at the stereo control center. The system is now ready for operation.

With this system, hum problems are reduced to those produced by inductive pickup in the cartridge. Other noise is below the surface noise on your best record. You can now stop worrying about preamplifier problems and concentrate your worry on other things.

APPENDIX

In considering the required load on the cartridge to give the RIAA high-frequency rolloff we must divide the cartridges into two general classes, those in which the output impedance is predominately inductive in nature, and those in which the output impedance is predominately resistive in nature.

Let us first consider those having an inductive characteristic. For these the resistive load on the cartridge should equal the reactance of the cartridge at 2100 cps.

$$X = 2\pi F L$$

so

$$R = X = 2 \times 3.1416 \times 2100 \times L$$

The answer comes out to be approximately

$$R = 13 \times L \quad (L \text{ in millihenries})$$

Thus if the cartridge has an inductance of 350 millihenries we will need a load resistance of 13×350 or 4550 ohms.

Since the bias network amounts to about 40,000 ohms we must find out what resistance in parallel with 40,000 ohms will result in a total resistance of 4550 ohms. The value of a resistor R_x which when placed in parallel with R_b will give a parallel resistance of R_t is given by:

$$R_x = \frac{R_t \times R_b}{R_b - R_t}$$

In the case under discussion

$$R_x = \frac{4550 \times 40{,}000}{35{,}450} = 5100 \text{ ohms}.$$

The nearest standard resistor in the 10 per cent tolerance series is 4700 ohms which should give satisfactory results. A 5-per cent resistor may be purchased in the correct value but after soldering and aging take place you may not be any better off than if you had used the 10 per cent value. Referring to *Fig.* 1 we see that if the whole cartridge output is to be fed into the preamplifier the value of R_1 goes to zero and R_2 takes the value calculated for R_x.

Let us consider another case where we want to use a cartridge with an inductance of 500 millihenries and an output voltage high enough that we would desire to divide it by a factor of two.

$$R = 13 \times 500 = 6500 \text{ ohms}$$

To impress on the input of the preamplifier only one half of the output voltage of the

Fig. 7. Completed units ready for mounting under motor board.

(Continued on page 26)

Designing A Transistorized Preamp

R. J. BOSSELAERS

On the surface, transistors are ideally suited for use in preamplifiers—which makes one wonder why they are not widely used. If the reason is lack of familiarity with their properties—this design example may help shed some light on those transistor procedures which differ from tube procedures.

THE TRANSISTOR has advantages over vacuum tubes especially for an application such as the hi-fi preamplifier because:

1. There is no hum problem.
2. The transistor is non-microphonic.
3. The noise figure of low noise transistors is better than that of the average vacuum tube.

However, the behavior of a transistor is different from that of a vacuum tube in many respects. This means that one cannot simply replace the tubes in a circuit by transistors and have a good design. Therefore designers not familiar with the peculiarities of a transistor are faced with problems they cannot solve. This may be the reason why the transistor has not yet acquired the place it deserves in the audio field.

These notes are written to provide guidance for the design of a high-quality transistor preamplifier.

Distortion and Feedback

To keep distortion within limits suitable for hi-fi purposes large feedback factors are required—together with low levels to maintain a low modulation percentage. With a 20-volt supply the maximum level that is practical is 1 volt rms.

To understand what feedback factors are required we can calculate the second harmonic distortion caused by the exponential relationship between base voltage and collector. (See Appendix.) The result is:

$$d_2 = \frac{q\hat{V}}{kT} \times \frac{1}{4N^2} = \frac{10\,\hat{V}}{N^2}$$

Where

$\hat{V}$ = peak input voltage
N = feedback factor

Since this is not the only cause of distortion we will set it at .2 per cent.

For $\hat{V} = .1$ volt and $d_2 = .002$ we find $N = 22.4$. If the feedback is obtained by an unbypassed resistor in the emitter lead as in *Fig.* 1, the d.c. voltage drop will be $N \times kT/q = .55$ volts. The drop in the collector resistor for an amplification of 10 times will consequently be 5.5 volts. With 3 volts between collector and emitter, the required supply voltage is 9 volts.

Block Diagram

From the calculations above it follows that the maximum input level at the base is about 100 *mv*. The minimum level is determined by the required signal-to-noise ratio. The noise figure of a transistor is minimum when it is connected to a source with a resistance in the order of 500 ohms. If the input resistance is considerably higher the noise level will increase with the square root of this resistance. Therefore input levels of 30 *mv* (maximum program peak) may not cause noise problems at the low impedance equalizer input, but the same level at the high-impedance tone-control stages could indeed give noise problems. For this reason the minimum required input level for a high impedance stage is also about 100 *mv*. Apparently a change of level means either distortion or noise and consequently there is no place for a volume control anywhere but across the output.

Since the required maximum output voltage is 1 volt and the maximum available level is also 1 volt, the tone control cannot immediately precede the volume control. An amplification of 20 db is required to make up for the loss in the tone control. Furthermore the input impedance of a tone control is generally not high enough and an extra amplifier stage is necessary to obtain this. These considerations lead to the block diagram of *Fig.* 2.

The circled voltages indicate the level for maximum program peak. Because of the distortion we cannot allow the level to be higher than 1 volt rms consequently the volume control will have to be turned up all the way for full power output of the power amplifier and any differences in input levels will have to be compen-

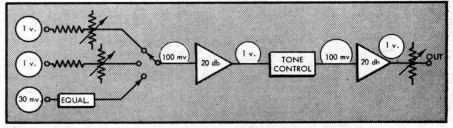

Fig. 2. Block diagram of preamplifier.

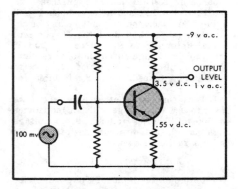

Fig. 1. Feedback obtained by an unbypassed resistor in the emitter lead.

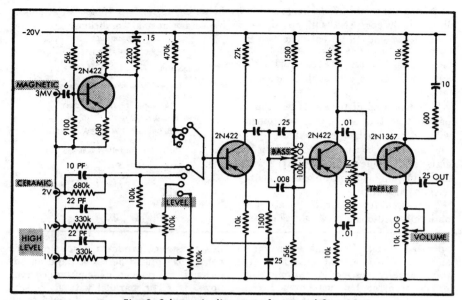

Fig. 3. Schematic diagram of preamplifier.

sated by level controls at the input. These level controls can be either fixed or separately variable for each input or as a common level control mechanically coupled to the volume control.

Tone Control

The conventional tone control circuit is connected to a low impedance source and is loaded only by a grid leak resistor. In a transistor circuit the situation is different. The base bias resistor is usually of the same order of magnitude as the collector resistor. Consequently the load for the tone control circuit is of the same order of magnitude as the impedance of the source.

A first thought might be to use an emitter follower and lower all impedances. This would be correct, but on second thought, if an extra transistor has to be used anyway it might as well be used as an extra amplifier that separates bass and treble controls. This way the design is more flexible. This approach is used in the final circuit of *Fig. 3*.

Equalizer

An equalizer circuit for a magnetic cartridge is also included in *Fig. 3*. The function of this circuit is:

a) to raise the output of the cartridge to a 100 mv level
b) to provide frequency correction for a RIAA recording characteristic

To keep the distortion low the equalizer has to be designed for the cartridge with which it will be used. For a universal input that will accept any cartridge a two transistor equalizer with level control is required. Examples of such equalizers can be found in literature. To help the reader to redesign the first stage for his particular cartridge an example that results in the values used in *Fig. 3* will be worked out.

Example

Sensitivity of the cartridge
    2.2 mv/cm/sec.
Inductance of the cartridge
    500 mh
Maximum recorded velocity at 1000 cps
    14 cm/sec.
$$\hat{V} = 14 \sqrt{2} \times 2.2 = 40.5 \ mv$$
For $d_2 = .002$ the feedback factor is:

$$N = \sqrt{\frac{10 \hat{V}}{d_2^2}} = \sqrt{\frac{405 \times 10^{-3}}{2 \times 10^{-3}}} = 14.2$$

The feedback is obtained by an unbypassed resistor, R_3, in series with the emitter.

$$R_3 = (N-1) \ r_e = (N-1) \left(\frac{2}{I_e}\right) (I_e \ in \ ma)$$

I_e is the d.c. emitter current. For a good noise factor it is best to make I_e approximately 0.5 ma. With this value R_3 becomes:

$$R_3 = \frac{13.2 \times 26}{0.5} = 680 \ ohms$$

The collector circuit is determined by the available d.c. voltage drop and the requirement for RIAA. equalization. This circuit is the same for all cartridges.

The sensitivity at 1000 cps can be calculated as follows:

$$\frac{R_3 + I_e}{2200} \times 100 \ mv = \frac{680 + 52}{2200} \times 100 = 33 \ mv$$

This is reasonably close to the 40.5 mv peak level.

To keep the output level constant at the high frequency end the cartridge should be loaded with the proper resistance. At the same time this resistance will reduce the noise of the transistor so that the signal-to-noise ratio is not affected too much.

The load resistance can be found from the 75 microsecond time constant for equalization:

$$\frac{L}{R} = 75 \times 10^{-6}$$

$$R = \frac{500 \times 10^{-3}}{75 \times 10^{-6}} = 6670 \ ohms$$

In *Fig. 3* the load resistance for the cartridge is R_2 in parallel with R_1 and the input impedance of the transistor which is

$$Z_i = \beta (R_3 + r_e)$$

For $\beta = 50$ Z_i would be 36,600 ohms with a total input impedance of 6450 ohms for the values in *Fig. 3*.

The required bias voltage at the base is

$$T_e R_3 + 120 \ mv = 460 \ mv$$

This is obtained with $R_1 = 6.6 \ R_3$.

The nearest standard value for $R_1 = 56,000$ ohms.

High-Level Inputs

There are three high-level inputs in *Fig. 3*. Two have individually adjustable level controls with a maximum sensitivity of 1 volt. The third input is for ceramic or piezo-electric cartridges. It is designed for cartridges with a sensitivity of 0.1 v/cm/sec with a maximum output of approximately 2 volts. Since the sensitivity at the base of the transistor is 100 mv and the impedance at this point is 36,000 ohms, the required

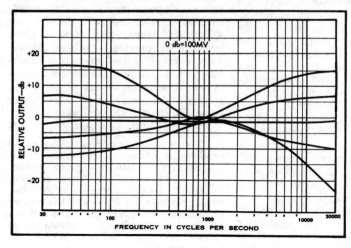

Fig. 4. Effect of tone controls.

series resistance, R_4, is $\left(\frac{2-1}{0.1}\right)36 = 685,000$ ohms. The nearest standard value is 680,000 ohms. To correct for the loss at the high-frequency end a small capacitor should be connected across R_4 such that $R_4C = 7$ microsecond. In this case 10 picafarads. The total input impedance is then 716,000 ohms. Since the recommended load for some cartridges is 2 megohms there will be some loss of bass which can be corrected with the bass control.

Measurements

The effect of the tone controls is shown in *Fig.* 4. There is 17 db boost and 10 db cut at 30 cps. At 10,000 cps there is 13 db boost and 13 db cut. The two intermediate positions shown reveal some interaction between the controls which was not found objectionable in listening tests.

The measured distortion is less than 1 per cent at 1000 cps for 1 volt output with volume control in maximum position and tone controls flat for both the high- and low-level inputs. At 2 volts the distortion is about 2 per cent, mainly second harmonics.

The signal-to-noise ratio is about 65 db in all positions of the volume control. Listening tests show that this is inaudible under normal conditions. Æ

Appendix

Let $i_c = I_E(1 + m\cos\omega t)$

and $V_{BE} = \frac{kT}{q}\ln\left(\frac{i_c}{I_{ES}}\right)$

The a.c. portion of V_{BE} is approximately:

$$\frac{kT}{q}(m\cos\omega t - 1/4m^2\cos 2\omega t)$$

so that $d_2 = 1/4\,m$

With feedback $d_2 = \frac{m}{4N}$ if N = feedback factor.

The peak input voltage is in this case:

$$\hat{V} = N\,m\,\frac{kT}{q}$$

or

yielding $m = \frac{q}{kT}\,\hat{V}\,\frac{1}{N}$

$$d_2 = \frac{q}{kT}\,\frac{\hat{V}}{4N^2}\,m = \frac{10\hat{V}}{N^2}$$

TRANSISTORS
(from page 10)

into the transistor. If we call it R_{out}, the current divider equation becomes

$$\frac{i_{load}}{\beta i_b} = \frac{1}{1 + \frac{R_L}{R_o}}$$

where R_{out} is usually very nearly equal to R_2.

Moving back to the input circuit, another current divider is seen. If a signal current, i_{in}, is applied to the circuit, the useful proportion that actually gets to the transistor is

$$\frac{i_b}{i_{in}} = \frac{1}{1 + \frac{r_{in}}{R_1}}$$

where r_{in} is the *transistor* (not stage) input impedance and in this instance is equal to $r_b + \gamma r_e$. If the last two equations are combined, the ratio of useful output current to input current becomes

$$\frac{i_{load}}{i_{in}} = \beta\left(\frac{1}{1 + \frac{r_{in}}{R_1}}\right)\left(\frac{1}{1 + \frac{R_L}{R_o}}\right)$$

The maximum gain is β. It cannot be achieved because biasing resistors must be used. The only way to minimize the loss is to make the biasing resistors as large as possible. Since there is one basic universal law which says "you can't get something for nothing," the maximum sizes of the bias resistors turn out to be discouragingly low. We'll return to this after a brief pause for impedance matching.

The maximum power transfer theorem states that, if the maximum power is to be transferred from a generator to a load, the resistance of the load must equal that of the generator. This theorem is, like all mathematical theorems, quite true. Unfortunately, it is also irrelevant. One seldom matches the input impedance of a vacuum tube. One never matches the output impedance of an amplifier to a loudspeaker. A 60-cps power generator is never matched to its load. Thus there is no good precedent or reason for impedance matching in transistors. The transistor is a current-operated device. You put a current in and you get a current out. Thus, the thing usually desired is not maximum *power* transfer but maximum *current* transfer. This point cannot be stressed too strongly. Do *not* match impedances. There is no mystic benefit to be gained from so doing, and in most cases gain will actually be reduced. In the circuit of *Fig.* 7, people have been known to add a resistor in series with C_1 to match to the output impedance of the preceding stage. On other occasions, R_2 has been lowered to match the input impedance of the following stage. *Don't do it!*[1]

Leakage Currents

Transistors will be destroyed if excessive voltages are applied to them. In most cases, a collector supply of 30 volts or less is required. To get enough bias current, this alone implies relatively small bias resistors.

The input impedance of a grounded-emitter transistor is almost equal to the current gain times the emitter resistance and since the internal emitter resistance is roughly equal to $26/I_e$, a high bias current reduces the input impedance considerably. But this again implies small bias resistors.

Most important, however, is the leakage current, I_{co}. This current flows at all times when the transistor is operating, but is defined as the current which flows in the collector-base circuit when the proper operating bias is applied but the emitter lead is open. I_{co} is relatively independent of the voltage across the collector-base terminals, but it increases with time, temperature, hard usage, and various other factors. And, if the resistance in the base circuit of (A) in *Fig.* 2 is high, I_{co} causes an additional component of collector current to flow. This component is equal to βI_{co} and, if I_{co} increases very much, can produce a voltage drop in R_2 large enough to saturate the transistor. I_{co} will often start out at 5 microamperes or more. In a high β transistor, this can be fatal.

Even in a circuit like (C) of *Fig.* 2, troubles can develop. The equivalent base resistance is R_1 in parallel with R_3 as far as I_{co} is concerned. I_{co} through this resistance produces a voltage. This voltage is applied across R_4, and an additional component of emitter current flows. If R_1 in parallel with R_3 is say, 10,000 ohms, and I_{co} goes from 5 to 50 microamperes, the base is raised by nearly half a volt. If R_4 is 225 ohms, I_e increases by 2 ma. This can shift the collector voltage by a considerable amount. It is apparent, therefore, that although the circuit of (C) in *Fig.* 2 greatly reduces the effects of β variations, it does not necessarily solve all biasing problems. To reduce I_{co} troubles, the current in R_1 and R_3 must be considerably larger than the largest anticipated I_{co}. For most purposes, 50 microamperes is a good estimate for I_{co} maximum in germanium transistors. Thus the R_1R_3 current should be at least 0.5 ma. It is a good rule of thumb to make the R_1R_3 current at least this large and preferably as large as the current in R_4 if that current is greater.

It should be noted that silicon transistors have much lower I_{co}'s and bias resistors can be ten times larger than those for germanium without causing trouble.

Conclusions

If Ohm's Law is applied with confidence, transistor circuits are no harder to understand than circuits containing vacuum tubes. This does not mean that complete understanding can come from arguments such as those used in this paper. High-frequency effects, direct-coupled circuits, feedback, transient response and many other items of great practical importance have been deliberately omitted in the interest of establishing a feeling for transistor operation based on a minimum of physical and mathematical reasoning. A little experience will show both the value and the limitations of this approach. Æ

---

[1] Note that impedance matching *is* of considerable importance in certain applications such as terminating transmission lines. In addition, if *transformer* coupling is available between a source and load and neither the load nor the source has zero or infinite impedance, it can be shown that the optimum turns ratio for maximum power transfer also gives maximum voltage and current transfer.

A Transistorized Stereophonic Control Unit

RICHARD Y. MOSS

Stereo controls and their functions have been a source of confusion for the audiofan. The following analysis dispels some of the confusion and leads to the design of a high quality preamp.

STEREOPHONIC SOUND arrived suddenly, perhaps unexpectedly, transported by the media of compatible disc recordings, FM multiplex, and simultaneous AM–FM and AM–TV transmissions, and multitrack magnetic tape recordings. While each of these techniques requires a different method of conversion from electromagnetic or acoustical information to an electrical signal, all have the common characteristic that two similar but separated audio channels are required for compensation and amplification. This article will discuss the design of a stereophonic control unit by examining the problem in two parts: first, the philosophy of stereo reproduction and the functions of the various controls which are necessary to such reproduction; and second, the actual circuit design and construction of such a control unit, incorporating transistors for increased reliability, low hum and microphonics, and compactness.

After a brief examination of the stereophonic control units on the market, one would be forced to the conclusion that the transition from monophonic to stereophonic reproduction entails an increase in the complexity of controls by a factor of at least two-to-one, and perhaps greater. This observation is supported by the initial approach of most manufacturers of stereo equipment, namely, to provide two of everything plus some peculiar additional functions. While "two-of-everything" is certainly a straightforward solution, it does not reflect very favorably upon the engineering "know-how" of these manufacturers, since such a solution does not imply anything more than a very superficial analysis of the problems of stereophonic reproduction. In lieu of any further criticism of what has not been done, let us proceed instead with a first-order analysis of what should be done, and reflect this analysis in the design of a control unit intended for stereo, not merely a double-barreled approach to monophonic sound.

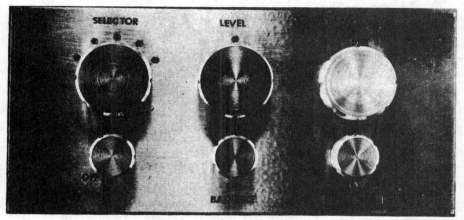

Fig. 1. Front panel of control unit.

Audio Controls in General

Audio control functions in a stereophonic system may be divided into two general categories: first, those functions which are common to all types of audio reproduction in general, and hence can be controlled in both channels simultaneously; and second, those functions which are peculiar to stereo, and which may require some definition before they can be appraised. In the first category we find: (1) input selector switching, (2) phonograph record and magnetic tape playback equalization, (3) master level control, (4) loudness contour compensation, (5) tone control, (6) scratch and rumble filtering, and (7) power switching.

It is often convenient to incorporate the first two functions, input selector switching and equalization compensation, into a single control, especially since most modern monophonic disc recordings and virtually all stereophonic discs are recorded with the RIAA characteristic, and the NARTB recording curve has become standard for tape recording at the 7½ and 15 ips speeds. The selector-equalizer should thus control at least three low-level inputs: a high-impedance microphone input, a magnetic phono input which incorporates RIAA equalization, and a magnetic tape head input with NATRB equalization. At least two high-level inputs are also necessary, one for a tuner, and one to be used for pre-equalized tape playback or a high-level phono cartridge. While level control of each input would be a pleasant luxury, it is desirable to limit the number of level pots to one per preamp channel, so as to be able to adjust the output of the low-level preamp to approximate the signal at the high-level inputs, thus avoiding sudden changes in volume when switching from any of the low-level inputs to a high-level signal.

Level control needs little discussion, except to point out that such a control should be ganged to both channels, avoiding the inconvenience and added complexity of separate concentric controls and unreliable mechanical clutches which are supposed to permit operating the two knobs in a ganged fashion.

Loudness contour compensation and tone controls may usually be considered simultaneously, since in the ordinary listening environment either will produce a satisfactory result, and in installations where both are present it is common that the loudness control produces such satisfactory results that the tone controls soon become dusty with disuse. For the purposes of a compact installation, then, the tone controls may be eliminated from the control unit and made a part of the power amplifier, wherein they are adjusted to compensate for unusual room acoustics and then left set; in many cases tone controls may be eliminated from the system entirely. In any case, the loudness contour is a necessity, and should be capable of reproducing the Fletcher-Munsen equal-loudness contours from the 0 db, or full room volume, curve to a fairly low listening level, say -35 db. It is also possible to consider scratch and rumble filtering under the general heading of tone control, but logical consideration of this topic soon leads to the conclusion that the audiophile who is interested in building his own control unit is probably using a professional turntable or high quality changer, and is concerned almost entirely with the reproduction of high fidelity program sources, thus predicating the proposition that the inclusion of filters designed to compensate for the shortcomings of lower quality equipment and low fidelity sources, is a waste of effort in an ostensibly high fidelity system. Even the enthusiast who may wish to play badly worn or scratched discs which are collector's items isn't likely to find much utility in such controls. He is more apt to dub these irreplaceable recordings onto tape using special filters as a part of the recording system, and then play the tape to save wear and tear on the originals. The conclusion to be drawn from this argument is that the elimination of rumble and scratch filters from a stereophonic control unit intended primarily for the reproduction of new, high quality recordings with professional quality equipment will hardly impose any hardship by limiting the flexibility of the control unit.

Power switching is a subject which could be expanded into a volume by itself, but in this control unit such switching will be confined to controlling the d.c. power for the unit itself, plus one additional circuit which may be used to control the coil of a power relay, and this in turn controls all the a.c. power to amplifiers, tuner, turntable, and tape or other auxiliary equipment. This is a more desirable situation than trying to mount a switch to handle as much as half a kilowatt within the control unit itself, and if the d.c. power is used in the relay control circuit, then there will be no a.c. power within the control unit except signals, with attendant advantages in hum pickup reduction.

The preceding analysis has thus reduced the original seven monophonic functions to four: (1) a combined selector-equalizer, (2) a master level control, (3) a loudness contour control, and (4) power switching. In each case, where there was doubt about the inclusion of a function, assumptions of high quality equipment and logical function were advanced in order to determine whether such a function was a necessity or merely a luxury.

Stereo Controls in Specific

A second set of control functions becomes necessary with the advent of stereophonic reproduction. Such terms as "balance," "separation," "phase reversal," and "channel reversal" have begun to appear in the profusion of recent literature on the subject of stereo, and the consequent confusion about their meaning warrants a brief definition of each, as well as an examination of their importance.

Balance means simply a comparison of the relative volume levels of the two audio channels; if the inputs are assumed equal, as is usually the case, the term "balance" describes the gain ratio of the A and B channels. Since the overall gain of each channel must necessarily include everything from the cartridge to the acoustic output as it reaches the listener's ears, it seems obvious that even if all the reproducing elements and amplifying elements are exactly balanced, the position of the listener within the room may be such that the sound does not seem balanced to him. Moreover, even the assumption of balanced inputs is unrealistic, thus the necessity for a control to adjust balance is compelling. Such a control usually operates so as to vary the gains of the two channels differentially; that is to say, to increase the volume in one channel while simultaneously decreasing that in the second channel.

"Separation" is a term which defines a more subjective phenomenon of stereo; that is, the separation of the two halves of the apparent source of sound. The two limiting cases of this effect are easy to imagine: on the one hand, a total lack of separation would cause the apparent source of sound to seem to be located at a point midway between the two speakers; the other extreme would occur when the speakers were located so far apart that there seemed to be a "hole-in-the-middle" of the sound emanating from the left and right hand speakers, giving an exaggerated stereo or "ping-pong" effect. Since this effect depends upon the electrical channel isolation, on speaker placement and room acoustics, and, in fact, on the microphone placement in the original recording studio, it would seem almost a necessity that some means of continuous adjustment of the separation should be included. One means by which this adjustment can be achieved electrically is by adding a portion of the signal from one channel to the signal in the other, and vice-versa. Each channel would then contain some information which is common to both and hence is monophonic in character, serving to fill the "hole-in-the-middle"; each channel would also contain some information which is peculiar to that channel and hence retain its stereophonic character. Using this control it would be possible to attain a "curtain-of-sound" spread across the space separating the two speakers, regardless of variation between recordings, room acoustics, and so on. It is scarcely necessary to mention, however, that the adding type of control can only reduce the separation of the two channels, since if it could increase the separation beyond that of the recording it would be possible to make monophonic recordings sound stereophonic! There have been circuits designed which use a matrix to take advantage of the separation already inherent on the stereo recording and improve upon it, but such circuits are nearly impossible to construct, except on paper, because of the extreme precision required.

The third term mentioned was channel reversal, which means simply the reversal of the two channels in left-right orientation, and is a purely aesthetic effect which assumes prior knowledge of the subject arrangement. An illustration of this effect would be a recording of a symphony orchestra, which has standardized so that the strings are to the left of the conductor, the percussion to the right, and so on. In the event that this is reversed in the reproducing system, it is a simple matter to correct, since the two channels may be interchanged at any point in the system, from the pickup leads to the loudspeakers. In view of this fact, a separate control to accomplish this function is not felt to be important enough to merit inclusion on any save a very elaborate control unit, and such a control will not be included in this case.

The fourth term, "phase reversal," is perhaps the most difficult to describe. The need for attention to this function arises because of the relative youth of the stereo art, so that there is no universally observed standard as to what direction of excursion of the phonograph stylus shall produce a signal of a given polarity; or in the case of magnetic tape, what state of magnetization of the oxide layer on the tape shall produce a signal of the same given polarity. In a purely stereophonic system, (i.e., the channels are assumed to have perfect isolation), the result of improper phase orientation will be cancellation of some of the signals. In the case where there is electrical adding of the signals, as in a separation control, the out-of-phase components will cancel and reduce the overall signal level; and in the case where the signals are not added electrically there will still be acoustic interference of the same sort which arises in an improperly phased public address system. The remedy is simple in principle: just reverse the direction of excursion of the speaker cone in one channel for a signal of a given polarity, providing that no electrical adding takes place. In the case where there is adding, then the phase of the signals in one channel must be reversed before the addition, either by interchanging the pickup leads for one channel, or by inserting a stage with a voltage gain of -1 in the voltage amplifying stages of the preamp; and since the introduction of a switch to reverse pickup leads is more than likely to introduce hum, and may not be possible with a three-lead system, a stage with a gain of -1 is by far the better solution.

The Control Unit

We have thus arrived at a realistic complement of controls for a flexible yet compact stereo control unit, see *Fig. 1*. The front panel contains six controls, four of which operate upon both channels simultaneously to cover the important monophonic functions, and the remaining three control the essential stereo functions. (This totals seven, but the phase reverse and power switch functions are combined in one control.) The remainder of this article will describe the design and construction of the control unit, incorporating all the electronic refinements which are consistent with high quality, reliable performance. This design takes advantage of the freedom from hum and microphonics which characterize the transistor, an attribute which provides a convenient solution to the problem of the low-level output of most stereo cartridges. The small size and weight, and the lack of heat generation of the transistor will also serve us in good stead so far as a compact design is concerned, and the reliability of a properly designed transistor circuit is such that it should practically never require service.

DESIGN DATA FOR THE TRANSISTOR CONTROL UNIT

The first portion of this article has discussed the philosophy of a simple but adequately flexible stereo control unit. After examination of the requirements for such a control unit, a set of functions was specified which would meet the needs of nearly any audiofan. The design problem now becomes one of electrical realization of the functions thus specified, and of high-quality, high-reliability audio equipment design in general. In considering each channel separately we must observe the performance criteria which have become standardized in spe-

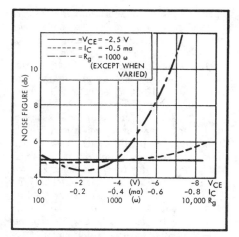

Fig. 2. Typical noise characteristics of a low-noise germanium transistor.

cifying the performance of monophonic equipment; the design is specialized only insofar as the peculiarities of stereo are concerned.

At the time that the idea for this control unit was conceived, it was decided that such a unit should be transistorized. Many reasons for such a decision may be advanced, and several of these merit examination. First, when a transistor is operated within its conservative ratings, it has an extremely long lifetime, or "mean-time-to-failure." For a high-quality transistor this lifetime approaches that of a passive device such as a resistor or capacitor, and even for the lower-priced transistors which we shall use, months or years will elapse before a typical transistor will require replacement. Second, the transistor and its associated circuitry are small and generate almost no heat. This means that the advent of stereo, requiring almost twice the equipment that monophonic reproduction entails, will not necessarily increase the size of the system by a factor of two to one. Transistor equipment can be made very compact, it does not require much power to operate, and the lack of heat generation means that such equipment may be mounted in the cabinetry without much thought to ventilation. Third, the transistor is virtually free from hum and microphonics, and this means that the hum level in the preamplifier—a problem because of the extremely low output voltages of some of the popular stereo cartridges—will be almost unmeasurable. In addition, we shall see that "transistor noise," which has manifested itself as a hiss or "frying" sound in early units, can be reduced to a vanishingly low level if proper design techniques are applied.

The discussion to follow will be somewhat different from the ordinary construction article. Because of the novelty of transistor circuits compared to their vacuum-tube counterparts, some detailed information will be included concerning the operating points of the transistor stages, and the equations used to calculate stability vs. temperature change, feedback resistors, and gain will be presented. While such calculations are necessarily tedious, it is hoped that by their presentation the reader will gain a better understanding of the approximations and computations associated with the design of transistorized equipment.

The circuitry of the control unit, *Fig. 3*, may be divided into two sections: the low-level preamplifier, where noise figure and exact frequency compensation are the important factors, and the high-level control stage, where signal-handling capabilities are important. The low-level preamplier circuit consists of two low-noise transistors Q_1 and Q_2, connected in the grounded-emitter configuration. The first transistor, Q_1, is operating at a low collector voltage and very low collector current to minimize the noise figure of the transistor, and the load resistor for this stage is a deposited-carbon type for the same reasons. The bias configuration is the "H" type, with the temperature-stability factor (defined as $S = \Delta I_c / \Delta I_{co}$) chosen to have a numerical value of less than 4. This value for S is chosen mostly from experience, which has shown that this is a reasonable criterion for reliable class A operation over the rated temperature range of a germanium transistor. The "H" bias configuration depends upon a large resistance in the emitter circuit to achieve stability, and then this resistor is bypassed with a large capacitor to maximize a.c. gain. In proceding with the design of this stage, the value of the collector-to-emitter voltage, V_{ce}, and the collector current, I_c, are first chosen from a graph of the noise characteristics of a typical transistor versus V_{ce}, I_c, and the generator impedance R_g. (*Fig. 2*) the values thus selected are:

$$V_{ce} = -8 \text{ vdc.}$$
$$I_c = -0.35 \text{ ma.}$$

Before resistor values can be calculated from the operating point, there are several restrictions on the parameters of the stage which must be taken into account. First, the voltage gain, G_v, and the input impedance Z_{in} must conform to external considerations. A survey of the popular stereo cartridges shows that an input impedance of 50,000 ohms is sufficient for all the cartridges requiring high-impedance, low-level inputs, and a lower Z_{in} may always be achieved by loading the cartridge with a resistor. It must be remembered that high values of input impedance are difficult to achieve in transistor stages which also have voltage gain, and for this reason we shall not attempt to make Z_{in} any higher than necessary. The compensated closed-loop gain of the low-level preamplifier should be about 40 db ($G_v = 100$) so that a 10-mv input will produce a 1-volt output. Since about 20 db for bass boost is required for the RIAA curve, the preamplifier will require an open-loop gain of 40 plus 20, or 60 db ($G_v = 1000$). This means that each stage within the loop should have a gain of $60/2 = 30$ db ($G_v = 31.5$).

We are now in a position to calculate some resistor values for the first stage. With the "H" bias configuration and the large amount of feedback present, the input impedance of the first stage will be primarily determined by the parallel combination of R_2 and R_3; we shall call this equivalent resistance R_b', and we can then say that approximately:

$$Z_{in} \cong R_b' \quad (1)$$

The next restriction on the values of the resistances in this stage is the S factor. For a grounded-emitter stage using a transistor with a high grounded-emitter current gain Beta (β) and small internal base and emitter resistances compared to the circuit values, S can be given by:

$$S \cong 1 + \frac{R_b'}{R_e'} \quad (2)$$

where R_e' is the total external emitter circuit resistance, $R_4 + R_5$. But we have specified R_b' at 50,000 ohms, and S can have a maximum value of 4, so we can easily calculate the minimum value of R_e' necessary to achieve this stability factor, and this turns out to be 16,000 ohms. Allowing an ample margin of safety, we will let R_4 equal 22,000 ohms, and the voltage drop across R_4 is $I_e R_e = 7.7$ volts. Making one additional assumption, that there is a small and nearly constant voltage difference between the base and emitter terminals of the transistor, amounting to about 0.3 volts, we know that the voltage across R_2 is $7.7 + 0.3 = 8$ vdc. We can similarly deduce that the voltage across R_3 is the supply voltage to the stage less 8 volts. In our case the power supply voltage is 30 volts, selected because experience has shown that this is a convenient voltage for audio work in transistors. To make certain that there is no cross-coupling between stages, there are two decoupling resistors in the supply line, R_7 and R_{19}, and the supply voltage to the Q_2 and Q_3 stages is actually about 28 volts, and that to the Q_1 stage about 25 volts. This shows that there are $25 - 8 = 17$ volts across R_3, and now that we know that the ratio of R_2 to R_3 is 8 : 17 (assuming that he current in the base of the transistor is negligibly small) and that the parallel equivalent of R_2 and R_3 is 50,000 ohms, we can easily calculate their values, which are 73,000 ohms and 156,000 ohms, respectively. Increasing these to the nearest 10 per cent EIA values gives: $R_2 = 82,000$ ohms, and $R_3 = 180,000$ ohms, and recalculation of R_b' now yields 56,000 ohms, still sufficiently close to the target figure of 50,000 ohms.

We have only two resistors left to calculate, the load resistor R_L (R_6), and the small unbypassed emitter resistor R_5, which controls the voltage gain. R_6 is actually already determined since we know the collector current (which equals the emitter current, approximately) and the voltage at the collector (which is the supply voltage less the V_{ce} and $I_e R_e$ drops), and the calculated value is 26,600 ohms, the closest value in a deposited-carbon resistor being 26,100 ohms. Once the load resistor is known, we can calculate the value of the feedback resistor R_5 from the equation for gain with feedback:

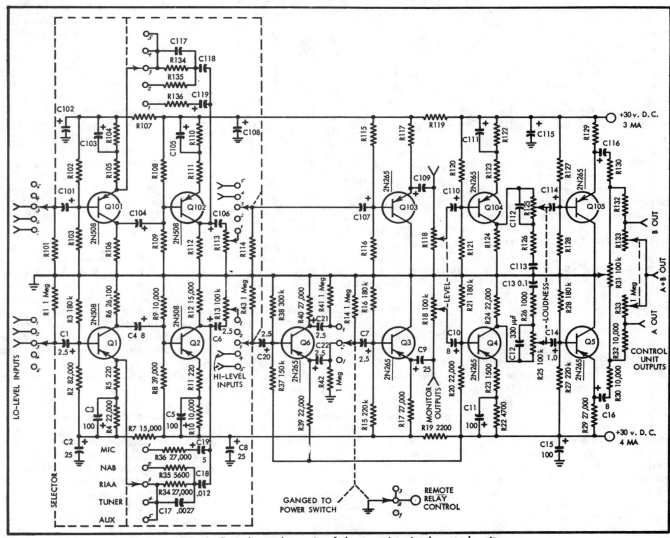

Fig. 3. Complete schematic of the transistorized control unit.

$$G_v \cong \frac{(\beta R_L')}{(h_{ie} + \beta R_5)} \quad (3)$$

where R_L' is the equivalent load imposed by the load resistor and any subsequent circuitry. H_{ie} is the grounded-emitter input impedance of the transistor, as specified by the manufacturer. It is interesting to note that this is the first time the parameters of the transistor have entered the equations explicitly; all the previous calculations have simply assumed some range of values.

Choice of Transistors

At this point, then, it will be necessary to select the specific transistor to be used in the circuit. A survey of the inexpensive, commercially available germanium "p-n-p" transistors with low noise-generation characteristics results in the selection of any of two or three types, all of which are pretty much interchangeable. The GE 2N508, the Philco 2N535B, (a newer version of the 2N207B), and the Raytheon 2N422 will all perform well in the preamplifier, and from this point forward we shall assume the characteristics of the 2N508. This transistor has a β of 100 and an h_{ie} of 3000 ohms, and a typical noise figure of less than 6 db (0 db = 1 μv.). Using this data, and assuming a loading by the following stage of 15,000 ohms, we can calculate from Eq. (3) that the G_v will be approximately 33 for an R_5 of 220 ohms. For the remaining control unit circuits, where noise is not so critical but larger signals entail larger collector voltages, we select a transistor with the same general characteristics but an increased maximum V_{ce}; such a transistor is the GE 2N265 or the Philco 2N534.

The preceding discussion has covered all the computations which are peculiar to the design and stabilization of a transistor stage with a specified voltage gain and input impedance. The values of coupling and bypass capacitors are calculated in the same manner as those for vacuum tube stages, that is, by making the time constant of the equivalent circuit small compared to the period of the lowest frequency we wish to pass or bypass. The technique just illustrated may now be repeated for the Q_2 stage, using the conditions and approximations just discussed, and assuming the loading of subsequent circuitry to be 50,000 ohms. Once this stage is completed, we are ready to calculate the values of the feedback and compensation resistors and capacitors, R_{34-36} and C_{17-19}.

When the selector switch S_1 is in position 1, R_{36} and C_{19} are the elements in the feedback loop. This corresponds to the "microphone" position, and we are desirous of a flat frequency response and a closed-loop gain of 4 db ($G_v' = 100$). The value of R_{fb} (R_{36}), in the condition where the presence of this resistor does not upset the open loop gain excessively, is:

$$R_{fb} \cong -\frac{(R_5)(G_v)}{(K-1)} \quad (4)$$

where K is the ratio of the open loop gain G_v to the closed loop gain G_v'. This equation gives a value of 24,000 for R_{36}, but since this would load the output of the second stage and change the open loop gain slightly, (we assumed a loading of 50,000 ohms) the corrected value of R_{36} becomes 27,000 ohms. In this case, C_{19} is merely a coupling capacitor with

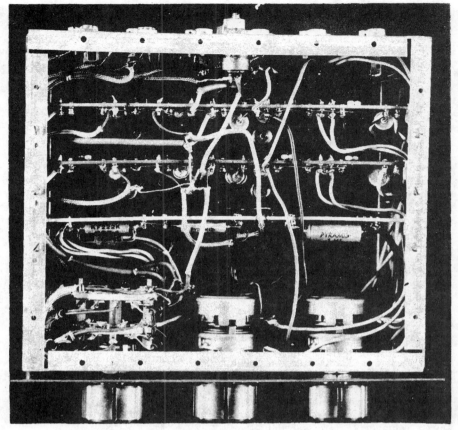

Fig. 4. Top view of the completed preamplifier unit. The upper two resistor boards are identical; the lower contains the equalization networks, loudness-control components, and the phase-reversal stage.

PARTS LIST

*$R_1, R_{14}, R_{41}, R_{42}, R_{43}$	1 megohm, ½ watt
R_2	82,000 ohms, ½ watt
$R_3, R_{16}, R_{21}, R_{28}$	180k ohms, ½ watt
R_4, R_{20}	22,000 ohms, ½ watt
R_5, R_{11}	220 ohms, ½ watt, 5%
R_6	26,100 ohms, ½ watt, 1% deposited carbon
R_7	15,000 ohms, ½ watt
R_8	39,000 ohms, ½ watt
R_9	120k ohms, ½ watt
R_{10}, R_{30}, R_{32}	10,000 ohms, ½ watt
R_{12}	15,000 ohms, ½ watt, 5%
R_{13}, R_{31}	100k ohms, ½ watt
R_{15}, R_{27}	220k ohms, ½ watt
R_{17}, R_{29}	27,000 ohms, ½ watt
R_{18}, R_{25}	100k-ohm dual potentiometers, linear, 10%
R_{19}	2200 ohms, ½ watt
R_{22}	4700 ohms, ½ watt
R_{23}	1500 ohms, ½ watt, 5%
R_{24}, R_{39}	22,000 ohms, ½ watt, 5%
R_{26}	1000 ohms, ½ watt
R_{33}	1-megohm potentiometers, linear, 10%
R_{34}, R_{36}, R_{40}	27,000 ohms, ½ watt, 5%
R_{35}	5600 ohms, ½ watt, 5%
R_{37}	150k ohms, ½ watt, 5%
R_{38}	300k ohms, ½ watt, 5%
*$C_1, C_6, C_7, C_{20}, C_{21}, C_{22}$	2.5 μf, 25 v, electrolytic
C_2, C_8, C_9	25 μf, 25 v, electrolytic
C_3	100 μf, 6 v, electrolytic
C_4, C_{10}, C_{16}	8 μf, 25 v, electrolytic
C_5, C_{11}	100 μf, 10 v, electrolytic
C_{12}	330 μμf, 300 v, mica
C_{13}	0.1 μf, 100 v, paper
C_{14}	1 μf, 6 v, electrolytic
C_{15}	100 μf, 50 v, electrolytic
C_{17}	.0027 μf, 100 v, paper
C_{18}	.012 μf, 100 v, paper
C_{19}	5 μf, 12 v, electrolytic
S_1	6-pole, 5-position switch, shorting
S_2	2-pole, 3-position switch, shorting
*Q_1, Q_2	"p-n-p" transistor, (GE 2N508)
Q_3, Q_4, Q_5, Q_6	"p-n-p" transistor, (GE 2N265)

* Values given for subscripts from R_1 to R_{36} (except R_{31}) apply also to resistors with subscripts R_{101} to R_{136}; values given for capacitors C_1 to C_{19} apply also to capacitors with subscripts C_{101} to C_{119}; transistor types listed for Q_1 to Q_6 apply also to transistors Q_{101} to Q_{106}.

All resistors 10% tolerance unless otherwise specified.

negligible reactance at the frequences from 20 to 20,000 cps.

The RIAA phonograph curve is realized by a low-frequency boost with a characteristic time constant of 300 microseconds, and a high-frequency rolloff time constant of 75 microseconds. Since the midfrequency gain is still supposed to be 40 db, R_{34} is 27,000 ohms. The bass-boost capacitor C_{18} will have a value of 0.012 μf, the boost time constant divided by R_{34}. Similarly, the rolloff capacitor C_{17} will be the rolloff time constant divided by R_{34}, and this yields a value of 0.0027 μf.

The NAB tape curve is realized by a single low-frequency boost with a time constant of 67 microseconds, where the high-frequency gain is considered the closed loop gain G_v'; thus the computation in this caste will be somewhat different. The output from a tape reproducing head is generally lower than that from a phono cartridge, so that it would be desirable to have the highest over-all gain possible with the equalization required. The total equalization between high and low frequency is 32 db, and since the open loop gain is 60 db, this means the maximum high-frequency gain (in this case the closed loop gain), will be 28 db ($G_v' = 25$). For this value and the known value of C_{18}, we calculate the corrected value of R_{35}, which is 5600 ohms. If we now calculate the gain at 500 cps (which is the crossover frequency for the NARTB curve), this midfrequency gain is 150, or about 43 db. This means that a tape head output of only 3.5 mv will produce a 0.5 volt output, while a 5-mv phono or microphone input is required to produce the same output.

At this point, it is timely to interject a note concerning the behavior of the large values of capacitance necessary in R-C coupled transistor stages. In order to maintain flat frequency response down to the lowest audio frequencies, the values of coupling capacitors tend to become very much larger than those in vacuum tube circuits, with values as large as 25 μf not uncommon. These are usually low-voltage midget electrolytics, and as such as subject to leakage currents considerably larger than those in the typical paper tubular capacitor. Therefore, it is important that d.c. return paths be provided for these leakages at inputs and outputs, and that the variable controls be arranged in such a way that the operation of the control does not change or reverse the d.c. voltage on the electrolytic, lest the resulting nonlinear discharge generate an audio signal not unlike a Bronx cheer.

The over-all characteristics of the low-level preamplifier stages may now be tabulated:

(A) *Flat Equalization:*

Frequency response flat ± 0.5 db from 20 to 20,000 cps.
Sensitivity 5 mv for 0.5 volts output.
Intermodulation Distortion 0.25 per cent at 1 volt output.
Harmonic distortion not measurable at 1 volt output.
Noise level more than 65 db below 1 volt output.

(This noise is characteristically below 30 cps.)

(B) *RIAA Equalization:*

RIAA equalization ± 0.5 db from 20 to 20,000 cps.
All other specifications equalled where applicable.

(C) *NAB Equalization:*

NAB equalization ± 0.5 db from 20 to 20,000 cps.
Sensitivity 3.5 mv for 0.5 volts output.
All other specifications equalled where applicable.

High-Level Section

The input to the high level section is made through the selector switch S_1, resulting in the choice of either the preamplifier output or a high-level input. The signal is then fed to the base of the emitter-follower Q_3. With the emitter-follower configuration we are able to realize a high input impedance and low output impedance at the cost of unity voltage gain, much in the same manner as the cathode follower in vacuum-tube circuitry. The output of Q_3 is labelled the "monitor output," and may be used to drive a tape recorder for dubbing, headphones, or almost any impedance greater than a few hundred ohms; this output is equalized to have a flat frequency response and constant amplitude, independent of the settings of the level and loudness controls R_{18} and R_{25}.

There is little to describe in the design of the emitter-follower stage because it is an extremely simple configuration, and there is no gain calculation to make. Suffice, then, that the operating point of the transistor is chosen so that:

$$V_{ce} = -12.5 \text{ vdc.}$$
$$I_c = -0.5 \text{ ma.}$$

and the S factor is again less than 4.

The Q_4 stage, like the Q_1 and Q_2 stages, is connected as a grounded-emitter amplifier with voltage gain determined by emitter degeneration. The smallest input anticipated at the high level jacks is 0.1 volt, and a 1-volt output of the control unit is sufficient to drive most power amplifiers to full output, so that an over-all voltage gain of 10 will satisfy the needs of the high-level portion. If each of the emitter followers has an over-all gain of 0.95, the gain of the Q_4 stage must be 12. Making the assumption of 50,000-ohm loading by the subsequent stages, and using exactly the technique described for the Q_1 stage and an operating point of:

$$V_{ce} = 11.5 \text{ vdc.}$$
$$I_c = 0.7 \text{ ma.}$$

the result is a gain of 12.5 for a load resistor of 22,000 ohms and an unbypassed emitter resistance of 1500 ohms.

The output of the fourth stage drives the loudness control R_{25} which, despite its simplicity, will approximate the Fletcher-Munsen loudness contours reasonably closely from the 0 db to the −35 db curves. The output of this control is connected to the fifth transistor, which is an emitter-follower circuit identical to the third stage. The output of Q_5 drives the balance control, R_{31}, and the separation control, R_{33}. The balance circuit is capable of an infinite range of attenuation, but the important range near balance is spread out so that it occupies a large portion of the range of rotation of the control. With this arrangement the balance can be adjusted very finely for nearly equal signals, or one channel can be eliminated from the output completely, by a 150-deg. rotation of the control. The separation control is likewise nonlinear in operation, making only about 6 db of variation in the separation through the first 100 deg. of rotation, and then decreasing the separation rapidly to zero at full rotation.

The output impedance of the control unit is about 20,000 ohms as a consequence of the isolating resistors R_{30} and

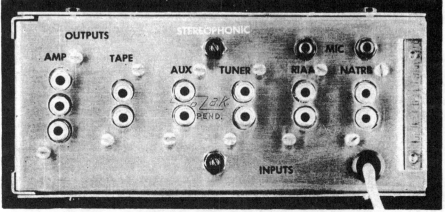

Fig. 5. Rear view of the completed transistorized preamplifier.

R_{32}, so that the unit should work into an impedance of at least 50,000 to 100,000 ohms. If a really low output impedance is desired, and this may often be the case, the present outputs may simply be followed by emitter-followers, one for each channel, identical in design to the Q_3 or Q_5 stages. For a typical installation where long cable runs are not necessary, the unit is quite acceptable without additional emitter-followers, and the treble loss and hum pickup will be negligible.

The phase-reversal feature, as discussed in part I, has been retained for separate treatment. The circuit is a "split-load" or "matched-load" phase splitter of a configuration which is common in vacuum-tube power amplifiers. When this circuit is inserted in ONE channel just ahead of Q_3 then the output at the collector of the reverser corresponds to an inverted signal with the same amplitude as the input, and the output at the emitter is a signal identical to the input. The Z_{in} of this circuit is the same as that of the Q_3 stage, namely 100,000 ohms. The design of this reverser is identical to that of a grounded-emitter stage with a voltage gain of 1, taking into account the fact that the equivalent load and unbypassed emitter resistors change, depending upon which output is being used. The operating point of Q_6 is:

$$V_{ce} = -10 \text{ vdc.}$$
$$I_c = -1.1 \text{ ma.}$$

The section of S_2 which selects either output, is one deck of the power switch. Thus S_2 is a 2-pole, 3-position switch, and the three positions are: (1) OFF, (2) ON, PHASE NORMAL, and (3) ON, PHASE REVERSED.

One additional feature bears a passing examination; the "third-channel" output. Because of the construction of the separation control R_{33}, there is always a signal available which is the sum of the signals in the two channels, and hence is monophonic, representing the sound from the "middle." This signal may be used to provide a "phantom" or third channel by the application of a third amplifier and speaker; however, the impedance at this point is very high, and any loading of less than 1 megohm will reduce the sum-signal level and disrupt the operation of the separation control. Therefore, if this third channel is to be used, it would be advisable to follow it with a very high input impedance emitter-follower or a cathode-follower, and to implement this, a schematic of such an isolation amplifier is appended as *Fig.* 7.

The performance of the high-level portion of the control unit may now be tabulated:

(A) *0-db Loudness contour:*
Frequency response flat ±1 db. from 20 to 20,000 cps.
Sensitivity 100 millivolts for 1 volt output.
Intermodulation Distortion 0.6% for 1 volt output.
Harmonic distortion not measurable at 1 volt output.
Noise level more than 65 db. below 1 volt output.
(This noise is characteristically below 30 cps.)

(B) *−35 db Loudness contour:*
Follows Fletcher-Munsen contour ± 3 db.
All other applicable specifications are equalled.

Construction

The circuitry of the author's control unit was laid out on three 2¾ by 6¾ in. perforated bakelite resistor boards; one channel on a board, except that the equalization networks and loudness-control components, and the entire phase-reversal stage, were placed on a third board. It is interesting to note that printed circuits would be ideal for this application, and could probably be made using some of the kits which are presently available. The author's three resistor boards were mounted inside a 3 × 6 × 7 in. aluminum chassis as shown in *Fig.* 4, with one 3 × 7 in. panel serving as the front and the other as the terminal panel, *Fig.* 5. To achieve a professional-looking unit, a 3½ × 7½ in. brass front panel was added, rubbed to a satin finish with crocus cloth and then steel wool, sprayed with Krylon clear plastic and then letter decals added, followed by about six more coats of plastic spray until the surface was smooth over the decals. *Figure* 1 shows the finished appearance. The author used machined Dural knobs to further enhance the appearance, but these are inordinately expensive to purchase, and any knob would suffice.

The schematic of a simple remote power supply, *Fig.* 6, is included without lengthy explanation. The supply is switched on and off by controlling the current in the coil of relay K_1, so that there is no 115-volt a.c. power inside the control unit at all. This supply will furnish + 30 volts d.c. at the total current drain of approximately 7 ma required for the control unit; the output voltage is adjustable by R_1, and the total ripple is quite low, well below 1 mv rms. The supply voltage could be obtained in any of a number of alternative ways, the only restrictions being that the ripple must not exceed 1 mv, and the supply voltage must never exceed + 35 volts, on pain of a damaged transistor. These requirements are so slight that the voltage could be obtained by a simple bleeder from the power supply of a power amplifier, or even from a battery.

The author's control unit has been in operation for twelve months at the time of writing, and has performed entirely satisfactorily There is a complete absence of hum, hiss or microphonics, and no audible distortion. The controls have justified the time spent in selecting them, in that the system has proved entirely satisfactory for both stereophonic and monophonic applications. The over-all cost of all parts, including the power supply described, was about $75, which is certainly comparable in price with the stereo preamplifier kits on the market—and there is no doubt that the kits could be assembled in quantity for a smaller price. Whatever the cost, the convenience of a high-performance control unit with only six knobs is one which cannot be bought on the market, at any price.

Æ

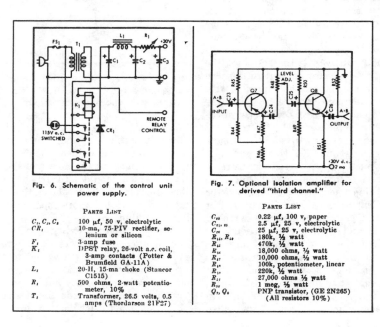

Fig. 6. Schematic of the control unit power supply.

PARTS LIST
C_1, C_2, C_3 100 µf, 50 v, electrolytic
CR_1 10-ma, 75-PIV rectifier, selenium or silicon
F_1 3-amp fuse
K_1 DPST relay, 26-volt a.c. coil, 3-amp contacts (Potter & Brumfield GA-11A)
L_1 20-H, 15-ma choke (Stancor C1515)
R_1 500 ohms, 2-watt potentiometer, 10%
T_1 Transformer, 26.5 volts, 0.5 amps (Thordarson 21F27)

Fig. 7. Optional isolation amplifier for derived "third channel."

PARTS LIST
C_{23} 0.22 µf, 100 v, paper
$C_{24, 25}$ 2.5 µf, 25 v, electrolytic
C_{26} 25 µf, 25 v, electrolytic
R_{43}, R_{48} 180k, ½ watt
R_{44} 470k, ½ watt
R_{45} 18,000 ohms, ½ watt
R_{46} 10,000 ohms, ½ watt
R_{47} 100k, potentiometer, linear
R_{49} 220k, ½ watt
R_{50} 27,000 ohms ½ watt
R_{51} 1 meg, ½ watt
Q_7, Q_8 PNP transistor, (GE 2N265)
(All resistors 10%)

A Transistor Protector

GEORGE FLETCHER COOPER

Thermal runaway is the culprit which will destroy a transistor unless the circuit is properly designed. The circuit protector described here takes advantage of the high collector impedance of a transistor to maintain relatively constant current.

IN THE CHANGEOVER from tubes to transistors most of us have found it necessary to get used to the hard and bitter fact that transistors are not indestructible. Even though you may not have to pay for them yourself this can be rather annoying—and I would imagine that when it is your own money which vanishes in a few milliseconds you would feel that "rather annoying" is not sufficiently strong. Total failure may be the result of doing something quite stupid, of course, in which case there really is nothing more to be said but there are some circuits which start off behaving quite normally and then . . . well then it is just too late.

Thermal Runaway

The basic effect which can land you round at the bank asking for a loan is, as you no doubt realize, thermal runaway. In general, this is a danger when the circuit you are using is one which is designed to let you get as much power as possible from a particular size of transistor, which very often means either a class-B amplifier or an inverter. These circuits are not easily made inherently stable because, for one reason, making them stable usually means a severe limitation of their output. Some of them have the rather unpleasant characteristic that they are most dangerous at some particular signal level, not when they are first set up.

Thermal runaway is the result of the fact that the collector current in the zero bias condition changes very quickly with junction temperature. Typically the grounded-emitter current with base open-circuited may increase by 8 times in going from 25-deg. to 45-deg. Centigrade. When your luck is out you may get the junction up to a temperature at which there is so much current that more heat is being generated in the transistor than is leaking away, so it gets hotter and . . . When you open up a wrecked transistor and look at the size of the germanium die you will see why the whole process takes place very quickly.

Users of power transistors take it for granted that they must mount the transistor on a large heat sink to help to keep it cool. For small transistors I have found it convenient to use two different types of heat sink. One is a simple radiating flag, made by cutting out a rectangle of aluminum, say 1/32-in. thick, and about 1½-in. by ½-in., and rolling one end of this into a cylinder—a drill forms a suitable mandrel—which will slip over the transistor. This is the form shown by some transistor manufacturers.

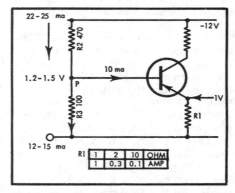

Fig. 1. The transistor protector drawn as an amplifier.

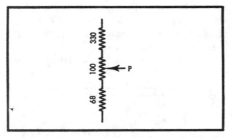

Fig. 2. To deal with a range of beta values, the bias chain R2 R3 includes a potentiometer as shown.

The other is more massive and consists of short sections of aluminum bar into which holes are drilled. Some silicone grease is put into the hole and the transistor, which should fit fairly closely, is then pushed in. These have the advantage of providing more mass and slowing everything down.

A recent painful experience in which two flagged transistors in a finished piece of equipment proceeded to "walk away briskly" so that the meter could be seen swinging firmly but, as it turned out, inexorably towards the destruction point led me to decide that prevention was better than retardation. Oddly enough we already have prevention up in the 15–25 amp class by the use of a special trigger circuit so that I chose, rather arbitrarily, a limit of 1 amp for the protector. There is nothing really special about this limit, however, and you can extend the principle up to higher currents if you wish, though each protector circuit will only have a limited range of currents over which it will work. The reason is that you need to use a big transistor if you mean to pass 25 amps and you cannot safely or wisely use this size of transistor down in the milliamp region.

Mostly we work with 12-volt supplies, with the battery permanently connected to the charger so that the terminal volts are up around 13–14 volts. This is only material if you are anxious to use the protector on 24–28 volts, when you must use a transistor with a higher voltage rating. Fortunately this is one of those simple circuits in which you do not have any difficulty in choosing a suitable type of transistor. I happen to have chosen one of the Clevite Spacesaver units (CST 1739-43) which will pass 3 amps and stand 20 volts in any condition you care to name.

The Basic Circuit

The basic principle of the circuit is very simple. The current through a transistor is almost independent of the collector voltage. Examining the data sheet through a magnifier it looks as though, for the CST 1742 and 1743, the collector current rises from 0.9 amps at 0.5 volts collector-emitter voltage to 1.0 amps at 15 volts collector-emitter voltage when the base current is 10 milliamps. If, then, we start with a 13-volt supply and a load which wants to take about 1 amp connected in the collector, we shall only drop about half a volt in the transistor and we shall get 12.5 volts across the load. Suppose, however, that at about 12 volts the load attempts to take 2 amps. This is not permitted, and the control transistor offers it only 1 amp, which means that there will only be 6 volts across the load while the control transistor drops the remaining 7 volts. And if the load is a short-circuit, well, the control transistor will still only pass 1 amp although it has 13 volts applied to it.

Perhaps we should look at the circuit diagram in *Fig. 1*. It is arranged here to look like a conventional amplifier because this makes it easier to follow. The emitter resistance R_1 is chosen to give 1 volt drop at the rated current so that for 1 amp we have $R_1 = 1$ ohm and for 100 milliamps we have $R_1 = 10$ ohms. You can, I hope, work out other values for yourself though I do not recommend you to go beyond $R_1 = 100$ ohms, corresponding to a current of 10 ma. A suitable set of resistors is then assembled on a switch.

Consulting the maker's curves I find that V_{be} is likely to range from 0.2 to 0.5 volts, from which I see that I must hold the base, point P on the R_2R_3 chain, at around 1.2–1.5 volts. Obviously I do not want any thermal runaway trouble with this transistor, so I make R_3 fairly low, and I have chosen initially a value of 100 ohms. The current through R_3 is then about 12–15 milliamps. For the CTP 1742 and 1743 there will be around 10 ma flowing into the base when the collector current is 1 amp, so that the total current in R_2 will be 22–25 milliamps. For a supply of around 13–14 volts the drop across R_2 must be about 12 volts and this will give us as the nearest standard value $R_2 = 470$ ohms.

This simple circuit is just a straightforward amplifier without very much negative feedback and is therefore rather dependent on the beta of the transistor. I have accordingly modified the base bias circuit to the form shown in *Fig. 2*, with P now connected to the slider of a potentiometer which can be used as a continuous control for the current. This is treated rather as a trimmer and can be left alone in the early stages of operating a breadboard, or else you can use the operating procedure I shall describe later. You can see immediately that if the current could increase by about 20–50 per cent of its selected value the transistor would be reverse biased by the emitter drop, so that this circuit will probably hold the current to around 10–20 per cent of the wanted value. It is not worth while calculating or measuring this exactly because all we are after is something to hold down a device which will take ten or more times the prescribed current if it gets into trouble.

A refinement to this system which I should introduce if I were designing it to sell is a simple indicator circuit. The circuit of this is shown in *Fig. 3*. Under normal conditions the collector of the control transistor is bottomed so that the voltage is around −1.5 volts. The Zener diode in the coupling to the indicator transistor is chosen to have a drop of around 4 volts although this is not frightfully important. With everything operating properly then, the Zener diode will not have enough volts across it to make it conduct and the second transistor will be cut off. When the conditions are abnormal the control transistor must drop enough volts to keep the current limited and the collector voltage will move towards −12 to −14 volts. As soon as it reaches −4 volts the Zener diode can pass current and thus current can be driven through the limiting resistor into the base of the indicating transistor.

The collector load of the indicating transistor is a 12-volt lamp bulb taking as little current as possible. I am not sure if it is easy to get anything below 50 ma. As the indicator transistor is made to conduct, this bulb will light up to show you that things are not what they should be. For a 50-ma bulb the

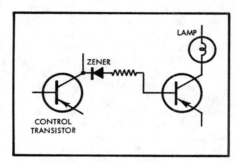

Fig. 3. An indicator added to the control transistor is a refinement.

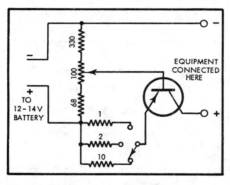

Fig. 4. The circuit as it appears to the user.

control transistor needs to be able to pass this current, of course, and it must also be able to dissipate 150 mw when halfway on. It will be prudent to put in 100–200 ohms emitter resistance, while the base resistance will probably be around 4700 ohms though this depends on the beta of the indicating transistor. This section of the device is, as I have said, strictly for amusement only and should be tailored to whatever transistor and lamp you find the most convenient.

Building the Protector

As usual I have been more interested in how the system works than how you make it. Now we must turn to the hardware. If we limit ourselves to 2 amps, since that is getting towards the limit of the transistor I have considered, the worst condition we can encounter is when the load is a complete short circuit. The internal dissipation in the control transistor will then be around 25 watts. The thermal resistance of the Clevite Spacesaver transistors averages 1.4° C/W but may be as high as 2.5°C/W. We cannot expect to get the external heat sink down below 1.5°C/W, so the thermal resistance may be 4°C/W. In an ordinary room at 25-deg. C, the 25 watts we are putting in under emergency conditions will then bring us up to 125-deg. C at the junction. This is far too high.

I carried out this exercise because it is very tempting to try to take 3 amps through a transistor rated at 3 amps. In fact we must limit this particular circuit to something around 1 amp, giving us 12 watts dissipation in the worst condition and a consequent temperature rise to around 75 deg. C. A check of the leakage current-temperature curves shows that we shall not be in any trouble here, as the leakage current will be less than 20 ma.

The heat sink for 1.5°C/W is nominally a 6-in.-square sheet of aluminum 1/16-in. thick. The transistor is best insulated from this and the edges of the sheet bent round to protect the shell from accidental connection to ground. I used an available box, about 4-in. cubed, and folded the edges of the transistor plate down about 1-in. all round so that it fits into the bottom of the box and leaks some more heat away into this skin. Ventilating holes allow the air convection currents to flow over the main plate and the box lid carries the switch for selecting values of R_1, the trimming potentiometer and the terminals. The circuit as it looks to a user is shown in *Fig. 4*. It is exactly the same circuit as *Fig. 1* but redrawn to emphasize the construction rather than the fact that it is basically an amplifier. Somehow I find that by arranging a circuit in a conventional form it is easier to understand.

Using the Protector

When using this protector the first step is to connect the ammeter or milliammeter in series with one of the output terminals and then short circuit the pair of wires which will go to the apparatus. This will enable you to trim the base potentiometer to the design current you want. Again, you could fit meter and switch to the unit, but it is rather a matter of taste which I leave to you. Having set up the device there really is no more to it except to put a voltmeter across the terminals and get to work. So long as the voltmeter shows around 12 volts you know that the equipment is not trying to take too much current.

Of course, you cannot expect the protector to think for you. Suppose you have two transistors in the equipment, one taking 10 ma and the other 900 ma. You just must not hope that this system will prevent the 10 ma transistor running up to 20, 30, 40, or 100 ma. That ought to be looked after in the circuit design anyway. Usually it is the big transistor which is the problem and it is not easy to think of normal circuits where the special difficulties arise.

There is one extremely important point to be watched when using a protector of this kind. The whole object of the exercise is to provide a constant current by using the very high collector impedance of the control transistor. If, therefore, you attach a simple amplifier to the terminals you have the protector trying to keep the current constant while the object of having an amplifier is to produce a varying current, the amplified signal. You will see that the results you get can only be thoroughly confusing. I do not think that this is at all a bad thing because I find that transistor circuits get designed on the bench with short connections to a good battery and used with long wires to an aging battery. This produces in a mild form the troubles which the protector highlights. We had all this long long ago with tubes when power supply units first came in and the magazines were full of articles on motorboating. When did you last see one of those? The answer is, as you well know, proper decoupling by means of a good big capacitor across the supply terminals. Just how big depends on the circuit and the easiest procedure is to put an oscilloscope across the capacitor and check that the ripple voltage is small enough. The other simple test is to double the value you have and see if there is any effect: if there is, you have not enough capacitance. If there is no effect you can try a smaller value.

Using the Spacesaver transistor we saw that overheating would limit us to about 1 amp at 12 volts, less at higher voltages and more at lower voltages in proportion. If you need more current you must use a bigger transistor. I do not think you can rely on going to better than 2.5° C/W between junction and air and I should like to allow more than 50°C temperature rise at the junction. The limit is then 20 watts of dissipation which with a 12 volt system and a short-circuited output, which puts it at its worst point, means you can allow 1.6 amps. To handle more current in this arrangement you must use more transistors in parallel.

One way round this limitation is to modify the indicator circuit so that it has a relay in it. Once the indicator transistor starts to take current the relay pulls in and locks through one of its own contacts. At the same time another contact opens the supply to the device which is demanding too much current. In this arrangement you need a capacitor across the relay to make sure that it has time to lock in before it disconnects the supply. I would expect the first hook-up of this arrangement to give you the usual relay chatter trouble. My own needs for high current systems are met rather differently, which is why I cannot give details of the relay circuit. When you are using transistors a good deal and taking large currents it is worth-while changing from batteries to power supply units. Stage two brings in transistorized smoothing and regulation and in no time at all you reach stage three, with an automatic overload trigger which can be set to cut off the supply on overload. You put this in originally to protect the power supply but it also works well to protect the equipment. The design of a big power supply unit with full protection is something I hope to describe in a later article.

Æ

STEREO PHONO PREAMP
(from page 13)

cartridge we should use the circuit of *Fig.* 8.

$$R_x = \frac{3250 \times 40000}{56,750} = 3550 \text{ ohms}$$

Again we have our choice between the nearest 10-per cent tolerance resistor, 3300 ohms, and the nearest 5-per cent tolerance resistor, 3600 ohms. In this second case there is even less reason to use the 5-per cent resistor since the error here has more effect on the voltage division than on the frequency response.

To illustrate the other general case let us consider a cartridge where the output impedance is essentially resistive. A typical cartridge of this class may have an inductance of 3 or 4 millihenries and a resistance of 600 ohms. Under these circumstances we may neglect the effect of the inductance and consider the output impedance of the cartridge to be a pure resistance. To achieve the proper rolloff in this case we must shunt the input of the preamplifier with a capacitor, the reactance of which is equal to the output resistance of the cartridge at a frequency of 2100 cps.

The value of the necessary capacitor is given by:

$$C \text{ (in } \mu f) = \frac{75}{R}$$

For the values given $C = \frac{75}{600} = 0.125 \ \mu f$

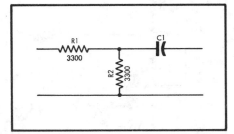

Fig. 8. Method of connecting input resistors to provide correct load and reduce input signal.

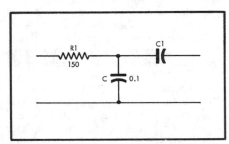

Fig. 9. R_2 of Fig. 1 is replaced by the 0.1-μf capacitor.

The nearest standard value is 0.1 μf. This is quite an error which may require either two capacitors in parallel to give the correct value of a 150-ohm resistor in series with the input lead from the cartridge as shown in *Fig.* 9. This extra resistance will have a negligible effect on the output of the cartridge at low frequencies.

Here we see that R_2 of *Fig.* 1 is replaced by the 0.1-μf capacitor.

Parts List

R_1, R_2	See appendix
R_3, R_{13}	470 k ohms, ½ watt
R_4	47,000 ohms, ½ watt
R_5	33,000 ohms, ½ watt
R_6, R_{11}, R_{16}	1500 ohms, ½ watt
R_7, R_{14}	27,000 ohms, ½ watt
R_8	270 k ohms, ½ watt
R_9, R_{18}	15,000 ohms, ½ watt
R_{10}, R_{15}	1000 ohms, ½ watt
R_{12}	18,000 ohms, ½ watt
R_{17}	1 megohm, ½ watt
R_{19}	68,000 ohms, ½ watt
C_1, C_2, C_4	10 μf 25 v, electrolytic
C_3, C_7	.05 μf 200 v, paper
C_5, C_{11}	100 μf 25 v, electrolytic
C_6	50 μf 6 v, electrolytic
C_8, C_9	100 μf 15 v, electrolytic
C_{10}	500 μf 15 v, electrolytic
Q_1	2N104
Q_2, Q_3	2N405
Q_4	GT34
D_1, D_2	1N91
T_1	Filament transformer, 6.3 v, 1a.

Æ

Amplifiers with Positive and Negative Feedback

CHARLES P. BOEGLI

Contrary to a widely held belief, this author discovered that the cathode-coupled phase inverter ("long-tailed pair") introduces a significant amount of distortion. By including this stage in the negative feedback loop he achieved an unusually low-distortion amplifier.

SEVERAL YEARS AGO, the writer had two articles[1] published on the design and construction of audio amplifiers utilizing over-all negative feedback with internal positive feedback. A number of readers constructed these amplifiers and satisfaction was the general result.

Those who are interested in the details of these amplifiers should refer to the original articles. Several difficulties were encountered with the circuits, primary among which were:

1. The output transformer was not designed for the manner in which it was operated.

2. The output transformer secondary was at a small d.c. potential above ground.

3. The inverter (the first stage of the amplifier) was not included in the negative feedback loop, so that the distortion introduced by this stage appeared undiminished in the output.

Both amplifiers used ordinary output transformers with the secondaries connected in unusual fashion. The speaker lines were connected to the 0- and 16-ohm taps of the secondary and the 4-ohm tap was grounded (for a.c.), so that a balanced output was being drawn from a transformer intended for unbalanced operation. The output transformer was carefully specified, and those who were foolhardy enough to construct their amplifiers with other transformers usually paid the penalty of instability or oscillation. For some time, the reason why one transformer worked well while another did not remained a mystery, but it was thought that unbalanced capacitances between each end of the winding and ground might be responsible.

---

[1] Boegli, Charles, A 35–watt "Infinite-Feedback" Audio Amplifier, *Radio and Television News*, July 1954, p. 39.

Boegli, Charles, A 13-watt All-Triode "Infinite-Feedback" Amplifier, *Radio and Television News*, November 1955, p. 68.

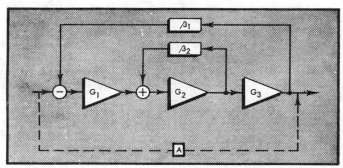

Fig. 1. Amplifier with over-all negative and internal positive feedback.

One hundred per cent negative feedback was obtained by connecting the ends of the secondary directly to the cathodes of the driver tubes. Internal positive feedback was brought from each driver plate to the grid of the other driver. Bias for the drivers was obtained by inserting a bypassed resistor between the center tap (that is, the 4-ohm tap) of the output transformer secondary and ground, so that the entire secondary was at a d.c. potential equal to the bias on the driver cathodes. If a speaker line became shorted to the chassis of the amplifier, the bias was disturbed and oscillation usually occurred. Nevertheless, speaker lines are usually not grounded, and this did not prove to be a very great shortcoming.

The inverter was not included within the negative feedback loop because of the desire to minimize the number of stages in the loop. A cathode-coupled inverter ("long-tailed pair") was used for inversion. This circuit was widely considered to be quite linear because of degeneration, so placing it outside the negative feedback loop was not expected to cause a significant increase in distortion.

Although the performance of the amplifier was quite good, distortion proved to be somewhat higher than had been anticipated. Since the only real source of distortion could be the inverter, a detailed study was made of the long-tailed pair with results that were, in some respects, surprising:

1. If the gain is defined to be the total plate-to-plate output divided by the grid-to-grid input, then the gain of the long-tailed pair is the same as that of a single tube operated under comparable conditions.

2. Because of the large d.c. voltage at the cathodes of the tubes, the output signal is somewhat limited in magnitude before distortion becomes excessive.

3. The distortion curve for this stage has a shape typical of a circuit without degeneration, suggesting that there is no improvement in linearity because of degeneration.

It was now certain that the inverter was the principal source of distortion in the amplifier. To reduce this distortion, the inverter would have to be included in the negative feedback loop, but it was not known what effect positive feedback around one stage would have on distortion arising in earlier stages. This lack of knowledge, coupled with the desire to explain the instability experienced in the amplifier when small changes were made in certain components, indicated need for further work on amplifiers using combined positive and negative feedback.

These circuits have now been rather thoroughly investigated. This paper details the work that has been done, and describes the resulting improved amplifier.

Analysis of Negative-Positive Feedback Amplifiers

Figure 1 is a block diagram for a gen-

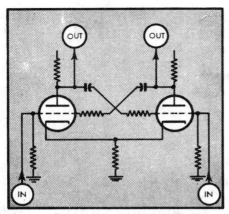

Fig. 2. Push-pull stage with positive feedback.

eral amplifier using over-all negative feedback and internal positive feedback. Each amplifier block can be assumed to consist of any number of stages, and in general the terms G and β, which designate gain and feedback factor, can be considered complex variables.

A straightforward analysis shows that the amplification of the circuit is

$$A = \frac{G_1 G_2 G_3}{1 - G_2 \beta_2 - G_1 G_2 G_3 \beta_1} \quad (1)$$

In the present case, β_2 is positive feedback and β_1 is negative feedback; when values are substituted into Eq. (1), the appropriate sign must be used (+ for positive feedback and − for negative.)

In the customary analysis of amplifiers of this type, the positive feedback is considered to be adjusted so that $G_2 \beta_2 = +1.0$, so that if the negative feedback is removed, the stage represented by G_2 just oscillates. If $G_2 \beta_2 = +1.0$ is substituted into Eq. (1), the resulting expression is

$$A = -\frac{1}{\beta_1} \quad (2)$$

This is the result that would be obtained with an ordinary feedback amplifier if the gain were infinite.

The effect of positive feedback upon the distortion introduced by each stage can be found by assuming a disturbing voltage δ to be injected at the outputs of the various stages, and finding the voltage produced at the same point by an analysis similar to that used to obtain Eq. (1). The following results are obtained:

(1) For a disturbance δ introduced at the output of G_1 or G_3, a voltage is produced at the same point amounting to

$$\frac{G_1 G_2 G_3 \beta_1 \delta}{1 - G_2 \beta_2 - G_1 G_2 G_3 \beta_1}$$

If $G_2 \beta_2$ is adjusted so that it just equals +1.0, this expression simplifies to −δ showing that the distortion introduced anywhere except in the stage around which positive feedback is brought is completely canceled.

(2) For a disturbance δ introduced at the output of G_2, a voltage is produced at the same point of magnitude

$$\frac{(G_1 G_2 G_3 \beta_1 + G_2 \beta_2)}{1 - G_2 \beta_2 - G_1 G_2 G_3 \beta_1}$$

If $G_2 \beta_2$ is adjusted so that it just equals +1.0, this expression becomes

$$-\delta \frac{(G_1 G_2 G_3 \beta_1 + 1)}{G_1 G_2 G_3 \beta_1}$$

and when this is added to the injected disturbance δ, the sum is

$$-\delta \frac{1}{G_1 G_2 G_3 \beta_1}$$

which represents a reduction in distortion by a factor equal to the loop gain in the absence of positive feedback. The loop gain should be as high as possible even when positive feedback is used, because the distortion arising in the stage around which positive feedback is introduced is reduced in this manner.

One very interesting observation about Eq. (1) can be made immediately. Suppose the internal feedback network is actually arranged so that $G_2 \beta_2 = +1.0$ at all frequencies from zero to infinity. Equation (1) then simplifies ideally into Eq. (2); that is, the over-all gain is completely independent of the individual G's (even though they might be complex quantities) and is simply the inverse of the negative-feedback characteristic. The β_1 can, however, be held constant over an exceedingly wide frequency range; certainly, if it should be desired, down to zero frequency and also up into the radio frequencies. It should then be possible to obtain a flat and level response from the entire amplifier even with internal stages and, in particular, an output transformer, of very poor characteristics. No problems of instability would enter except when β_1 became zero which, as we have seen, can be at frequencies very much higher than those of interest in audio work. At such frequencies, of course, rather drastic steps could be taken to insure stability.

Unfortunately, the problem of making $G_2 \beta_2 = +1.0$ over a wide frequency range itself appears insoluble. No matter how great the precautions, G_2 will drop off at high frequencies because of interelectrode and stray capacities; actually, this drop sometimes occurs at frequencies considered to be of interest in audio work. At the low end, it is quite possible by means of conductive feedback to keep $G_2 \beta_2 = +1.0$ down to d.c. The results of the attempt are, however, rather peculiar.

Making $G_2 \beta_2 = +1.0$ has the effect of causing the stage represented by G_2 to oscillate (in the absence of other factors), so this stage can be considered, for the moment, a wide-band untuned oscillator. Now, the effect of keeping $G_2 \beta_2 = +1.0$ down to d.c. is to extend the range of oscillation down to d.c.; that is, the stage becomes not only an a.c. but also a d.c. oscillator. A d.c. oscillator, by extension of the definition of an a.c. oscillator, is a device that will generate a d.c. voltage with no external input. Experiments have shown that this is precisely what happens with conductive positive feedback, and the d.c. voltage which then saturates the stage causes it to be ineffective in amplifying a.c. signals.

For example, a typical push-pull stage incorporating positive feedback is shown in *Fig.* 2. Its similarity to a multivibrator is at once evident. In fact, the only difference is that the positive feedback is controlled at the point where oscillation just begins, while in a multivibrator the positive feedback is very much greater than that. Now, if the positive feedback is extended down to d.c. by elimination of the blocking capacitors, the stage becomes in effect a flip-flop circuit quite incapable of passing an a.c. signal.

The inference, of course, is that in all practical cases, $G_2 \beta_2$ must drop off at low and high frequencies, becoming less than +1.0. An investigation must therefore be made of the frequency response of G_2 with a.c. positive feedback, so that this response can be controlled to prevent instability in the final amplifier.

Stages with Positive Feedback

Consider an amplifier with positive feedback (*Fig.* 3) and let β be constant with frequency from zero to infinity while G drops off at 6 db per octave below some frequency $\omega_1 = 1/T_1$ and above another frequency $\omega_2 = 1/T_2$. Then β may be considered a real quantity while

$$G(s) = \frac{GT_1 s}{(T_1 s + 1)(T_2 s + 1)} \quad (3)$$

The T's are time constants, and s is the Laplace transform argument. By ordinary feedback analysis, the closed-loop transfer function is

$$A = \frac{GT_1 s}{(T_1 s + 1)(T_2 s + 1) - \beta G T_1 s} \quad (4)$$

We now make the following substitutions: $T_2 = T$, $T_1 = aT$, and $G\beta = K$, which transform Eq. (4), after expansion of the denominator, into

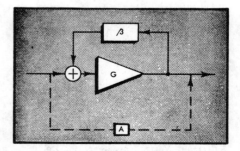

Fig. 3. Amplifier with positive feedback.

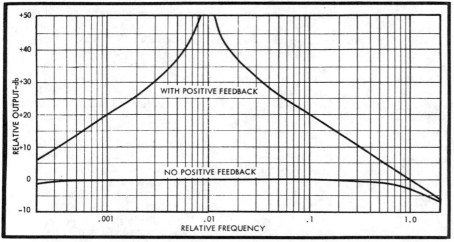

Fig. 4. Response of a single stage with and without positive feedback.

$$\frac{A}{G} = \frac{aTs}{a^2T^2s^2 + (1+a-Ka)Ts+1}$$

In the usual case where positive feedback is considered, K is about 1.0 and a is much larger than 1, so that as an approximation,

$$\frac{A}{G} = \frac{Ts}{T^2s^2 + (1-K)Ts + 1/a} \quad (5)$$

To find the shape of the response curve, $j\mu$ is substituted for s, and the magnitude of the resulting expression is calculated. This turns out to be

$$\left|\frac{A}{G}\right| = \frac{\mu}{\sqrt{\mu^4 + \mu^2\left(B^2 - \frac{2}{a}\right) + \frac{1}{a^2}}} \quad (6)$$

in which $B = (1-K)$. Since the expression for amplification, Eq. (6), contains only B^2, it remains unchanged if $-B$ is substituted for $+B$; that is, the response will be the same if K is 0.95 or 1.05. The fact that the phase shift in the two cases is different is of no consequence in the present study.

Now, when $\mu \neq 0$, the amplification, Eq. (6), approaches infinity as the denominator of the expression goes to zero. To find the value that B must have to make the amplification go to infinity at some frequency, we set the denominator equal to zero and obtain

$$\mu^2 = -\frac{1}{2}\left(B^2 - \frac{2}{a}\right) \pm \frac{B}{2}\sqrt{B^2 - \frac{4}{a}}$$

which is real and positive only when $B = 0$, or $\mu^2 = 1/a$.

In other words, when positive feedback is led around an amplifier and adjusted so that oscillation just begins, that oscillation occurs at a single frequency located at the geometric mean of those represented by the two time constants of the amplifier. Thus, what has ordinarily been assumed to be a wide-band untuned oscillator is in reality a tuned oscillator. Equation (2) was derived from Eq. (1) by substituting $G_2\beta_2 = +1.0$. It is now evident that this condition, and the conclusions drawn from it, can exist only at a single frequency in an a.c. amplifier, unless $\beta_2 = \beta_2(j\omega) = 1/G_2(j\omega)$ for more than one frequency.

The shape of the response curve for a typical amplifier with positive feedback adjusted so that $K=1.0$ is shown in *Fig.* 4. The response rises for $f < \sqrt{f_1 f_2}$ and falls for $f > \sqrt{f_1 f_2}$ at 6 db per octave except for a region at the geometric mean where the gain rises to infinity and the slope consequently increases.

It can be shown that if the amplifier transfer function has a different shape, e.g. so that the response at each end drops off 12 db per octave, the response with positive feedback remains unaffected except for regions near where the relative gain is 1.0. For this reason, it is not possible to alter the shape of the response significantly by changing the transfer function of the amplifier.

If another case is considered in which the amplifier is flat from zero to infinite frequency and the feedback factor drops off in the same manner as was assumed for the amplifier in the first example, the over-all response is unchanged except for regions above $1/T_2$ and below $1/T_1$. Once again, the important part of the curve remains unchanged.

These are valuable properties. By using them correctly, it is possible to design an amplifier with negative and positive feedback which shows greater stability and lower distortion than can possibly be attained with negative feedback alone. The logic behind the development of such an amplifier is presented in the next section.

General Amplifier Design

In a three-stage amplifier with negative feedback around the entire unit including the output transformer, oscillation is certain to occur at some frequency where the loop phase shift is 180 degrees if the loop gain at this frequency is equal to or greater than 1.0. The problem in the design of such an amplifier has always been to control the response at the extremes of frequency in such a manner that the loop gain is less than 1.0 when the phase shift is 180 degrees. This may be accomplished in several ways. One is to introduce phase-correcting networks. Another is to stagger the time constants of the stages so that two of them are flat out to extreme frequencies and the third has a response that drops off gradually (6 db per octave) toward low and high frequencies. In this manner, the phase shift can be kept at 90 degrees until the loop gain is less than 1.0.

The first method suffers from the fact that the response of an audio amplifier is apt to depend upon the nature of the load, so a fixed phase-correction network may work satisfactorily for one load and fail to prevent oscillation for another. All loudspeakers are not identical. The second method, however, can be used to produce an amplifier completely stable with any previously assigned range of loads. Positive feedback is an ideal way to apply the second method to amplifier design.

The steps in the practical design of a completely stable audio amplifier with

Fig. 5. Rear view of amplifier.

over-all negative and internal positive feedback are:

1. An amplifier with at least three stages (input, driver, and output) is constructed. The over-all negative feedback loop is closed, and the loop gain is adjusted (by controlling the gain of one of the stages) until the amplifier shows no trace of instability when the worst desired load is connected to the output. This load will generally consist of a resistor of the correct size shunted with a capacitor as large as is likely to be encountered in the use of the amplifier.

In stabilizing this amplifier, a small amount of network phase-correction may be used, but the amplifier should not oscillate even in its absence. Stability may be considered adequate when the high-frequency peak is no higher than 2 db.

2. The positive-feedback loop is now closed, and the time constants of the feedback network are controlled so that the high-frequency peak remains no higher than 2 db. (Remarks relative to high-frequency peaks also apply to low-frequency peaks). Under these conditions, the amplifier is necessarily no less stable than it was with the negative feedback alone.

When the loop gain is adjusted to prevent instability in Step 1, it will generally be found that the final loop gain is quite small. In fact, if the maximum treble drop of the open-loop amplifier is 18 db per octave, it may be shown that the loop gain can generally not exceed 1.8 before the high-frequency peak exceeds 2 db. A completely-stable feedback amplifier with more than two stages must often have a quite-low *loop* gain. This places no restriction upon the gain of the amplifier, for if G is the gain of the open-loop amplifier and β is the feedback factor, then the loop gain is $G\beta$ while the gain is

$$A = \frac{G}{1 - G\beta}$$

If $G\beta$ is fixed, therefore, any value of A may still be obtained by adjusting G. For example, if the loop gain is set at 1.8, the over-all gain will be $0.36G$. If the over-all gain is to be 25, G must be set at around 70. With such a completely stable negative-feedback amplifier, the distortion introduced in the various stages will obviously be reduced only a small amount (in the example, by a factor of 2.8). By the converse of this reasoning, the conclusion is reached that negative-feedback amplifiers having high loop gains and in which distortion is brought to a low value tend to be unstable.

That the introduction of positive feedback need now cause no additional instability is evident from the typical response curve of *Fig.* 4. A 6 db per octave slope in response corresponds to a phase shift of 90 degrees, which is insufficient to cause oscillation. To control the positive feedback so as to prevent the entire amplifier from being unstable therefore simply means that the relative gain of the positive-feedback loop must be 1.0 at the point where the response of the negative-feedback amplifier is just beginning to rise because of phase shift.

The work that has so far been presented indicates that the reason for the instability difficulty with earlier amplifier circuits lay in the failure to control the high-frequency cutoff point for the positive-feedback network. This difficulty might have been remedied by various expedients, but the problem of inverter distortion would still have remained.

Practical Amplifier Design

There are many possible designs for amplifiers incorporating both negative and positive feedback. Before proceeding with a circuit, the designer must lay down ground rules which, according to his own experience and study, lead to what he believes is good performance.

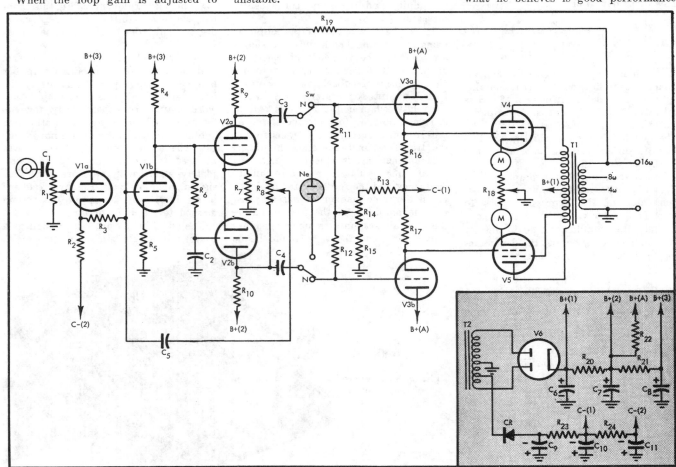

Fig. 6. 50-watt power amplifier.

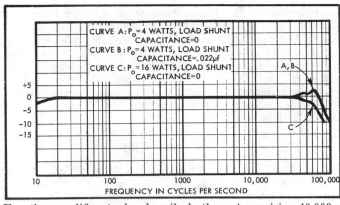

Fig. 7. Response of amplifier.

For the amplifier to be described, the ground rules were:

1. The output stage should be push pull.
2. The output tubes should be driven by cathode followers.
3. The output transformer secondary should be operated single ended.
4. Feedback should be achieved entirely by means of a resistive network.
5. Performance of the amplifier should not be affected by reasonable variations in load or in signal-source impedance.

With the exception of Rule 4, reasons for most of these rules are obvious enough. The reason for Rule 4 is that feedback amplifiers are quite sensitive to distortion introduced in the feedback connection. Where feedback is returned from the output transformer to the cathode of the input tube, nonlinearity in the grid-cathode voltage appears as distortion in the amplifier output. It was for this reason that the present amplifier combines the feedback with the input signal in a resistive network.

Using these criteria, the writer constructed the amplifier shown in *Fig. 5* according to the circuit diagram of *Fig. 6*. The output stage consists of push-pull fixed-bias 6CA7's in a distributed-load circuit with a Triad HSM-189 25-watt output transformer. The grids of the 6CA7's are directly coupled to the cathodes of the 12AT7, and the fixed bias is applied to the 12AT7 grids. This connection eliminates the large coupling capacitors that would otherwise be required for the 6CA7's because of the low permissible grid resistances. It also reduces the loading on the phase inverter, permitting it to operate with minimum distortion.

The inverter, a long-tailed pair, is directly coupled to the preceding amplifier. Anode-follower feedback is brought to the grid of the amplifier from two sources: negative feedback from the output-transformer secondary through a 1-megohm precision resistor, and positive feedback through a .047-mfd capacitor from a 5-megohm control shunted across the phase-inverter plates. A precision 40,000-ohm resistor in series with the input grid of the amplifier completes the feedback network.

The effects of signal-source impedance variations on the feedback are eliminated by a cathode-follower input stage, which is outside both feedback loops. The cathode resistor is returned to a well-filtered negative voltage rather than to ground, which allows use of a large cathode resistor and minimizes the distortion of this stage.

The positive feedback is initially set by breaking the negative-feedback loop and adjusting the 5-megohm control until the amplifier-phase inverter combination begins to oscillate. A push-button switch connects a neon bulb from one plate of the inverter to the other, and simultaneously disconnects the output tubes to prevent damage to them and the loudspeaker during adjustment of the positive feedback.

The experimental circuit uses a 50-ohm control in the output-tube cathode circuits for cathode-current balancing. Balancing is facilitated by two 100-ma meters permanently inserted into the circuit. The fixed-bias control permits the 6CA7 cathode currents to be varied between about 20 to 60 milliamperes each.

With the exception of the positive feedback, no particular pains are taken to achieve high gain in any of the amplifier stages. On the contrary; the absence of a bypass capacitor across the single-ended amplifier cathode resistor insures a low gain.

The power supply is conventional, utilizing a 5V4 rectifier with minimum R-C filtering. The negative fixed-bias voltage is obtained from the 70-volt tap on the power transformer by a silicon rectifier and an R-C filter. It has been found necessary to filter the C- to minimize the hum in the amplifier output.

Amplifier Adjustment and Performance

When the positive-feedback control was centered (resulting in zero positive feedback) and the negative-feedback loop was closed, measurement of the response at low and high levels showed no evidence of a high-frequency peak. The push-button switch disconnecting the output stages and inserting the neon lamp into the circuit was depressed, and the positive-feedback control adjusted until the neon lamp indicated that the stages around which positive feedback was connected were oscillating. The push button was released, the output tubes balanced at 50 ma each, and the amplifier was placed in service for a month to insure that all stages were properly aged and stabilized. After this period, the amplifier was subjected to a series of tests to determine its quality.

Tube interchangeability

The first three stages of the amplifier are directly coupled. One might well be anxious about the effects of tube replacement upon the performance of these stages. The first 12AT7, which comprises the cathode follower and first amplifier, is in the most sensitive position. A series of seven randomly selected 12AT7's was tried in this posi-

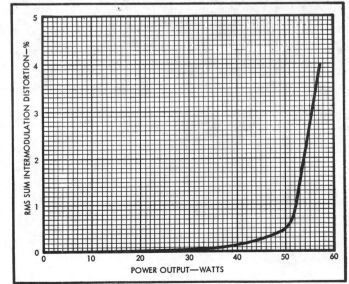

Fig. 8. Rms sum intermodulation distortion of amplifier (60 and 3000 cps, 4:1).

tion, and the inverter plate voltages were checked.

Variation in plate voltage was found to be less than ±15 volts. The initial positive-feedback setting was correct for five of the tubes; an extremely small adjustment sufficed for the remaining two. Evidently, it is within the capability of the amplifier to accommodate a wide range of 12AT7 tubes in the sensitive input position.

The same series of tubes was tried in the cathode-follower driver position. Only slight changes in output-tube currents resulted; they were easily corrected by adjusting the bias potentiometer.

It is worth noting that the amplifier operated satisfactorily with 6L6, 5881, and 350B tubes in the output sockets. The maximum power output was, however, diminished. A considerable adjustment of the bias control was required to achieve the recommended cathode currents with these tubes.

Frenquency response

The response of the amplifier into a 16-ohm resistive load was measured at 4 and 16 watts. The resulting curves are shown in *Fig.* 7. They show that the effect of connecting an 0.022-µf capacitor across the output terminals is negligible.

The drop in response at about 10 cps was deliberately introduced by the input capacitor (C_1 in *Fig.* 6). Without this capacitor, the amplifier tended to maintain a flat response down to d.c. Because the output transformer was not able to handle extremely low frequencies, the amplifier tended to overload severely. Prior to the introduction of this capacitor, objectionably large variations in output-tube plate currents resulted from minor eccentricities in 33⅓-rpm records.

Distortion

The rms sum intermodulation distortion (60 and 3000 cps, 4:1) was unmeasurable at low levels. It rose to one per cent at a power output of 52 rms watts, as shown in *Fig.* 8.

Evidently, the power output of the amplifier was limited by the output tubes rather than the output transformer. Because of the 25-watt rating of the output transformer, the frequency response of the amplifier is doubtless quite restricted at a 50-watt output.

Whether or not a wide frequency response is required at power levels which the amplifier will, in home use, be required to deliver only during unusual transients, is a question each individual must answer for himself. At any rate, the 50-watt frequency response can be improved, if desired, by the substitution of a larger output transformer for the one specified.

Hum and noise

The hum and noise at the speaker terminals amounted to 0.01 volt (= 6×10^{-6} watts) with the input shorted. An appreciable portion of this noise was contributed by the C− supply, which probably could have been better filtered. Even with the present circuit, however, hum is inaudible two feet from a highly-efficient speaker.

Output impedance

As is usual with amplifiers of this type, the output impedance is about zero ohms over the audible range. The infinite damping factor contributes to the cleanness of response by preventing hangover and undesired speaker-cone movements.

Sensitivity

An input signal of 1.1 volts drives the amplifier to 50 watts output, with the input gain control wide open.

Listening tests

Listening tests have been conducted, to date, by about a dozen critical listeners. Although the tests were separately conducted, most of the listeners used the word "transparent" to describe the reproduction. The amplifier must be heard to appreciate the relief with which the various instruments stand out in reproduced orchestral sound. Æ

PARTS LIST

R_1	250,000-ohm potentiometer	R_{16}, R_{17}	15,000 ohms, ½ watt	V_4, V_5	6CA7
R_2	22,000 ohms, ½ watt	R_{18}	50-ohm potentiometer	V_6	5V4
R_3	40,000 ohms, 5 watt, wire wound	R_{19}	1 megohm, 1 watt, deposited film	CR	diode, 200 PIV min., ½ amp (Sarkes-Tarzian F-2 or equiv.)
R_4	220,000 ohms, ½ watt	R_{20}, R_{21}, R_{22}	27,000 ohms, 2 watt	T_1	output transformer, 6600 ohms to voice coil, Triad HSM-186
R_5	6800 ohms, ½ watt	R_{23}	1000 ohms, 1 watt		
R_6	1 megohm, ½ watt	R_{24}	4700 ohms, ½ watt	T_2	power transformer, 700 v ct, 150 ma 70 v tap, Triad HSM-241
R_7	150,000 ohms, ½ watt	C_1	.047 µf, 400 v		
R_8	5-megohm potentiometer	C_2	0.33 µf, 400 v	M	0–100 ma meter
R_9, R_{10}	330,000 ohms, ½ watt	C_3, C_4	0.1 µf, 600 v	SW	d.p.d.t. switch, spring return
R_{11}, R_{12}	1 megohm, ½ watt	C_5	.047 µf, 400 v	NE	NE-10 neon bulb
R_{13}	12,000 ohms, ½ watt	C_6, C_7, C_8	40/30/30 µf, 475 wv		
R_{14}	25,000-ohm potentiometer	C_9, C_{10}, C_{11}	20 µf, 150 wv		
R_{15}	47,000 ohms, ½ watt	V_1, V_2, V_3	12AT7		

Feedback Techniques in Low-Level Amplifiers

DONALD L. SHIRER

Although more widely used than in previous years, feedback in low-level ampliers is still not used as widely as its' advantages would indicate. For those "on the brink," here's as persuasive argument for—**plus a large helping of** how.

IT IS CURIOUS in this day when the principles of negative feedback are almost universally used in audio power amplifiers, to find that there seems to be some reluctance to apply them to the lower-level stages of the amplifier chain as well. Perhaps there may be some justification for this, since one of the principal uses of negative feedback is to reduce distortion, and low-level stages do not distort the signal as much as amplifiers which must handle large signal voltages, but this is only part of the story. By judicious choice of the feedback paths and components, not only can distortion in the output signal be reduced, but the frequency response can be controlled, the effective input and output impedance of the amplifier may be changed and *at the same time* the circuit may be adapted to other functions such as mixing or tone control. The price that must be paid for this flexibility is a reduction in amplifier gain, which sometimes (but not always! ... see the discussion on equalizers) necessitates an additional amplifying stage at a slight increase in cost and complexity. However, this is usually far less significant than the advantages gained through the use of feedback.

Only so-called negative feedback will be discussed here, and its application to units which are primarily low-level current or voltage amplifiers. It is assumed that these units either themselves accept an audio-frequency signal from a microphone, pickup, or other transducer and deliver an amplified (and perhaps modified) signal at a higher level that can be fed to a power amplifier, or else that these separate units may be combined into a more complex preamplifier which performs this function. We thus regretfully ignore here the many interesting applications of feedback to oscillators, relaxation circuits, and to amplifiers designed to handle extremes of frequency at either end of the audio range, which are best treated in a more rigorous manner. At audio frequencies, we can gladly simplify our calculations to neglect such complications as circuit and tube capacitances, and treat transistors and vacuum tubes as "perfect" amplifiers having a gain independent of frequency. Although the consequences of such a sweeping assumption would usually be drastic in power amplifier design, I have rarely encountered stability troubles from this particular simplification in preamplifiers, and even if they arise, they may generally be eliminated without radical redesign of the circuit.

Basic Feedback Circuits

Two feedback circuits are shown in *Fig. 1*, in which the triangular block represents a "perfect" voltage amplifier whose voltage output is A times the input signal voltage, v_i, over the entire audio frequency range. This amplifier may be a single vacuum tube, several R-C coupled tube stages or even a "black box" exhibiting the specified raw gain A. In (A) of *Fig. 1*, a certain fraction β of the output voltage, v_o, is fed back to the input of the amplifier through the voltage divider indicated by the rectangular box which is assumed to have an impedance much greater than the load resistor, R_L, so as not to overload the amplifier. The feedback voltage, v_f, is then inserted in such a manner that $v_o = [A(v_i + v_f)]$ is less than the output (Av_i), which would occur if the feedback path were broken. It is clear that you should not connect the feedback voltage so that it *increases* the output, or the increased output would cause a larger feedback voltage, which in turn causes a still larger output, which causes a bigger ..., etc., and you soon have a dandy oscillator instead of an amplifier. With the proper connections, though, the amplifier soon settles down at its lower output voltage. Now $v_f = \beta v_o = \beta A v_i + \beta A v_f$, so evidently the condition that v_f oppose v_i (that is, be out of phase with v_i) requires that βA, the gain "around the feedback loop" be negative—thus "negative" feedback. If the voltage divider, β, consists of passive elements (resistors, capacitors, inductors, etc.) only, inversion of the signal cannot take place there, so either A must be negative, which is the same as saying that the amplifier contains an odd number of phase-inverting stages, or

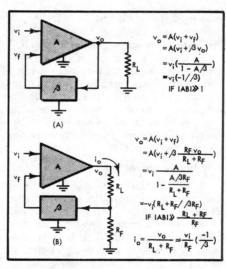

Fig. 1. Voltage amplifier.

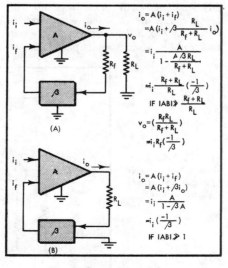

Fig. 2. Current amplifier.

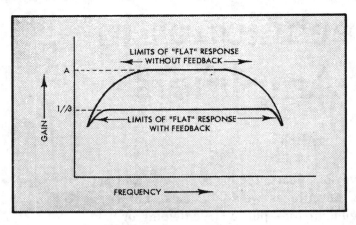

Fig. 3. Extension of frequency response with negafeedback.

else if an even number of stages are used, the feedback signal must be inserted at a point where it will produce the effect of an input signal of opposite sign, usually at the cathode of the first stage. (For this criterion to be applied, remember that cathode followers and grounded-grid amplifiers do not invert the signal.) Examples of both types of amplifiers will be given later. This connection is known as parallel, shunt, or voltage feedback.

Series or current feedback is shown in (B) of *Fig.* 1. Here the output current passes through a resistor R_f which converts it to a voltage signal (usually $R_f \ll R_L$), then through the voltage divider, β, to a point in the amplifier which will provide the proper negative feedback polarity. In either case, if the amplifier gain is sufficiently high, it drops out of the equation for the total gain when feedback is included, as does any mention of the load resistance, so that we have "stabilized" the gain against changes in vacuum tube μ or changes in the load resistance. The factor β can be rather small (if A is large enough) without violating the simplifying condition $A\beta \gg 1$, so that the gain, including feedback, which depends on 1/β, may be quite large. The difference between the two cases appears in the equations below each figure. It is evident that for (A) of *Fig.* 1 it is the output *voltage* which is a stabilized function of the input voltage. Voltage feedback has thus made it appear as though this amplifier has a very small effective output impedance, so that R_L may be varied without affecting v_o as long as it is not so small that greater output currents are demanded than the output stage in the amplifier can supply. On the other hand, the current feedback in (B) of *Fig.* 1 stabilizes the output current, so that now the amplifier behaves as though it had a very large output impedance. The change in output impedance in each case is in about the same ratio as the reduction in gain caused by the application of feedback. (See *Appendix* I.)

The same two feedback configurations can be applied to amplifiers incorporating transistors, or any "black box" in which the current output i_o is A times the current input i_i. In (B) of *Fig.* 2, a certain fraction, β, of the output current is extracted by means of the low-impedance current divider indicated by the rectangular box, and fed back to the amplifier in such a manner as to oppose the input current.

In (A) of *Fig.* 2, the output voltage is converted to a current by placing R_f between the amplifier output and the low-impedance input of the current divider, then a fraction, β, of this current is fed back to the input. Again voltage feedback stabilizes the output *voltage* ((A) of *Fig.* 2) and lowers the output impedance of the amplifier; current feedback stabilizes the output *current* and increases the effective output impedance as in (B) of *Fig.* 2.

Stability Considerations

If the divider network, β, contains only resistive elements, the adjusted gain of the amplifier including feedback will be flat over a range greater than that of the amplifier itself. This is so because even though A is decreasing at the extremes of the frequency range, the adjusted gain will not seriously decrease until A becomes comparable to 1/β. The adjusted gain will then drop off with further fall in A itself somewhat as shown in *Fig.* 3. Phase shifts in the output voltage always accompany this decrease in A at high and low frequencies, but should not contribute to instability in properly-designed amplifiers. Proper design in this case usually amounts to little more than making only *one* R-C coupling element contribute to the amplifier droop at low frequencies and *one* other network cause the high-frequency rolloff.

In the typical flat amplifier shown in *Fig.* 4 for instance, the time constant R_3C_3 is made, say, five times less than R_1C_1, R_4C_4, or R_5C_5 so that the first coupling network contributes most of the low-frequency droop in the amplifier response. The high-frequency loss is limited here by all the tube and circuit capacitances. To ensure its proper "ruin" by only one R-C network, a small capacitor, C_2 may be tied across R_2 to start drooping the amplifier's gain above 20,000 cps or so. These simple expedients will generally suffice in the great majority of preamp circuits.

Adding capacitances or inductances to the current or voltage divider, β, will enable you to produce an amplifier having a frequency-varying gain which is dependent only on the feedback network. Several voltage dividers are shown in *Fig.* 5, accompanied by sketches of the behavior of their division ratio, β, and the total amplifier gain, 1/β, as a function of frequency. Remember that in order for the adjusted gain to be given by 1/β, A must be several times the highest value of 1/β over the desired frequency range (if you measure gains in db, A must be approximately 6 db more than the maximum value of 1/β for the approximation to hold). Since the impedance of the divider circuit will probably also change with frequency, be sure that your voltage dividers have high impedances and the current dividers low impedances compared to the impedance level of their associated circuits. A single R-C network will give slopes up to 6 db/octave on the gain curve, and the responses can be easily sketched by joining the flat portions and the 6 db/octave slopes with smooth curves near the crossover frequencies.

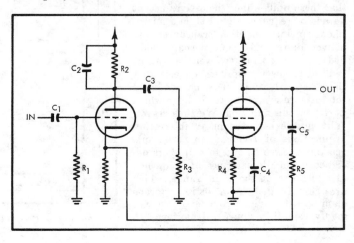

Fig. 4. R-C coupling networks in typical feedback amplifier.

Networks of this type can obviously be combined to form quite complicated frequency response curves which are useful in equalizers and tone-control networks. Inductances are not usually used in feedback networks because of their size, weight, cost, and tendency to pick up hum if not well shielded, but R-L-C circuits can give much sharper slopes than a single R-C network. For instance, the feedback divider shown in (D) of *Fig.* 5 was found to be useful in designing an equalizer for a tape recording amplifier needing considerable treble boost around 15,000 cps. Several other types of feedback networks are considered later.

So far, nothing has been said about distortion reduction, though it is fairly easy to show (see *Appendix* II) that the distortion produced by the amplifier without feedback is reduced by the same factor $(A\beta - 1)$ as the gain when negative feedback is applied. It is true that some additional higher-order harmonic distortion may arise because the feedback allows the initial distortion products to pass back through the amplifier, but the usual low-level amplifier has only a small degree of distortion to begin with, until overloaded, so that the higher-order distortion products are nearly always negligible.

"Flat" Amplifiers

Dynamic microphones and many types of vibration transducers produce signals in normal use of about 1 mv or smaller, and demand "flat" amplification up to a level of at least 0.2 volts before the signal can drive most power amplifiers to full output. While part of this amplification may be provided later, generally it is wise not to insert level or tone controls until the signal level is brought up to 0.1 v or so, to avoid hum and noise pickup in an excessive number of low-level stages. It is usually worthwhile to expend considerable effort to avoid noise in the input stages so that the amplifier may handle as wide a dynamic range as possible.

A transistor circuit which has shown considerable adaptability as a low-level amplifier for medium-to-low impedance inputs is shown in *Fig.* 6. Without the feedback path (F-F in the figure) it has

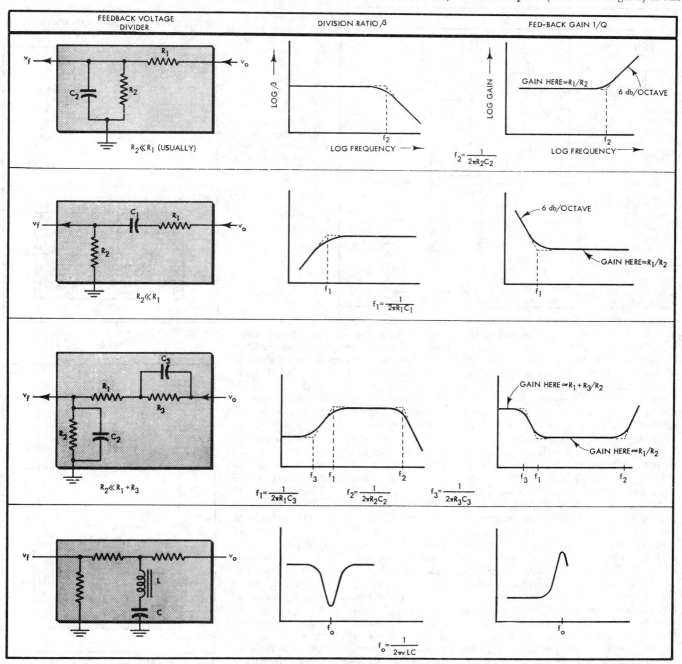

Fig. 5. Gain curves produced by sample feedback dividers.

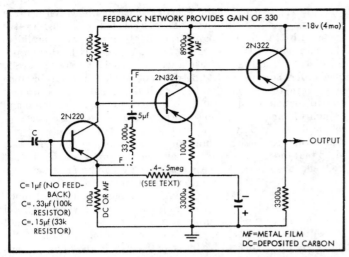

Fig. 6. Low-level transistor preamplifier.

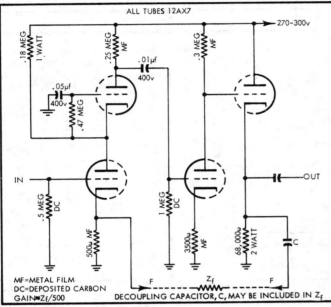

Fig. 7. Low-level vacuum-tube preamplifier.

a gain of 69 db. About 6 db of this is used for local feedback, by means of the unbypassed emitter resistors in each stage, which acts to improve the stability of the bias levels as well as providing some distortion reduction. D.c. coupling after the input capacitor leaves only the R-C time constant in the second emitter circuit to provide the low-frequency cutoff, but also demands an inner d.c. feedback path back to the first base to stabilize the bias potentials on the two grounded-emitter stages. The inclusion of three local feedback paths in addition to the over-all feedback loop makes this circuit rather insensitive to different transistors, although for optimum results the first base bias resistor (marked 0.4–0.5 megohms) should be adjusted to provide exactly 0.58 ma emitter current in the input transistor. The operating points are chosen for low transistor noise, although this will also depend on the impedance level of the input signal. An input impedance of 1000 ohms will produce the best signal-to-noise ratio in the 2N220 stage, but a 250-ohm microphone can be directly connected to this stage with a loss of only 2.5 db in signal-to-noise ratio. This will entail no loss in gain since the effective impedance of the first base will be raised by the over-all feedback up to a value in the range 10–50,000 ohms. This amplifier can provide an output for almost four volts into a 10,000-ohm load without clipping, giving a great overload reserve and eliminating the need for a volume control at the preamp input. The frequency response is flat from below 15 cps (which is "all the further" my oscillator goes) to 15,000 cps, where it is 3-db down without the over-all feedback path—supplying as little as 6 db or more of over-all feedback produces flat response to beyond 20,000 cps. The emitter follower is not strictly needed for most purposes, but serves to keep the output impedance low over the entire frequency range if the preamp is used as an equalizer, and can drive output circuit impedances down to about 3000 ohms without too much restriction on the output level. Low-noise metal-film and deposited-carbon resistors are used in strategic locations, and the power supply was designed to be taken either from two 9-volt batteries in series, or dropped through a decoupling network from a higher-voltage supply.

Connecting a 33,000-ohm resistor between the points F-F effectively converts this amplifier into the circuit of (A) in *Fig. 2*. The divider consisting of the 33,000-ohm and 100-ohm resistors provides the negative voltage feedback which is properly applied to the first stage emitter since there are two (an even number) phase-inverting stages in the feedback loop. Neglecting the shunting effect of the first transistor's base resistance, the dividing ratio, β, is approximately 100/33,100, giving an adjusted gain ($1/\beta$) of *330*, or about 50 db. The gain reduction is thus 69-50, or 19 db, a factor of 90 times, which serves

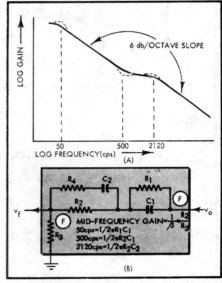

Fig. 8. RIAA playback curve with required network.

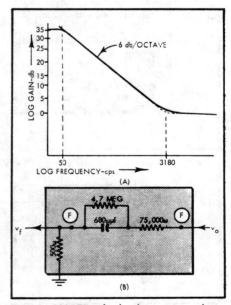

Fig. 9. NARTB playback curve and required feedback network.

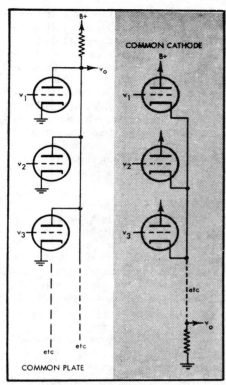

Fig. 10. Summing chains.

to reduce the output impedance and harmonic distortion figures by the same ratio. Many people will refer to this as "19 db of feedback"—a convenient shorthand phrase. In this amplifier it makes the harmonic distortion unmeasureable if the output is below 1 volt. Exceptionally low-level microphones may demand a gain of 60 db, which may be obtained by a division ratio of 1/1000, that is, by replacing the 33,000-ohm resistor with one of 100,000 ohms. There is still 9 db available for distortion reduction, although now the input impedance has come down to about 6000 ohms. This variation should not be used with high-impedance microphones without a step-down transformer.

A vacuum tube preamp which may be used with crystal or other high-impedance microphones is shown in *Fig. 7*. A double triode is connected in the grounded-cathode grounded-grid cascode arrangement to provide a low noise level and at the same time a considerable gain in the first stage. One unusual feature is the 120,000-ohm bypass resistor to the plate of the grounded-cathode half of the cascode tube. This increases the current drawn by this section, raising its g_m and giving an extra 5 db of gain over an ordinary cascode circuit. The second tube is R-C coupled with a cascode follower direct-coupled to its plate to provide a low impedance output. The slightly higher raw gain of this circuit (84 db) permits a higher effective gain than the amplifier of *Fig.* 6 with the same amount of distortion-reducing feedback. For instance, a 0.5 meg feedback resistor will provide a 60-db gain (the division ratio is 500/500,000) with a comfortable 24 db of feedback reserve. The feedback voltage is correctly applied to the cathode of the cascode stage since there are only two phase-inverting stages —the grounded-grid and cathode-follower tubes producing an output in phase with the input signal. To realize the low-noise possibilities of this circuit, it should be well decoupled from the power supply and d.c. should be used on the filaments.

Equalization

The previous amplifiers can easily be converted to phono preamps by adding frequency-sensitive elements to the feedback path to make the gain follow the RIAA playback curve, shown in (A) of *Fig.* 8. This has a 6 db/octave rise in the bass region with a turnover frequency of 500 cps, a treble turnover at 2120 cps followed by a 6 db/octave droop, and a bass plateau at 50 cps.

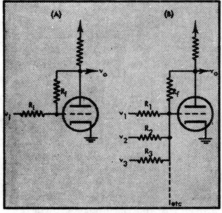

Fig. 12 Mixing amplifiers.

The divider network shown in (B) of *Fig.* 8 will produce this characteristic in a circuit with voltage-feedback. The elements between F-F would be inserted in place of the simple feedback resistor in *Figs.* 6 and 7. C_1 and R_2 produce the bass boost below 500 cps, C_2 shunts R_2 at high frequencies to produce the treble rolloff, and R_1 limits the rise of the impedance in series with R_2 to produce the shelf below 50 cps. R_4 is a safety factor to reduce excessive phase shift and possible oscillations at very high frequencies. If it is chosen about $R_2/20$, then it only affects the treble response above 40,000 cps, well out of the audio range. Values for these components which seem to work well in the transistor amplifier of *Fig.* 6 where R_3 is the 100-ohm emitter resistor, are, for a mid-frequency gain of 40 db:

$R_2 = 10,000$ ohms
$R_1 = 0.1$ megohms
$R_4 = 500$ ohms
$C_1 = 0.0032$ µf
$C_2 = 750$ µµf

It is easy to buy one per cent, deposited carbon, low-noise resistors in these values, but capacitors may be as much as 20 per cent off their nominal ratings. To produce accurate equalization then, you must buy several capacitors of the nearest EIA tolerance and select the ones reading closest to the desired value on a capacitance bridge.

The specified amount of gain (40 db at 1000 cps) will only bring extremely low-level signals from the stereo cartridge up to about a 0.2 volt level, but additional gain may be provided in the following stages. The gain at low frequencies is approximately 60 db, since the rise in impedance of C_2 makes the effective feedback ratio 100/100,000, but the amplifier still has about 9 db of reserve which is useful in reducing distortion. Do not worry if you remember that the input impedance is not as great with less feedback; magnetic pickups have inductive reactances and so their impedance only becomes large at high frequencies where the feedback is much greater, thus providing the droop in the RIAA curve and at the same time increasing the effective input impedance.

You can obtain more gain with the circuit of *Fig.* 7 if you are willing to put up with a slightly worse signal-to-noise ratio. If we pick

$R_2 = 0.15$ megohms
$R_1 = 1.5$ megohms
$R_4 = 7500$ ohms
$C_1 = 0.0021$ µf
$C_2 = 500$ µµf

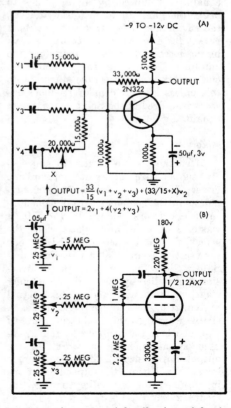

Fig. 11. Plate-to-grid feedback and feedback summing circuits.

(R_g is, of course, the 500-ohm cathode resistor), the mid-frequency gain will be about 50 db. Even at low frequencies where the actual gain is nearer 70 db, there will still be 14 db of feedback reserve, and at mid-frequencies, 24 db or more of feedback is available to reduce distortion.

NARTB equalization for tape playback heads has a single bass turnover at 3180 cps for the 15-ips speed, with a shelf appearing below 50 cps, and the same curve is used at 7.5 ips by general adoption. A feedback network such as is shown in *Fig.* 9 will provide the equalization for the preamp of *Fig.* 7, with a high-frequency gain of 44 db ($\beta = 500/75,000$). Although the extreme bass boost demands an extra 35 db of gain at 50 cps, this amplifier still has 5 db of reserve gain at this point, with much more feedback at higher frequencies. If additional distortion-reducing feedback is desired at the low frequencies, some gain must be sacrificed for this privilege.

Other equalizers may be similarly designed, choosing the feedback network arrangement to give the proper shape to the response curve. The amplifier must have a gain without feedback equal to, or greater than, the desired gain at the highest point on the gain curve, plus the minimum amount of distortion-reducing feedback desired at this point. There will be a greater amount of feedback at all other frequencies as a result. While in most cases, the same response curve may be obtained with losser-type equalizers followed by an amplifier to bring the gain up to the desired level, the advantage of the feedback equalizer is that the gain which is "thrown away" in the losser pad is used to reduce the amplifier distortion in the feedback circuit. It takes no more tube stages to do this unless a great deal of feedback reserve is desired at the point of maximum equalized gain.

Mixers

Often the occasion arises to add several signals together, as in a microphone mixer. The use of a triode for each signal, as in the common-plate or common-cathode networks shown in *Fig.* 10, is wasteful of tubes and space, when the same job can be accomplished by one tube section connected as a feedback amplifier. Since the necessary phase inversion is supplied by the single amplifying stage, we must insert the feedback signal at the grid, instead of the cathode. A simple way to do this is shown in (A) of *Fig.* 11 where the resistors R_f and R_i are used as a summing network. If the output impedance of the tube is much lower than R_f, and the impedance of the input generator is smaller than R_i, the feedback ratio is just $R_i/(R_i+R_f)$ so that the gain from grid to output = $1/\beta = (R_i+R_f)/R_i$. But the input signal is reduced at the time it reaches the grid by a factor $R_f/(R_i+R_f)$ by the dividing network acting backwards as it were, so that the total gain from input to output is

$$\frac{R_f}{R_i+R_f} \cdot \frac{R_i+R_f}{R_i} = \frac{R_f}{R_i}.$$

Another way of looking at it is that since the stage gain is quite high, for any reasonable output, the voltage at the grid will be very small, essentially zero. Then the current flowing into this virtually grounded summing junction (v_i/R_i) must be equal to the current flowing out ($-v_o/R_f$) by Kirchoff's Law, so that the gain $=-v_o/v_i=R_f/R_i$ again.

Increasing the number of inputs will not affect this result. In spite of the fact that the additional input resistors shunt some of the input signal to ground, the feedback ratio is also reduced, and by the same amount, so that the total gain from, say, input point no. 2 in (B) of Fig. 11 to the output is still R_f/R_2. A limit on adding inputs comes as usual when the feedback factor is so small that $A\beta \approx 1$. But until that point is reached, the summing junction acts like a virtual ground, there is no interaction among the signal sources, the gain in each channel may be adjusted individually, and the load impedance presented to the generator feeding channel N is just R_N.

Figure 12 shows two amplifiers utilizing this method of mixing circuits. The transistor amplifier in (A) of *Fig.* 12 mixes three inputs of equal current level, amplifies each by a factor of about 2, provides a fourth input of adjustable current gain from 1 to 2 and still has over 15 db of distortion-reducing feedback. The vacuum tube feedback mixer in (B) of *Fig.* 12 shows a volume control added to each channel, with channel 1 amplified by a factor of 2 and the other two by a factor of 4. The grid and base resistors used as bias returns do not alter the design significantly. Note that the feedback resistor is used as part of the base bias network in (A) of *Fig.* 12 to reduce the number of circuit elements at the expense of some additional current drain, but that a blocking capacitor must be used to prevent the vacuum tube plate voltage from affecting the grid bias.

It is entirely possible to use frequency-sensitive networks for the input and feedback impedances and combine the job of mixing and equalization in one stage. The same method of feedback may be used with a three-stage amplifier providing mixing, equalization, higher gain and the possibility of increased distortion reduction in one package. Many times, though, if R-C coupling is used between all the stages in a high-gain three-stage amplifier, instability and even oscillation may result the first time the feedback loop is connected. If some of the stages may be d.c.-coupled without causing bias instabilities, there will be less possibility of inadvertently designing a feedback oscillator. Seldom are more than three low-level stages needed in a preamp, and if they are, it is better to provide a combination of several one-, two-, or three-stage feedback amplifiers than to attempt over-all feedback from the output to the input, unless, of course, you are either an expert or lucky.

Tone Controls

The configuration just discussed,[1] the odd number of amplifying stages, voltage feedback, and the combination of the feedback and input signals at a virtually-grounded summing junction, is characteristic of the "operational" amplifiers having ubiquitous application in electronic analog computers. A more prosaic use in audio amplifiers is as a tone-control stage in preamps. *Figure* 13 shows four R-C networks which will produce high and low cut and boost characteristics. One possible adaptation of this circuit would be to fix the input and feedback resistors for the desired mid-frequency gain and to select the capacitors by a multipoint switch to give the proper response and turnover frequency. This produces the constant-slope, variable-turnover characteristic many experts think desirable for music tone-controls; a different arrangement is necessary for the older fixed-turnover, variable-slope controls.[2] Some ingenious feedback arrangements, notably the Baxendall-type illustrated in *Fig.* 14, provide an adjustable lift-cut characteristic by adjusting one separate potentiometer for the treble and bass frequencies, at the cost of some interaction between the functions, which is usually not noticeable in practice. This configuration can also be applied to a two-stage amplifier.[3]

Potpourri

Before concluding this article, I would like to point out several applications of the feedback techniques which may be a little beyond the well-travelled path of most audio experimenters.

In *Fig.* 15, the symmetrical conduction characteristic of back-to-back Zener diodes is used to provide an amplifier which limits sharply at an output level

---

[1] Sometimes called an anode follower. See Charles P. Boegli, "The Anode Follower," AUDIO, Dec. 1960, p. 19.

[2] Barhydt, "A Feedback Tone Control Circuit," AUDIO, Aug. 1956, p. 18.

[3] Dynaco Preamp—Described in *Audiocraft*, March 1958, p. 16—See "Errata," *Audiocraft* July 1958, p. 33.

equal to the Zener breakdown voltage of the silicon diodes. These can now be obtained with Zener voltages ranging from about 4 to 200 volts. Feedback limiting requires no special 4-grid tubes or large output power capabilities and the gain below the clipping level is practically unaffected.

An operational amplifier with a twin-T network as an input "impedance" is shown in *Fig.* 16. The sharp notch may be useful as a rumble filter or harmonic distortion analyzer in some cases. The notch may be sharpened by including the twin-T network *inside* the amplifier feedback path, or if the network is used as a feedback "resistor," the notch becomes a peak and we have a frequency-selective amplifier.

Thermistors, strain gages or other resistive transducers may be inserted into the feedback path of a high-gain amplifier instead of the input, to provide an output which is the inverse of their normal response, which may be an advantage in some cases, and to eliminate their elaborate bridge-bias networks.

At audio frequencies, the application of negative feedback techniques can be generalized to a few simple practices. Its benefits—reduction of distortion to almost unmeasurable amounts, extension of frequency response, hand-tailoring of response curves, multiple use of stages —so far offset the additional gain it requires, that there is no reason why it should not be an integral part of the design of every low-level amplifier. A careful follow-through of the small amount of arithmetic used in this article, particularly in the examples, will provide most of the mathematical sophistication (not much!) needed for a reasonable grasp of the feedback principle. Æ

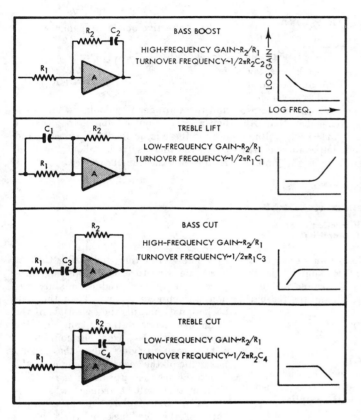

Fig. 13. Feedback networks suitable for tone controls when used with high-gain inverting amplifier.

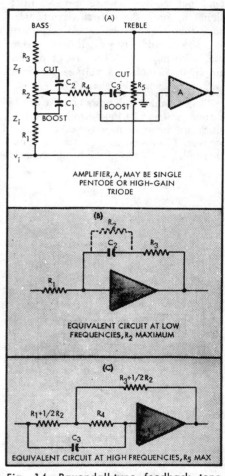

Fig. 14. Baxendall-type feedback tone control.

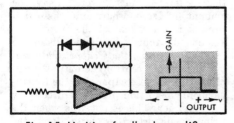

Fig. 15. Limiting feedback amplifier.

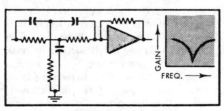

Fig. 16. "Notched" amplifier.

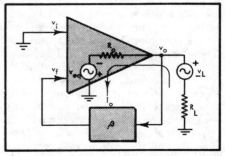

Fig. 17. Equivalent circuit.

Appendix I
Output Impedance of a Feedback Amplifier

The output of an amplifier may be thought of as coming from a zero-impedance voltage amplifier in series with the resistance R_o, the output resistance, if no feedback is used. Since the load resistor R_L will drop the output voltage by a factor of $R_L/(R_L+R_o)$ from the output supplied by the equivalent generator v_{eq}, the generator output must be $v_o(R_L+R_o)/R_L$, and as defined previously, $v_o = A(v_i + v_f)$. Suppose we temporarily short the input to the amplifier to eliminate extra distractions and pretend that the load tries to impose a little signal v_L onto the output circuit. Perhaps this might represent noise pickup or a signal induced by feedthrough from the following stage. The circuit of our amplifier, including a voltage feedback loop, now looks like *Fig.* 17.

The current which flows because of the action of v_L is just the output current i_o, since we have eliminated other signal sources. It passes through the combination of $R_L + R_o$ and the internal generator, which inserts a voltage $v_{eq} = (v_i + v_f) A(R_L + R_o)/R_L$ into series "aiding" with v_L. Even though v_i is zero, a voltage $v_f = \beta v_o$ is fed back to the amplifier input in proper phase to produce the polarities shown. The current must then be bigger than that which would normally flow, for now by Ohm's Law:

$$i_o = \frac{v_o - v_o A\beta (R_L + R_o)/R_L}{R_o}$$

and the effective output impedance is

$$\frac{v_o}{i_o} = \frac{R_o}{1 - A\beta(R_L+R_o)/R_L}$$
$$\approx \frac{1}{(-A\beta)} \frac{R_L R_o}{(R_L + R_o)}$$

if $|A\beta|$ is much larger than 1, or a factor of $1/(-A\beta)$ smaller than the parallel combination of R_o and R_L, which is what we would expect to find perhaps, without feedback. (Remember that $A\beta$ must be negative.) Thus in a voltage amplifier the output impedance is reduced by the same factor as the gain.

We may use the same method to analyze a current amplifier to avoid the use of a more unfamiliar equivalent circuit if we change the value of the equivalent generator slightly and take a feedback *current* βi_o back to the input. Since

$$i_o = v_{eq}/(R_o + R_L) = A(i_f + i_i),$$
$$v_{eq} = (i_f + i_i) A(R_o + R_L) = \beta A(R_o + R_L) i_o$$

if the input is shorted. Then by Ohm's Law,

$$v_o = i_o R_o - \beta A(R_o + R_L) i_o$$

and the effective output impedance is:

$$\frac{v_o}{i_o} = R_o - \beta A(R_o + R_L) = R_o \cdot [1 + (-A\beta)(R_o + R_L)/R_o]$$

and the amplifier now looks as though it had a *greater* output impedance than it does if there is no feedback present, and again the increase is by about the same factor that the gain is reduced.

Appendix II
Distortion Reduction in a Feedback Amplifier

No real amplifier produces an output which is an exact multiple of its input; let us suppose that for a certain pure sine-wave input signal, S, producing an output voltage (or current) $S \cdot A/(1-A\beta)$, there also appears on the output a small distorting signal, D. In most cases the amount of distortion is primarily dependent on the level of the signal at some point inside the amplifier, being caused by curvature in tube characteristics, etc., so that the larger the signal, the greater the distortion.

Perhaps the gain of the amplifier actually changes at different signal levels, providing a limiting (or expanding) effect. This would correspond to a distortion signal of the same frequency as the sine wave but with varying phase. It is also possible that the pure sine-wave appearance of the output signal will be altered, which would be caused by distortion components of the second, third, or higher-order harmonic frequencies. The latter case is reported and discussed most often, but the gain-changing effect of a fundamental frequency-distorting signal is present in every amplifier to some extent (and can be dramatically seen when the amplifier overloads!).

In either circumstance, when a fraction, β_{v_o}, of the output voltage is fed back to the input, it contains a part of the output distortion signal, β_D. This is amplified and appears at the output as $A\beta_D$, so that the distortion signal D actually appearing at the output is the combination of the distortion produced inside the amplifier, D', and the fed-back voltage, so:

$$D = D' + A\beta_D$$

and hence

$$D = D'/(1 - A\beta).$$

Thus the distortion signal measured at the output is reduced by the same factor as the gain through the application of feedback.

The main part of the distortion produced by the amplifier is cancelled by feeding a portion of the distortion signal back to the input and amplifying it out-of-phase with the originally distorted wave, but a small amount remains. Unfortunately, since the fed-back signal looks just like an input signal to the amplifier, it will be distorted too as it is amplified. Thus there will be an additional bit of residual distortion because the distorted correcting signal cannot eliminate all the original departure from a sine-wave output. If the distortion is small in the first place, these additional terms will be much smaller, but they have an effect disproportionate to their size.

No longer will the output look like the signal as the amplifier originally distorts it, but it will not be quite identical to a sine-wave either, indicating that its harmonic structure has changed. Suppose the amplifier produces mainly second-harmonic distortion. Then the feedback signal contains the essentials of the difference between the output and the input, that is, the second-harmonic term. When re-amplifying this, the second harmonic of the second harmonic, or the fourth harmonic, appears in the output. This in turn will be reduced when it is returned out-of-phase to the input through the feedback network, *but it will not be entirely eliminated,* and a further very small eighth-harmonic signal is generated.

Human ears are much more sensitive to *equal amounts* of fourth harmonic distortion than to second, so it is fortunate that its level is much lower. If, say, the amount of second-harmonic distortion present is 1 per cent of the output, the amount of fourth harmonic distortion *created by adding the feedback path* will only be .01 per cent of the output—usually insignificant compared to the fourth-harmonic distortion produced directly by the amplifier. This is not true when the primary distortion is large, and the sudden increase in high-order distortion products near the overload point causes a feedback amplifier to actually sound worse than an amplifier without feedback operating in this region.

Æ

Universal Feedback Amplifier Circuit

ARNOLD J. KAUDER

A simple amplifier of exceptional performance which should be adequate for practically any installation is the basis for this article, but its greatest value lies in the "universal" instructions for adjusting any feedback amplifier.

THE AMPLIFIER to be described has performed well with five different output transformers, which has led the writer to use the designation "universal." The amplifier has in each case been found completely stable with (a) no load, (b) 8-ohm resistive load, (c) 8 ohm loudspeaker load, and (d) a 0.1 µf capacitor load added to any of the load conditions of (a), (b), or (c) above. The feedback factor employed has been 20 db ±1 db.

Few of the "Williamson Type" and other amplifiers seen by the author have been capable of meeting such a stability test. Breathing of the loudspeaker cone, due to very-low-frequency oscillations, and supersonic oscillations readily seen on an oscilloscope are all too common. Either type of oscillation can produce negative charges on the grid sides of the output-tube coupling capacitors, with distortion and limited power output resulting. Marginally stable amplifiers have also been observed which do not normally oscillate, but are highly regenerative at extreme frequencies and do oscillate when audio signals with steep leading edges on the waveforms are fed to the input terminals.

A brief history of the development of the circuit is believed to be of interest and is as follows:

Development

The author was a "high fidelity" fan many years ago and is still not ashamed of the performance of a class-A push-pull 2A3 triode amplifier (power output of 7 watts) still on hand. After a lapse of 10 years, a reviewed interest in high fidelity led to a study of feedback and the present day amplifiers which have achieved recognition in the literature. The writer found to his annoyance that it was not possible to duplicate a published amplifier circuit and employ a different output transformer and a more compact layout—unless extensive redesign of the coupling and feedback circuits was carried out.

The author then made an analysis of the problems in the design of feedback amplifiers and established the following principles for his own amplifier.

1. It should have as few stages as possible to achieve the required gain. Extra stages contribute phase shifts at low and high frequencies which reduce the inherent stability of the amplifier.

2. The simplest possible circuitry should be employed. Additional components which are not necessary simply add to the cost, complication, and potential for failure.

3. The output stage should be biased for Class A operation. It is not generally realized that plate current cutoff in a power output tube biased to AB operation can cause ringing and oscillation, as well as other forms of distortion due to the steep wavefronts in the output-transformer current waveform.

4. Multiple feedback paths should be employed, rather than a single path from output to the input, to achieve maximum stability.

5. Although the author is of the school which believes that a power output of 6 watts is adequate for the home, a power output between 15 and 20 watts should be employed to assure near-perfect linearity at the normal maximum 5-watt home listening level and to allow a reserve margin of power to compensate for tube aging, inefficient or mismatched speakers and other variables.

Circuitry

The schematic of the resulting amplifier is shown in *Fig.* 1, and its power supply in *Fig.* 2. It consists of a pentode voltage amplifier directly coupled to a split-load triode phase splitter, which is resistance coupled to the push-pull output pentodes. Actually, the amplifier may be thought of as the simplest two-stage pentode resistance-coupled type, with the phase splitter added to convert the design to push-pull operation. This simple straightforward design is neither startlingly new or even original, but fulfills admirably the requirement of the least possible number of stages and coupling networks. This is the initial key to stability in a feedback amplifier. The final key to the high level of stability lies in the three feedback paths, which will be discussed in a later paragraph.

To satisfy the Class A and power design requirements, the output tubes had

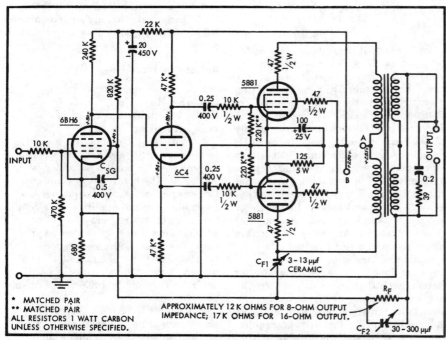

Fig. 1. Schematic of the author's "Universal" feedback amplifier.

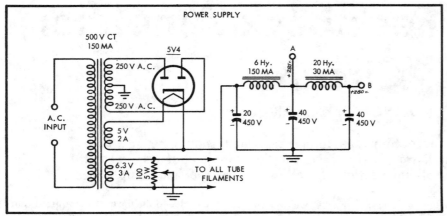

Fig. 2. Power supply for the "Universal" amplifier.

to be beam power tubes of the 6L6 type, with improved construction as exemplified in the type 5881 and 6L6GB. No available triodes are capable of providing either the high gain, or the power output in *Class A* operation which beam power tubes provide—and provide with modest power-supply and driver-stage requirements. Stabilizing resistors of suitable values are employed in the grid, screen grid, and plate circuits. The tubes are self-biased, since fixed bias does not appear to offer any advantage with beam-power tubes, while self bias permits the use of higher impedance in the grid circuits, with less loading of the driver stage.

The push-pull output stage is driven by a split-load or cathodyne phase splitter employing a type 6C4 miniature triode. After considerable study of phase splitters, this type is considered by the writer to be superior to all others, including the cathode-coupled inverter—actually the only other type deemed worthy of consideration. There has been much talk about unbalance of the cathodyne phase inverter at high frequencies, but after reference to the available literature, the author considers this to be much talk not based on adequate investigation.

Actually, the equivalent source impedance of either output channel is[1]

$$R_o = \frac{r_p R_L}{r_p + R_L (\mu + 2)} \quad (1)$$

Substituting a value of 56,000 ohms for R_L, 10,000 ohms for r_p and 19.5 for μ as applicable for the 6C4 tube employed in this amplifier,

$$R_o = \frac{10,000 \times 56,000}{10,000 + 56,000 \, (19.5 + 2)}$$

$$= 460 \text{ ohms approximately.}$$

The output capacitances C_o for each channel are, respectively[1]

Plate Channel

$$C_o = 2C_{pk} + C_{gp}\left(1 + \frac{1}{A}\right) \quad (2)$$

[1] "Radiotron Designer's Handbook." Edited by H. Langford-Smith, Fourth Edition, p. 330.

Cathode Channel

$$C_o = 2C_{pk} + C_{gk}\left(\frac{1}{A} - 1\right) + C_{hk} \quad (3)$$

where A = the channel gain of approximately 0.9

and C_{pk} = plate-to-cathode tube capacitance

C_{gp} = grid-to-plate tube capacitance

C_{gk} = grid-to-cathode tube capacitance

C_{hk} = heater-to-cathode tube capacitance

For a 6C4 triode

$C_{pk} = 1.3$ µµf
$C_{gp} = 1.6$ µµf
$C_{gk} = 1.8$ µµf
$C_{hk} = 2.5$ µµf

Substituting these values in Eqs. (2) and (3):

Plate channel

$$C_o = 2(1.3) + 1.6\left(1 + \frac{1}{0.9}\right) = 5.96 \text{ µµf}$$

Cathode channel

$$C_o = 2(1.3) + 1.8\left(\frac{1}{0.9} - 1\right)$$
$$+ 2.5 + 5.28 \text{ µµf}$$

If these output capacitances are added to values of 10 µµf for wiring and 10 µµf for the input capacitance of the power output tubes, the total shunting capacitance of the two channels are 25.96 µµf and 25.28 µµf respectively, a difference of 0.68 µµf or approximately 1.5 per cent. Referring back to Eq. (1), the value of 460 ohms equivalent source impedance R_o with a shunting capacitance of 30 µµf will result in an output within 3 db of mid-frequency output beyond 10 mc, well outside the range of any audio amplifier this citizen ever wishes to possess. Any unbalance within the range up to a few hundred kilocycles is considered negligible.

As employed, the cathodyne splitter contributes little phase shift, is virtually free of distortion itself, is easily balanced with two matched load resistors and readily meets the driving requirements of 30 volts peak-to-peak output for the two power output tubes. The phase splitter is direct coupled to the input amplifier tube, a high transconductance type 6BH6 miniature pentode. The design of the first stage is conventional. However, the screen grid potential is chosen to permit operation with a relatively low value of bias resistor in the tube cathode circuit, to reduce the impedance of the main feedback network.

While it is believed that other triode and pentode tubes of similar characteristics may be employed in the first two stages, the 6BH6 and 6C4 were chosen because they have low heater currents (150 ma), which have been found to minimize hum problems.

Reasons for Stability

The high degree of stability in this amplifier is achieved by means of three negative feedback paths of the simplest type as described below:

(a) The first stage has an unbypassed cathode resistor, which constitutes the first path providing negative feedback.

(b) A small adjustable capacitor C_{f_1} (3–12 µµf) connected between the plate of one output tube and the cathode of the input amplifier tube, provides the second negative feedback path. This path provides considerable feedback at the supersonic frequencies, eliminating the peak found in this region as the feedback factor is increased.

(c) The third feedback path is provided by a voltage divider connected across the output transformer, consisting of the feedback resistor R_f and the 680-ohm unbypassed cathode resistor of the input pentode amplifier tube.

An adjustable capacitor C_{f_2} (30–300 µµf) is connected across the feedback resistor to provide control of the feedback factor at the higher frequencies below the peak leveled by C_{f_1} discussed above.

Adjustment Procedure

The following procedure should be employed to adjust the feedback networks for stable operation of the Universal Amplifier.

1. With C_{f_1}, C_{f_2} and the main feedback resistor R_f, disconnected, run a response curve on the amplifier with a 4-, 8-, or 16-ohm resistor connected to the proper amplifier-output terminals. Depending on the quality of the output transformer, the response should be level within about 3 db (30 per cent approximately) of the midrange response (400 to 1000 cps) to about 10,000 cps or higher. Any reasonable audio oscillator covering from 20 to 100,000 cps is suitable.

2. Connect a 25,000-ohm variable resistor to maximum resistance as R_f and note whether the amplifier gain decreases. If it increases, reverse the out-

Fig. 3. Top view of the amplifier described.

put tube plate connections to the output transformer, or reverse the connections to the secondary winding of the output transformer.

3. Decrease the value of R_f slowly and look for a peak in the response of the amplifier in the high frequency ranges (usually found at about 40,000 cps or higher).

4. Connect C_{f_1} and adjust as required to eliminate the high-frequency peak discussed above.

5. Continue to decrease R_f and readjust C_{f_1} as required until the amplifier gain is reduced to one tenth of the value without feedback. This is a feedback factor of 20 db. With all but the highest quality output transformers a second response peak, in the frequency range somewhere between 17 and 30 kc will usually become evident. This peak will be considerably broader than the first response peak.

6. Connect C_{f_2} and adjust as required to level the second peak.

7. Check the amplifier frequency response again and make slight readjustments of C_{f_1} and C_{f_2} as required to level any rises in the response curve. Connect the loudspeaker in place of the load resistor and repeat this step.

8. The amplifier may now be given the acid test of connecting capacitors across the output terminals. If the amplifier is not stable with at least a 0.02 μf capacitor across the output terminals further adjustments of C_{f_1} and C_{f_2} should be made as necessary. R_f and C_{f_2} should now be measured and replaced with fixed value components.

If a peak in the low-frequency response of the amplifier becomes evident during the feedback adjustments, the value of the screen-to-cathode bypass capacitor C_{sg} of the input pentode amplifier may be reduced (in one case from 0.5 μf to 0.05 μf) to level the low-frequency response.

Feedback factors as high as 40 db (a reduction of 100:1 in gain) have been achieved with high-quality transformers. However, the amplifier then requires a 10-volt rms input signal for 15 watts output, whereas approximately one volt is sufficient for full output with 20 db of feedback. No difference in listening performance has been discernible with increase of the feedback factor beyond 20 db.

If any perfectionist (more power to such) does not accept the balance of the cathodyne phase splitter as being adequate, he may achieve theoretically perfect balance by adding a small adjustable capacitor across the cathode load resistor and check for perfect balance in the megacycle region. However, no test instruments should be connected to the cathodyne inverter itself, or the inherent added capacitance will invalidate the test results. The balance measurements must be made at the plate terminals of subsequent stages.

The author has tried the "ultra-linear" connection of the output tubes but has observed no advantage in performance to compensate for the reduction in gain. Those who prefer this method of output tube operation may change the bias resistor to a value approximating that for triode operation of the output tubes and follow otherwise the procedure outlined in this article.

This amplifier has been termed "Universal" since it is believed capable of

Fig. 4. Bottom view of the author's amplifier.

producing the optimum performance which is obtainable from any output transformer of suitable impedance ratios. There is no reason to believe that larger and smaller power tubes than the type 5881 cannot be used with equal success. However, the author takes a dim view of any type of operation of the output tubes except class A unless the output transformer is a top-grade unit and the screen grids of the output tubes are fed from a regulated power supply.

Additional Suggestions

The following notes are offered to prospective builders of the Universal Amplifier.

Note 1: The network across the transformer output terminals is intended to provide increased stability at high frequencies with no load. It is not needed with the best output transformers.

Note 2: It is recommended that all grounded leads be made to a common buss bar, which is isolated from the chassis except at the ground terminal of the input connector.

Note 3: No bias balancing control for the output tubes has been employed, since the 6L6GB and 5881 tubes seem to be quite uniform.

Note 4: The center tap of the 100-ohm balance potentiometer connected across the heater winding is grounded, since this connection makes hum inaudible with the ear at one foot from the loudspeaker. With different tubes or a different layout, it may be advisable to return the potentiometer center tap to a positive potential, which may be provided by a resistive voltage divider at the output of the power supply. Positive potentials up to about 55 volts may be investigated for minimum hum.

Note 5: The author's speaker system calls for an 8-ohm output impedance and this was provided on each output transformer used. Values of 16 or 4 ohms will require a different value of feedback resistor R_f. The primary impedance should be between 5000 and 7000 ohms for type 5881 and similar tubes.

Distortion and intermodulation measurements have been made, but will not be presented in this article. Suffice that the amplifier is essentially distortionless up to the overload point of the output tubes or transformer, whichever is reached first. The frequency response is flat from the low-frequency limit of the output transformer to somewhat higher than the resonant frequency of the output transformer, beyond which the response falls smoothly at a rate of 6 to 10 db per octave. The resonant frequencies of the transformers used have ranged from approximately 38 to 100 kc.

Figures 3 and 4 are top and bottom views of one version of the Universal Amplifier which provides plate and filament power for a large AM-FM radio phonograph console, and therefore uses two 5V4G rectifier tubes.

Figures 5 and 6 are views of a second version of the Universal Amplifier using silicon rectifiers in the power supply. Masking tape used to protect the paint finish of the transformers during construction is shown in *Fig.* 5 as a useful suggestion. Æ

Fig. 5. Another embodiment of the "Universal" amplifier—this one uses silicon rectifiers in the power supply section.

The Anode Follower

CHARLES P. BOEGLI

Possessing virtues far in excess of its namesake the cathode follower, the anode follower provides the audio engineer with a simple tool for achieving a wide range of high-quality audio circuits.

The "ANODE FOLLOWER" is so called because, in its simplest version, the anode of the amplifier tends to reproduce the input signal in much the same manner as does the cathode in a cathode follower. It is a tool of unusual versatility to the electronic designer. Although the cathode follower is more or less limited to a gain approaching 1.0, the anode follower suffers no such limitation; gains of more or less than 1.0 are easily attained. In common with the cathode follower, the anode follower may have high input and low output impedances. Again, it is not greatly restricted in this sense.

The term "anode follower" is generally applied to single stages of amplification supplied, in addition to the active element, with a series input impedance and a feedback impedance. In its strict sense, the term is applicable only when the input and feedback impedances are identical, so that under the usual conditions, the gain approaches 1.0. For want of a better term, the plate of a tube, the collector of an *npn* transistor, and the colector of a *pnp* transistor, are all called anodes, even though the last is a negative element. The versatility of this type of amplifier arises from possibilities of using nonidentical impedances; in such cases, although the similarity to the cathode follower ends, the term "anode follower" persists.

Circuits of this type find manifold uses in the entire field of electronics. They may be used as simple mixers, to add several inputs with very little interaction or loss of gain; by their use, signal filtering may often be accomplished with minimum loss of gain; as impedance-matching devices, they are far more versatile than cathode followers; and in the audio field, such stages provide good amplification with wide frequency response and notable lack of distortion.

Design of such a stage involves more, however, than simple addition of two impedances to an ordinary vacuum-tube or transistor stage. If satisfactory results are to be obtained, attention must be paid to the choice of impedance values, and unless the impedances are properly chosen, the circuit is apt to perform somewhat differently than expected. Unfortunately, treatments accorded the circuit in various texts make assumptions that tend to obscure the factors affecting the proper choice of impedances. For this reason, it has appeared worthwhile to consider the circuit at some length; this article summarizes the results of that work.

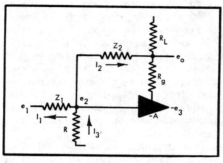

Fig. 1. Basic scheme of the anode follower.

ANALYSIS

Consider an amplifier (*Fig.* 1) in which a feedback element, Z_2, is connected between output and input. Z_1 is a series input impedance and R is an input shunt. In practical circuits, R may be very large (as in the case of tubes with a single input) or quite small. The latter situation arises when the amplifier has a low input impedance (e.g., a transistor) or when the amplifier is used for mixing a number of inputs, in which case R represents the paralleled resistances of all inputs other than the one whose behavior is being investigated.

The usual analysis of anode-follower circuits ignores the existence of R but, as will be seen, its effect upon the performance of the circuit may be quite profound.

With the currents as shown in *Fig.* 1,

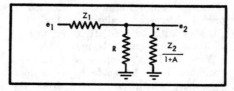

Fig. 2. Equivalent input circuit for the anode follower.

the following equation is obtained:

$$\frac{e_2}{R} = \frac{e_1 - e_2}{Z_1} + \frac{e_0 - e_2}{Z_2} \qquad (1)$$

which, rearranged, yields

$$e_2\left(\frac{1}{R} + \frac{1}{Z_2} + \frac{1}{Z_1}\right) = \frac{e_1}{Z_1} + \frac{e_0}{Z_2}. \qquad (2)$$

Equation (2) is the basic equation for the anode-follower circuit.

Gain

Let $e_2 = -e_0/A$, indicating that the amplifier has a phase shift of $n \cdot 180°$ where n is an odd number. Then, equation (2) becomes

$$-\frac{e_1}{Z_1} = e_0\left[\frac{1}{Z_2} + \frac{1}{A}\left(\frac{1}{R} + \frac{1}{Z_1} + \frac{1}{Z_2}\right)\right]$$

from which

$$\frac{e_0}{e_1} = -\frac{1}{\frac{Z_1}{Z_2} + \frac{1}{A}\left(\frac{Z_1}{R} + \frac{Z_1}{Z_2} + 1\right)} \qquad (3)$$

The exact implications of this equation depend upon the use to which the circuit is to be put. Examples of such uses will be treated subsequently.

Impedances

In equation (2), let $e_0 = -Ae_2$; then

$$\frac{e_2}{e_1} = \frac{RZ_2}{(A+1)RZ_1 + Z_1Z_2 + RZ_2}. \qquad (4)$$

An analysis of the voltage divider shown in *Fig.* 2 shows the voltage e_2 is related to the input e_1 by precisely the same equation (4) as was derived for the feedback amplifier. In *Fig.* 1, the current entering the circuit through Z_1 sees, at e_2, an impedance of R in parallel with $Z_2/(A+1)$. When viewed from the grid of the tube or the base of the transistor, therefore, the impedance Z_2 looks like $Z_2/(A+1)$.

In the expressions for gain and input impedance appears the term A for the amplification of the amplifier. This A is the gain that would be measured if the Z_2 were connected between the output and ground rather than between output and input. Calculations of A must therefore include the loading effect of the feedback impedance.

The output impedance may be found by letting $e_1 = 0$ and considering a signal to be applied at e_0. A certain fraction, β, of this signal will be fed into the amplifier input, resulting in an output $e_3 = -\beta A' e_0$ which will cause more current to flow in R_g than would be the case in the absence of the feedback. The current is, as a matter of fact, multiplied by the factor $(1 + A'\beta)$ so that the generator resistance, R_g, appears from the output terminals as

$$\frac{R_g}{1 + A'\beta}$$

In the general case, β is complex so the net generator impedance, R_g, will also be complex. A complete expression for Z'_g, applicable to *Fig.* 1, is derived simply by replacing β by its equivalent in terms of impedances:

$$\frac{Z'_g}{R_g} = \frac{1}{1 + A'\left[\frac{Z_1 R}{Z_1 R + Z_2 R + Z_1 Z_2}\right]}$$
$$= \frac{Z_1 R + Z_2 R + Z_1 Z_2}{(1 + A')Z_1 R + Z_2 R + Z_1 Z_2}. \quad (5)$$

It must be remembered that the actual output impedance of the circuit is Z'_g in parallel with R_L.

The gain A' that appears in equation 5 is the gain that would be realized from the amplifier with an infinite load resistance; it may be considerably larger than the quantity A that appeared in previous expressions.

Spurious inputs

The effects of spurious inputs such as noise, drift, and microphonics are generally expressed in terms of an equivalent signal to the grid of the tube or the base of the transistor. If the spurious input is of magnitude δ, then the output of the amplifier in *Fig.* 1 is not given by $e_0 = -Ae_2$, but rather $e_0 = -A(e_2 + \delta)$. The effects of δ are therefore added to e_2.

If the above expression for e_0 is solved for e_2 and this e_2 is substituted into equation (2), the result is

$$-e_0 = \frac{e_1}{\frac{Z_1}{Z_2} + \frac{1}{A}\left(\frac{Z_1}{R} + \frac{Z_1}{Z_2} + 1\right)} + \frac{A\delta}{1 + \frac{ARZ_1}{RZ_1 + RZ_2 + Z_1 Z_2}}. \quad (6)$$

When this is contrasted to the open-loop equivalent, consisting of the passive filter of *Fig.* 2 followed by an amplifier A, for which the output is

$$-e_0 = \frac{e_1}{\frac{Z_1}{Z_2} + \frac{1}{A}\left(\frac{Z_1}{R} + \frac{Z_1}{Z_2} + 1\right)} + A\delta \quad (7)$$

it is seen that the effects of the spurious signal are reduced in the anode follower by the factor

$$F = \frac{RZ_1 + RZ_2 + Z_1 Z_2}{(1+A)RZ_1 + RZ_2 + Z_1 Z_2} \quad (8a)$$

over that which would be observed with the open-loop circuit. This factor, of course, holds whether the spurious signal effects at the plate or the grid are under consideration.

In case where R is much larger than Z_1 or Z_2

$$F' = \frac{Z_1 + Z_2}{(1+A)Z_1 + Z_2}. \quad (8b)$$

APPLICATIONS

Amplifiers

The equations so far derived may be used to design a stage of amplification with predetermined characteristics, or to find the effects of certain uncontrollable factors on the performance of an existing stage.

When straight amplification is being considered, the object of the anode-follower circuit is usually one of the following: (a) to devise a highly-stabilized stage whose amplification is substantially unaffected by a reasonably small change in tube or transistor characteristics, (b) to provide an amplifier of low output impedance, (c) to accomplish control over the input impedance, or (d) to control the frequency response.

In the usual case, it is desirable that the gain be controlled by impedances Z_1 and Z_2, remaining substantially independent of A. With the substitution $A \to \infty$ in equation (3), the gain expression becomes

$$\frac{e_0}{e_1} = -\frac{Z_2}{Z_1}$$

and the problem is to determine the magnitudes of Z_1, Z_2, and A such that this condition can be closely realized. Now, the presence of the shunt input resistance R may greatly affect the performance of the stage. In the case of a vacuum tube, R is usually the grid-return resistor, around one megohm; but if a transistor is used R may be of the order of 2000 ohms. Because of its small input resistance, a transistor anode follower may not operate as expected unless attention is paid to the magnitudes of Z_1 and Z_2.

From equation (3) it may be seen that if the gain is to be determined principally by Z_1 and Z_2, then we must have

$$\frac{Z_1}{Z_2} \gg \frac{1}{A}\left(\frac{Z_1}{R} + \frac{Z_1}{Z_2} + 1\right).$$

(1) If $Z_1 \leqslant Z_2$ and $Z_1 \gg R$, the condition is $AR \gg Z_2$.

(2) If $Z_1 \geqslant Z_2$ and $R \gg Z_1$, the condition is $AZ_1 \gg Z_2$ or $Z_1/Z_2 \gg \frac{1}{A}$.

(3) If $Z_1 \gg Z_2$ and $Z_1 \leqslant R$, the condition is $1 \gg 1/A$.

In the discussion that followed equation (4) it was shown that, viewed from the input terminal of the amplifier, Z_2 looks like $Z_2/(A+1)$. Condition (1) above is tantamount to saying that $Z_2/(A+1)$ must be small compared to the shunt resistance, R. Since the other two conditions apply when $Z_1 \leqslant R$, it may be concluded that a condition for the proper functioning of an anode follower is that the shunt resistor, R, multiplied by the gain of the amplifier, must be large compared to the feedback impedance.

Anode-follower feedback may be used to overcome the detrimental effects of certain inalterable amplifier characteristics. Consider a grounded-emitter transistor amplifier with an input resistance of 3600 ohms, a gain of 100, and a collector-base capacity of 36 mmfd. Suppose further that this amplifier must be driven from a source with an internal impedance of 30,000 ohms. Viewed from the input, the collector-base capacity looks like $(101)(36) = 3636$ mmfd and this, together with the source resistance, leads to a 3 db drop at about 1500 cps. If, for audio work, a response out to 40,000 cps were desired, a resistance equal to the reactance of 36 mmfd at 40,000 cps could be introduced between collector and base. The resistor would be about 100,000 ohms and the voltage gain of the stage (= output voltage/generator open-circuit voltage) would be 3. This gain is, of course, quite low and it points up the unsuitability of this type of transistor in a grounded-emitter transistor stage for use with high-impedance sources, in cases where good voltage gain and wide response are desired.

The amplifier, A, may consist of a single stage of amplification, or any odd number of phase-inverting stages. Because of instability problems, the number of stages is usually limited to three, but even a three-stage amplifier may have properties unattainable with a single stage. For example, by loading a low-impedance tape-playback head with about 10 ohms, a constant-current output is obtained which requires very little equalization. Given a grounded-emitter transistor stage with a gain of 100, the feedback resistor would have to be 1000 ohms to attain this low input impedance, but such a heavy load on the transistor output is apt to lead to low gain and excessive distortion. If, however, three stages are used ($gain = 10^6$) then the feedback resistor may be 10 megohms for the same 10-ohm input impedance.

Mixing

An anode follower may be used to mix several inputs, with good isolation between the signal sources. The various inputs are all connected, through their own series input resistances, to the anode-return resistor. The series resistor of each input looks into a shunt resistance equal to the paralleled input resistor of

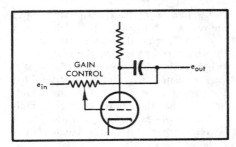

Fig. 3 Anode-follower gain control.

all other inputs. There is thus a limitation on the gain that may be obtained with a mixer.

Let it be desired, for example, to mix $(n+1)$ inputs, each having an internal resistance of R ohms, by means of an anode follower with a gain of about 1.0. For each individual input, Z_1 will be a resistance equal to R, while the shunt resistor will be R/n. Presumably, the feedback impedance Z_2 will also be R. In this case, obviously, $Z_1 \gg R$ (that is, $R \gg R/n$) so that the first condition of the previous section applies. Thus, we must have $AR/n \gg R$ which imposes a rough lower limit on A. If, for example, ten inputs are to be mixed, $n=9$, so we have $R \gg R/9$. If we let $AR/n = 9R$ (which introduces a sort of pseudo consistency) then $A = 81$.

The symbol "$\gg$" is, of course, much less definite than the symbol "=". If we wish to find how far the actual circuit departs from ideal performance, recourse must be made to the original equations. In the case of the 10-channel mixer we should find

$$\frac{e_o}{e_1} = -\frac{1}{1+\frac{1}{81}(9+1+1)} = -0.9$$

instead of the 1.0 that was expected. Were a gain of exactly 1.0 desired, an adjustment could be made to the value of Z_2 to obtain it.

Gain control

It is possible, by using the anode follower, to duplicate the gain characteristics of any combination of passive filter and simple amplifier. Although gain control is customarily carried out by means of resistive attenuators, it may also be performed by an anode follower (*Fig.* 3).

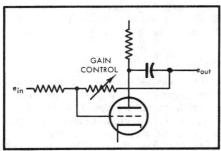

Fig. 4. Anode-follower gain control with fixed input resistance.

The anode-follower gain control has the advantage that the distortion of the stage is greatly reduced at low levels. For this reason, this type of gain control is particularly felicitous with transistors, which are generally operated with much larger ratios of signal-to-power-supply voltages than are tubes, and which are therefore more susceptible to distortion.

The input resistance of the gain-control stage varies, depending upon the setting of the gain control. A minimum input resistance may be set by introducing a small series resistor into the input; its effect is also to set the maximum gain of the stage. Where an approximately fixed input resistance is desired, an alternative arrangement can be used (*Fig.* 4) in which a fixed input resistor and a varying feedback resistor are employed. The maximum size of the feedback control is determined by conditions previously outlined.

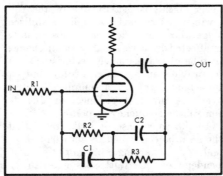

Fig. 5. RIAA equalizer stage.

Equalization

The anode-follower circuit permits a great deal of control over the frequency response of an amplifier. When equalization is the object, the impedances Z_1 and Z_2 may be complex, and their limiting values are determined by already-stated considerations.

For instance, let us consider the design of an amplifier to provide correct RIAA recording characteristic for playback. As is well known, the desired playback equalizer shows a flat response up to 50 cps; between 50 and 500 cps it drops with increasing frequency at 6 db per octave. Between 500 cps and about 2100 cps the response is again flat, while above the latter frequency the response again drops, as the frequency increases, at the rate of 6 db per octave.

In the design of an equalizer stage, the gain at one particular frequency may be specified. For a phonograph equalizer, it is usually wise to set the low-frequency (below 50 cps) gain, which is the maximum gain over the audio spectrum, at from one-fourth to one-tenth that of the open-loop stage. *Figure* 5 shows the design of an anode follower for accomplishing the necessary equalization. At low frequencies, the gain of this stage is determined by the feedback resistances $R_2 + R_3$ and the input resistance R_1. At 50 cps, C_1 begins to shunt R_2 and the response begins to drop; this drop continues until the reactance of C_1 is equal to R_3, which should occur at 500 cps. At 2100 cps, capacitor C_2 begins to shunt R_3 and the response again drops off above this frequency.

Most phonograph pickups operate well with a load resistance of 27,000 ohms or more. Thus, R_1 may be 27,000 ohms. If the tube has a gain of 150, the gain at 50 cps may be set at about 15. Within the ranges of commercial capacitors and resistors, the values $R_2 = 330k$, $R_3 = 33k$, $C_1 = 0.01\ mfd$, and $C_2 = 0.0022\ mfd$ meet the requirements quite accurately, and permit a gain at 1000 cps of about 1.22. R_1 may be reduced for pickups that can operate into lower resistances, with a resultant improvement in gain.

Tone control

A tone-control stage is a variable equalizer of rather simple characteristics. A tone control using the anode-follower circuit has been designed by Baxandall; a simplified and highly satisfactory version is shown in *Fig.* 6. The difficulty in its design is carrying out two control functions (i.e., bass and treble) independently. If it is remembered, however, that the bass-control capacitors present effective short circuits at high frequencies, then it can be seen that the components effective in the bass control are capacitors C_2 and C_3 along with resistors R_3 and R_4 and the bass control itself; the components effective in treble control are the capacitor C_1 and the resistors R_3 and R_4, along with the treble control.

This circuit is also very effective with transistors, provided suitable impedances are used. *Figure* 7 shows a transistor stage which permits as much as

Fig. 6. Simplified Baxandall tone control.

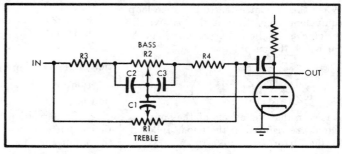

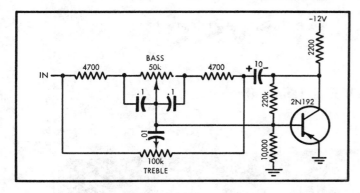

Fig. 7. Transistor anode-follower tone control.

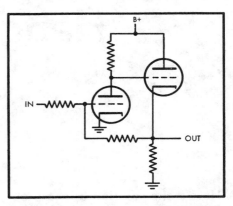

Fig. 9. Amplifier with extremely low output impedance.

15 db boost at 20 cps. Linear controls are used, and in the flat position the response is down only 1 db at 10 cps and 100,000 cps. The *IM* distortion (60/7000 cps, 4:1) with a 12-volt supply is less than 0.3 per cent at an output of 1 volt rms.

Capacity pickups

Capacity pickups are useful devices for the measurement of small displacements, particularly where it is important to avoid loading the unit being measured. An excellent phonograph pickup may be designed by causing the stylus to move a small metallic plate closer to, and farther from, a fixed plate. Often, the maximum permissible dimensions of the capacitor plates are very small, so that the capacity between them is minute, particularly in comparison to stray capacities that exist elsewhere in the circuit. For example, in a capacity pickup, the capacity between plates may be 2 mmfd. while the stray capacities between each side of the connecting line and ground may be over 200 mmfd. If an attempt is made to obtain an output from the pickup by polarizing one plate, grounding the other, and obtaining the signal from the polarized plate, the effect of the stray capacities is to attenuate the signal severely.

An anode follower may be used to overcome these effects. Here, the capacity pickup is, in effect, connected between the anode and input terminals of an amplifier. It has already been demonstrated that by this means the capacity is effectively multiplied by $(A + 1)$. At the same time, the strays are split, half of them being shunted between the anode and ground, and the other half between the input terminal and ground. By this means, the stray-capacity attentuation of the signal is reduced to a considerable extent.

D.c. amplification

A d.c. anode follower can be constructed by inserting a v-r tube in the feedback path of the conventional circuit to attain a favorable distribution of d.c. voltages (*Fig. 8*). An amplifier of this type has a large useful gain, reasonably low drift, low output impedance, and input and output terminals at approximately ground potentials in the quiescent state. If it is necessary that the input and output terminals be exactly at ground potential, a small resistor may be inserted in the cathode of the amplifier tube to adjust to this equality.

Because the plate-load resistor of the tube must carry not only the plate current of the amplifier, but also the v-r tube current, it is generally of somewhat lower resistance than it would have been in a similar a.c. amplifier. For this reason, high-perveance triodes are very useful in d.c. anode-follower amplifiers.

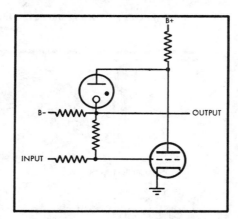

Fig. 8. D.c. anode-follower amplifier.

Amplifiers of extremely low output resistance

It will be remembered from equation (5) that the effect of anode-follower feedback is to reduce the effective generator resistance of the amplifier; that is, the plate resistance of the tube or the collector resistance of the transistor. The benefits of this reduction are much more noticeable with triodes, for which the plate resistance is usually lower than the load resistance, than with pentodes or transistors, for which the generator impedances are quite high.

The addition of a cathode follower to the amplifier, as shown in *Fig. 9*, permits amplifiers of extremely low output resistances to be obtained with great economy of parts. In this circuit, the open-loop generator resistance is the output resistance of the cathode follower —already a low value—and it is reduced appreciably by the anode-follower connection. For example, a 12AX7 cathode follower has an output resistance of some 500 ohms, while the same tube operated as a voltage amplifier may easily show a gain of 60. By using the two halves of a 12AX7 in the circuit of *Fig. 9*, and proportioning the resistors to yield a voltage gain of 1.0, an output resistance of some 17 ohms is obtained.

The connection of *Fig. 9* is also very useful with transistors. Output resistances lower than one ohm can be obtained in this manner. Thus, high-impedance techniques may often be brought to bear on circuits which are presently considered low impedance, such as 250- or 500-ohm audio circuits. Æ

Hanging Hi-Fi System

HAROLD C. MANGELS (and wife)

Here's one solution to the problem of where to place high-fidelity components—hang them!

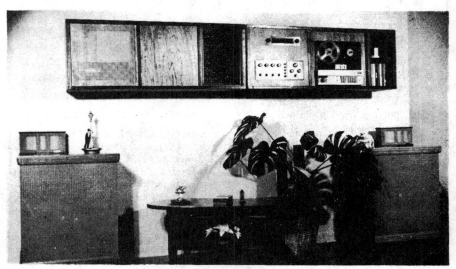

Fig. 1. Front view of the system.

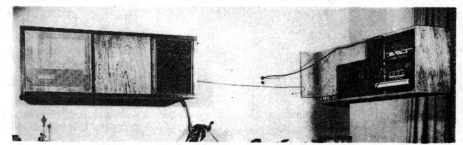

Fig. 2. The right cabinet swung out revealing the unusually easy access to the components.

THIS-DO-IT-YOURSELFER and his wife became interested in building a component high-fidelity system as a natural adjunct to their interest in doing-it-themselves and in music. In their own words, "We chose a hobby which we both could enjoy creating, planning, and finally listening" By planning each step carefully, they assembled a stereo system which consists of an EICO HFT90 tuner kit, a Harman-Kardon Citation I preamp kit, a Dynakit Stereo 70. a pair of Bozak 207-A's plus matching enclusures. and a Bell stereo tape deck They also have a "center" channel consisting of a Pilot amplifier and a Jensen speaker

Of course these selections are clearly related to their do-it-yourself orientation—almost all of the electronic components were kits and were built as a family project They also built the enclosures for the Bozak speakers Finally, they devised the ingenious method for enclosing the system shown on this page

They installed their system in two hanging cabinets, each one 4-ft long and 12-in deep. Instead of rigidly fastening these cabinets to the wall, they hung them on door hinges so that the components would be readily available for servicing

The left cabinet contains the Dyna anplifiers, a utility cabinet, the Pilot center amplifier, and the Jensen center speaker The right cabinet contains the EICO tuner, the Harman-Kardon Citation I preamp, the Bell tape deck, and provision for tape storage

The Mangels are true audiofans—they are already thinking about their next system. In their own words, "As happy as we are with our system, we are already planning improvements. Perhaps an improved tape deck, maybe multiplexing, or possibly the new Bozak midrange speaker" Æ

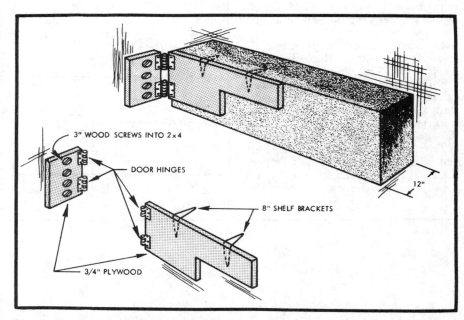

Fig. 3. Drawing showing how the right cabinet is mounted. The hinge for the left cabinet is mounted in a similar manner on the left end of the cabinet.

A Case for the Custom Console

F. H. JACKSON

A custom console to meet your specific equpiment needs is relatively simple to construct and requires only the most basic hand tools. The approach described in this article makes it possible for anyone to produce a furniture-quality console.

To paraphrase the noted sculptor's description of the genus homo sapiens' lifetime as the "Seven Stages of Man," one might as easily draw the simile anent the genus "audiofan"! Fortunately for the writer, and his acquaintances, fired with the desire for fine reproduction of music in the home, the stages are confined to three in number, whose total duration need not *necessarily* sum to a lifetime!

The first might prosaically be termed the planning stage, wherein the audio literature is combed not less avidly than are the shelves of the neighborhood emporium devoted to such wares (presided over by the ever-patient proprietor).

The second might be termed the stage of creation. Decision, ever procrastinated, has finally molded actions. Within short days passing our scrimped savings fast disappear into the maw of desire!

The third stage, for lack of more apt semantics, might be called the contemplative. This article is being composed at that most difficult of times. For the author, the planning and the creation are over. The satisfaction and enjoyment of a fine concert wafting through his home, is tempered by the gnawing query, "Has the final stage fulfilled the promise of the long past first?"

These stages through which the author's system developed are universal enough in nature as to cause the reader similar concern as he plods his way through that first stage!

One problem faced, and no doubt common to fellow devotees of the audio art, is the ever present problem of combining audio component quality with living room decor. Optimizing the equipment quality somehow always seems to outstrip the desires of the distaff side to decoratively house that which husband hath wrought!

Secondly, there is need for convenient access, coupled with easy component removability. Thirdly, it was desired to incorporate in the design the quality of adaptability, i.e., obsolescence conversion. Lastly, convenience features compatible with the design criteria were to be incorporated. How well the solutions to these problems wear is the continuing subject of both this article and the third stage of this audiofan's career in high fidelity.

The basic system chosen (see appendix) was to be that of a three unit stereo presentation of both high-fidelity broadcast and phonograph programming. The three units were to consist of separate left- and right-channel speaker cabinetry together with complete component console housing for all program and amplification sources. (*See Fig.* 1.) The cynic so easily states that given appropriate sums most problems are solved. To confound this crass viewpoint, the writer offers this system and in particular his solution to the task of creating a furniture setting for such a system at a total cabinet cost of under $100! This cabinetry cost was not predicated upon ownership of an extensive inventory of power tools. On the contrary, the author's sole claim to a power workshop is a vintage electric drill with a $12 sabre saw attachment!

How Were These Items Constructed?

At the risk of offending the power tool sellers, we must state that most local lumber yards, for a very nominal charge, will cut and mill any raw stock to your working drawings. On my unit the tolerances maintained were such that for ease of assembly, the method approached the so called "kit" type of cabinet construc-

Fig. 1 (below). The complete system.

Fig. 2 (right). Exploded view of the laminated case construction.

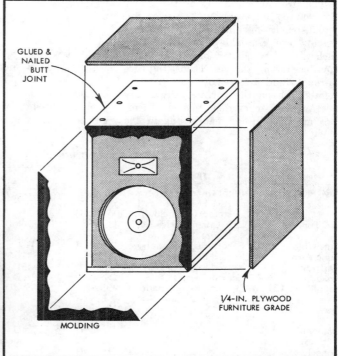

Fig. 3. Top edges of the console are covered with wood tape.

tion. The obvious advantage to this method is the wide latitude in design permissible, *versus* the justifiably limited number of styles available from kit manufacturers.

Another area of expense coupled with the construction of a console suitable for milady's living room is that connected with jointing and cementing of large scale furniture. Rather costly jigs and fixtures are usually required for such construction. Since these items were not readily available to the author, some substitution became a necessity. The resultant innovation, which is felt to be the prime one of those in this system, has been termed "laminated case construction". (See *Fig. 2*.) Cabinet makers may shudder within their professional stoicism to learn that this method uses common finishing nails and quick-setting casein cement! The method was first tried on the construction of the speaker cabinets and the encouraging results prompted the use of the same method, without modification, in building the console.

The procedure consists of two steps. Step one is to simply butt the sides and nail using casein glue (several national brands are available). Three-quarter-inch fir plywood is adequate. After counter-sinking the nail heads, the case is rough sanded to remove any gross irregularities on the surface. Previously cut and corner-mitered, ¼-inch, furniture-grade, plywood panels (again the friendly millwright at the local lumber yard was responsible) were coated on the underside with glue, as was the rough case. Carefully fitted, the panels were butted at all corner miters and weighted until the glue set. The end grain in this case was covered by the application of grille molding although an alternate method, easily applied, is to use plywood tape. This tape method was used extensively on the end grain covering of the console. Most observers believe the cases are of solid hardwood construction. (See *Fig. 3*.) This laminated case construction and the implementation of the Provincial motif in all three cabinets (which helped to create a unity of design) through use of ordinary builders molding, did much toward beautifying the ensemble. The grille frames are definitely enhanced by the graceful curves of the covering molding. The console inset doors and the drop

Fig. 4. The top is off! Note the laminated case construction in the end panels.

front have that added touch of detail, which would be noticealy lacking were these panels of plain surface. Mention should be made at this juncture, that a contribution to the over-all effect was gained by the hand fitting and mitering of these moldings on assembly to comply with the final tolerances existing on the several mating items.

The second major problem to be resolved by our design was that of accessibility. It is only within the last several years that the ogre of maintainence accessibility has been given notice, much less resolution, by the commercial package interests, and that primarily by the television receiver manufacturers. It is to be seen (*Fig. 4*) that this problem was resolved most easily by the inclusion of a lift-off top in the console design. This feature greatly eases the disassembly breakdown of the system into units for repair or transport. In addition the preamplifier and tuners are so inserted into the control panel, that their removal is accomplished by simply sliding them on their base toward the rear of the console for an inch or two. This is possible because the normal bezel-type mounting was not employed. Insteal, a matching rectangle was cut in the panel for each of the three units to be mounted. As each unit face protrudes about ¼ inch from the panel face, it is difficult to tell whether or not they are permanently affixed. This feature has already proved of value in the case of some minor repairs to one of the panel units. The operation took less than a half hour, including repair of the unit.

The third criterion for our design, in which it was desired to create solutions amenable to our other goals, is that called adaptability, or obsolescence conversion. One area in which this attribute was incorporated was the design of the speaker enclosures. Close perusal of *Fig. 5* will reveal the method used to allow for future horizontal placement of the speakers, should the need arise. The bases are entirely separate from the speaker enclosures, and the enclosures are finished on all four sides. They need only be lifted off their bases, placed on the side (after orienting the horn tweeter) for functional use as a bookshelf enclosure.

In line with solution to the problem
(*Continued on page 65*)

Fig. 5. The bases are separate and the enclosure finished on all sides to permit a variety of placements.

Appendix
System Parameters

Component	Function	Mfr.	Model
AM-FM Tuner	Right Channel Either AM-FM or FM-FM stereo (see below)	Pilot	FA-670
FM Tuner	Left Channel FM-FM Stereo	Heath	FM-3
Turntable	Record Reproduction	Fairchild	412
Tone Arm	Record Reproduction	ESL	S-1000
Cartridge	Record Reproduction	Shure	M3D
Preamp	Control Center	Dynakit	PAS-2
Amplifiers (2)	Left and Right Channel Power Output	Allied	83 YU 793
Speakers (2)		University	CUL 10 (kit form of 2 way SLOH system)
Cabinets (spkr)		Custom	To Univ. CUL 10 Specifications
Console		Custom	

Common-Bass Stereo Speaker System

FRANCIS F. CHEN

One speaker can handle the bass of both stereo channels if the crossover frequency is low enough to avoid directionality.

THERE ARE TWO COMMON PROBLEMS that the hi-fi hobbyist often faces. One is that his good wife, understanding though she may be in other ways, objects to the "Laboratory Look" in her living room. The other is that his good wife, music-lover though she is, thinks that one-half of the family budget is too much to spend on electronics. Now everybody who reads this magazine knows that you can't get good music reproduction without a certain minimal outlay of cash or without a certain minimal number of components and interconnecting cables, which may not all look beautiful. And *nearly* everybody does not have the inexhaustible capital and engineering time necessary to build one of the "ultimate" systems that so often appear in print.

The more realistic problem of getting highest quality and versatility (important to the hobbyist) within the ever-present restrictions of cost and decor is a challenging one. To solve it, one must give careful thought as to what is important and what is superfluous. To be sure, in the present case the problem of cost was solved not only by "Doing-It-Myself" and by careful choice of components, but also to some extent by the Principle of Infinitesimal Accretion ("You don't mind, dear if I get a couple of EL-34's this week?"); but the latter ploy is at any rate a useful one for the hobbyist to have at his command.

The Common Woofer

The first problem considered was that of the most expensive part of a hi-fi system. Why do good speakers cost so much? The answer is that they do not, as long as one sticks to 8-in. speakers. It is only the 12- and 15-in. speakers that bear the disheartening price tag. It is, however, an inescapable fact that good bass requires large piston area and low resonance. The obvious solution, then, is to use three speakers: two matched 8-in. units for the midrange and treble of the two channels, and a 15-in. woofer for the common bass. This costs considerably less than two 15-in. full-range units. The savings are partially offset by the need (to be explained later) for three amplifiers instead of two, and for a crossover unit. The latter can easily be constructed, however, using the two-tube circuit to be described later; and it turns out that three amplifiers, of which only one need handle the bass, do not necessarily cost more than two amplifiers, both of which must produce, say, 25 "clean" watts at 30 cps. Again the costliness of good bass has made itself apparent, this time in terms of the power handling capacity needed in the amplifier, and in particular in the amount of iron needed in the output transformer.

At this point I should note that in the choice of a woofer I considered using a battery of, say, 20 or 30 small, cheap speakers in a series-parallel array, instead of using a single 15-in. speaker. The piston area would indeed be large, and one might hope that the distortion would be small in spite of the cheapness of the magnets, because the cone excursion of each speaker would be small. However, besides being a somewhat clumsier arrangement, this method would also entail some risk in that the flimsy magnets may not properly damp the cone motions. Besides, 20 or 30 speakers are not exactly cheap. It was therefore decided to leave the multiple-speaker array to possible future experimentation, in spite of several favorable reports in the literature.

Crossover Frequency

The next problem to consider was the choice of a crossover frequency. Now it is a well-known physical principle that two signal sources cannot be distinguished directionally if they are separated by a distance of the order of magnitude of the wavelength emitted. Since the geometry of my living room requires the speakers to be about 13 feet apart, and since a wavelength of 13 feet corresponds to a frequency of 85 cps, it would appear that a crossover well below 100 cps would be necessary to avoid losing any stereophonic effect through the use of a single woofer for both channels. However, the dividing line is a fuzzy one: one cannot say that at 100 cps there is definitely no directional effect, while at 110 cps, or 150 cps, there definitely is. Aside from the vagueness of the physical principle cited above, the acoustics of the room and the psychology of hearing would also enter in; and indeed, common-bass stereo systems have been made with crossover well above 100 cps. The decision, therefore, was to make the crossover frequency as low as possible, consistent with other limitations.

In listening tests, it may sometimes appear that a low note, below 100 cps, has some directionality. This is probably due to the sudden onset of the note. This initial transient consists of higher frequency components and would be reproduced by the tweeter. For this reason, directionality should be tested only with steady tones.

The "other limitations" mentioned above are the ones imposed by the bass response of the 8-in. speakers. The units chosen were Wharfedale Super-8 FS/AL's, which have high flux density, good efficiency all the way to 15,000 cps, smooth response, and a soft suspension. The free-air resonance of these speakers is around 70 cps. Because of the decor of the living-room, these speakers had to be mounted in different types of cabinets; therefore, differences in cabinet resonances would cause the speakers to be mismatched in the low-frequency region. For this reason, it was felt that the crossover frequency should occur at least an octave above the free-air resonance of the speakers, and a frequency of 150 cps was chosen.

Type of Crossover

The usual type of crossover, an LC circuit inserted between the speaker and the amplifier, cannot easily be obtained at a frequency as low as 150 cps. The reason is that the capacitor has to be large—in the neighborhood of 100 µf. This would be most unwieldy unless one used an electrolytic. However,

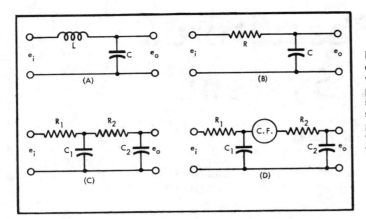

Fig. 1. Simple crossover networks. Only low-pass networks are shown: the corresponding high-pass networks would be similar with the elements interchanged.

electrolytics do not maintain their capacitance value very accurately and, moreover, would have to be used "back-to-back" (requiring double the capacitance in each) in order to "hold off" the a.c. voltage. When used this way the capacitance of the electrolytics can change suddenly in the middle of a loud passage! The inductor, of the order of 10 mh, would also be a large affair, since it would have to be wound with heavy wire to avoid a large insertion loss.

The alternative is to use an "electronic" crossover ahead of the amplifier, and this is the more sophisticated and the more soul-satisfying way of doing it. This would require a separate amplifier for each speaker, but, as mentioned before, this is not necessarily more expensive. By having separate amplifiers and speakers for the highs and the lows, one gains in addition a most attractive bonus: intermodulation distortion (except in the program source) is for all practical purposes completely eliminated!

An attenuation of 12 db per octave is generally recommended for the crossover network. One can achieve this by a half-section LC filter, as in (A) of *Fig. 1*, or by two simple RC filters, as in (C) or (D) of *Fig. 1*. The LC filter has a distinct advantage in that the phase shift is 0 deg. on one side of the crossover frequency and 180 deg. on the other. Thus one can bring the high- and low-frequency speakers into phase by merely reversing the leads on one speaker. However, the use of an inductor at the preamp level would be asking for trouble with hum pickup. This leaves the RC network, with its horrible phase shift characteristics, and leads us to a discussion of phase shift and frequency response in low-frequency RC crossovers.

The Two-Section RC Crossover

Consider first the one-section low-pass filter in (B) of *Fig. 1*, in which the resistor, R, and the capacitor, C, form a simple voltage divider (operating into infinite impedance) for the input signal, e_i. At any angular frequency, ω, the output voltage e_o will be given by

$$\left(\frac{e_o}{e_i}\right)_L = \frac{1}{j\omega C} \bigg/ \left(R + \frac{1}{j\omega C}\right) \quad Eq.\ (1)$$

If we define the crossover frequency, ω_o as $1/RC$,

$$\left(\frac{e_o}{e_i}\right)_L = \left(1 + \frac{j\omega}{\omega_o}\right)^{-1} \quad Eq.\ (2)$$

The corresponding high-pass network, with R and C interchanged, would give

$$\left(\frac{e_o}{e_i}\right)_H = R \bigg/ \left(R + \frac{1}{j\omega C}\right)$$
$$= \left(1 + \frac{\omega_o}{j\omega}\right)^{-1} \quad Eq.\ (3)$$

At frequencies which are low compared with ω_o, *Eq. (2)* shows that $(e_o)_L \approx (e_i)_L$, and at frequencies which are high compared with ω_o, *Eq. (3)* shows that $(e_o)_H \approx (e_i)_H$. At frequencies near ω_o, there is a phase shift, since there is a sizable imaginary part to *Eq. (2)* and *(3)*; and the question arises as to how the outputs from the high- and low-pass networks are to be added.

If the two signals are added together first and then fed into the same speaker, the acoustic output would be found by adding *Eq. (2)* and *(3)* and then squaring the result:

$$\left[\left(\frac{e_o}{e_i}\right)_L + \left(\frac{e_o}{e_i}\right)_H\right]^2$$
$$= \left[\frac{1}{1+\frac{j\omega}{\omega_o}} + \frac{1}{1+\frac{\omega_o}{j\omega}}\right]^2 = 1 \quad Eq.\ (4)$$

Even though the phase shift is 45 deg. in the critical region around $\omega = \omega_o$, the phases are such that the vector sum is always unity. Thus the frequency response is flat throughout. The same would be true, but not as exactly, if the two outputs were fed to two speakers right next to each other. However, if the two speakers were far apart and the crossover frequency fairly high, the total acoustic intensity would be the sum of those from each speaker; that is, the voltages would not add in phase and would have to be squared before adding:

$$\left(\frac{e_o}{e_i}\right)_L^2 + \left(\frac{e_o}{e_i}\right)_H^2 = \frac{1}{1+\left(\frac{\omega}{\omega_o}\right)^2}$$
$$+ \frac{1}{1+\left(\frac{\omega_o}{\omega}\right)^2} = 1 \quad Eq.\ (5)$$

This is also equal to *1*! From this standpoint, the single RC network is the ideal crossover; the frequency response is flat no matter how the signals are added.

Incidentally, you can easily see that for the LC circuit of (A) in *Fig. 1*, e_o/e_i is real, and there is no difficulty with phase shifts. However, just *because* e_o/e_i is real, the equations analogous to *Eq. (4)* and *(5)* cannot both be true; only *Eq. (4)* is true in the case of the LC circuit.

The trouble with the single RC section is, of course, that it rolls off at only 6 db per octave, which usually does not provide sufficient isolation of frequencies. In our particular case, with f_o at 150 cps, this means that the cabinet resonances of the 8-in. speakers at around 70 cps will be only 6 db down—not a very great difference to the ear. The next logical step would be to try two RC sections in cascade, as in (C) of *Fig. 1*. Here, if R_1 and C_1 are equal to R_2 and C_2, the first section would not be working into a large impedance and would not provide as fast a rolloff as it should. This can be improved if the second section is made higher in impedance than the first, or better yet, if the two sections are isolated by a cathode follower, as in (D) of *Fig. 1*.

In this case the response is found by applying *Eq. (2)* and *(3)* twice:

$$\left(\frac{e_o}{e_i}\right)_L = \left(1 + \frac{j\omega}{\omega_o}\right)^{-1}\left(1 + \frac{j\omega}{\omega_o}\right)^{-1} \quad Eq.\ (6)$$

$$\left(\frac{e_o}{e_i}\right)_H = \left(1 + \frac{\omega_o}{j\omega}\right)^{-1}\left(1 + \frac{\omega_o}{j\omega}\right)^{-1} \quad Eq.\ (7)$$

This has been written this way, without the exponent *2*, because by "squaring" we shall always mean multiplying a quantity by its complex conjugate, which

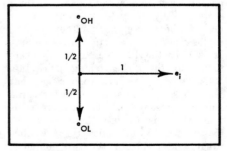

Fig. 2. Behavior of the dual RC network at the crossover frequency.

is not what is required here. Now if we square and then add, we will not get unity:

$$\left(\frac{e_o}{e_i}\right)^2_L + \left(\frac{e_o}{e_i}\right)^2_H = \left(1+\frac{\omega^2}{\omega_o^2}\right)^{-2} + \left(1+\frac{\omega_o^2}{\omega^2}\right)^{-2} \neq 1. \quad Eq. \ (8)$$

Moreover, the phase angle between the two signals will change with frequency. At the crossover frequency, $\omega = \omega_o$, each of the terms in the above equation is equal to $\frac{1}{4}$, so the total intensity is only $\frac{1}{2}$ of what it should be.

Most previous designers have gotten around this either by negative feedback to lower e_i in the flat regions of the response curves or by using different ω_o's in the low-pass and the high-pass sections, so that at the actual crossover point, $(e_o/e_i)^2$ for each section is down only to $\frac{1}{2}$. However, this still does not provide unity gain and zero phase shift at other frequencies. These considerations have been given in detail in two excellent articles by Norman Crowhurst[1,2], the latter of which, unfortunately, did not appear until my system was all finished. The point I want to make here, however, seems to have been missed in these articles, although Mr. Crowhurst touches on this in a more recent article[3] which begins to attack the most basic and difficult problems of stereophonic sound. And that point is, why should RC crossovers be designed so that the separate intensities of the two speakers add up to unity? Isn't it possible that under some circumstances the sound waves from the two speakers add in phase?

This depends on the frequency. At very high frequencies phasing cannot make any difference; there are so many reflections that phasing is all mixed up by the time the waves reach the ear anyway. At 6000 cps, the wavelength is only a couple of inches; and if phasing mattered, the cone of a tweeter would have to lie in the same plane as the cone of the midrange unit, within half an inch or so. This is impossible, since cones are deeper than that. In actual practice, I have been unable to tell the difference when the leads are reversed to a tweeter which crosses over at 8000 cps. At middle frequencies, which are important for the stereophonic effect, phasing makes a difference, but just how is a complicated business. Everyone knows, however, that if the phase were as much as 180 deg. off, the stereo effect is lost for two spatially separated speakers. Our interest now is in what happens near the crossover frequency to a midrange unit and a woofer which are not necessarily separated. In this case the effect of phasing is probably just as great, but easier to analyze. At very low frequencies, it seems to me, phasing must be correct and one must add the (complex) signals to the two speakers together first before squaring, as in *Eq.* (4) to get the total sound intensity. This must be true because the bass reflex principle is known to work; if only total intensity mattered, the back wave from the port of a reflex cabinet would add to the speaker resonance instead of reducing it.

If this is true, observe what the double RC crossover would do. If we add *Eq.* (6) and (7) without squaring them first, we would get (after multiplying numera-

[1] N. Crowhurst, "The RC Crossover Compromise," AUDIO, July 1957.
[2] N. Crowhurst, "Electronic Crossover Design," AUDIO, Sept. 1960.
[3] N. Crowhurst, "Audio Matrixing," AUDIO, Nov. 1960.

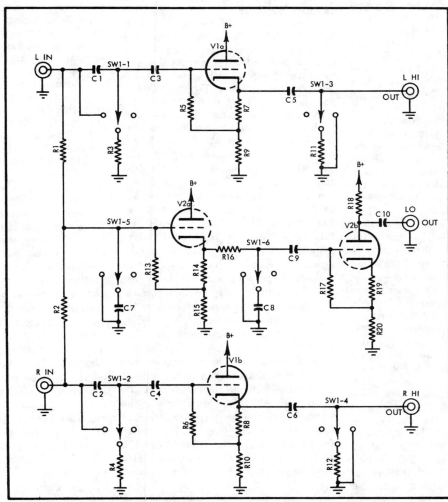

Fig. 3. Schematic diagram of the electronic crossover.

tor and denominator by ω_o^2 and $j^2\omega^2$ respectively),

$$\left(\frac{e_o}{e_i}\right)_L + \left(\frac{e_o}{e_i}\right)_H = \frac{\omega_o^2}{(\omega_o + j\omega)(\omega_o + j\omega)}$$
$$+ \frac{j^2\omega^2}{(j\omega + \omega_o)(j\omega + \omega_o)} \quad Eq.~(9)$$

Now let us reverse the leads to one speaker, so that the sign between the two terms becomes minus, thus cancelling the j^2, and then combine over the common denominator:

$$\left(\frac{e_o}{e_i}\right)_L + \left(\frac{e_o}{e_i}\right)_H$$
$$= \frac{\omega_o^2 + \omega^2}{(\omega_o + j\omega)(\omega_o + j\omega)} \quad Eq.~(10)$$

Upon squaring, each factor in the denominator becomes $\omega_o^2 + \omega^2$, and we get unity. Voila! No shifting of crossover points with the resultant mess in phase shifts; no complicated alignment procedure to perfect a feedback circuit.

For those who like vectors, this is what happens at the crossover frequency (*Fig.* 2). The voltages from the high-pass and low-pass filters have magnitude ½ and are shifted ±90 deg. in phase relative to the incoming signal, so that they are 180 deg out of phase with each other. If we square each separately, each becomes ¼, and the sum is only ½. However, if we reverse the leads to one speaker to bring the high and low signals into phase and then add them before squaring, we get 1. Things work out equally nicely at all other frequencies, as long as such simple addition of phased signals occurs.

How low a frequency must one have before this occurs with actual acoustic signals? To determine this, a simple test was performed in the living room. Equal low-frequency signals were fed from an oscillator, through amplifiers, to two speakers. In this test the speakers were separated, but this is unimportant, since the assumption is that the crossover frequency is so low that there is no directionality. The intensity of the sound at the opposite end of the room was observed both by ear and by a microphone feeding an oscilloscope. As the leads to one speaker are reversed, the intensity should go from 0 (complete cancellation) to 4 times the intensity of one speaker alone, if the acoustic waves added in phase. Of course complete cancellation does not occur in actual practice, but at 100 cps, there was a large change in loudness as the leads were reversed. At 150 cps the change was much less pronounced. The conclusion was that the simple circuit in (D) of *Fig.* 1, with two identical RC sections in cascade, should be used with crossover frequencies below 100 cps, and that at our previously chosen crossover frequency of 150 cps, the frequency response of this circuit is only approximately flat but that it should be as good as that of the fancier circuits usually used.

The Two-Tube Stereo Crossover

Before building the circuit of *Fig.* 3, we considered the two electronic crossovers on the market. One, the Marantz, seemed to be carefully designed with feedback; but it was prohibitively expensive. The other, by Heathkit, was unnecessarily bulky for our purpose and seemed to have been designed to provide a peak at the crossover frequency.

The circuit of *Fig.* 3 serves the functions of dividing the frequencies above and below 150 cps for each channel and of adding the low frequencies from the two channels together. Cathode followers are used both to provide high input impedance and to isolate the two sections of the cascaded RC networks. Only two tubes, each a twin triode, are necessary; and the whole circuit can be enclosed in a 3-in. × 5-in. × 7-in. aluminum utility box, which also contains two octal sockets for distributing B+ and filament power to two preamps. A four-prong Jones plug receives power from the bass amplifier.

The circuit is exceedingly simple. In the left channel (the right channel is identical), C_1 and R_3 form one RC network with $\omega_o = 1000$, corresponding to a crossover frequency of about 150 cps. The cathode follower V_{1A} then lowers the impedance level so that the second RC section, consisting of C_5 and R_{11}, with the same ω_o, can be made of such low impedance elements that no output cathode follower is necessary to drive the cable to the amplifier. In the bass channel, R_1 and R_2 serve both to add the left and right signals together and to form the first RC section with C_7. The second RC section, R_{16} and C_8, follows the cathode follower V_{2A}. An amplifier stage, V_{2B}, with a gain of approximately 2 is necessary because the adding network cuts the bass gain by 2. R_{18} and R_{20} may be varied, keeping their sum constant, to change the gain of the bass channel, depending on the gain of the bass amplifier. Here the gain has been made slightly less than 2 because my bass amplifier has higher gain than the treble amplifiers. The final adjustment, of course, is to be made with the level controls on the amplifiers. When testing with a signal source in only one channel, be sure to short the other channel input to ground, or the adding network will not halve the bass gain the way it would in actual use.

The other two positions of the switch SW_1 provide crossover frequencies of 0 and ∞—that is, with the entire program going straight through to the tweeters alone or to the woofer alone. This frill may be omitted, if desired, but is quite useful for checking the system as well as for having music even when one of the speakers or amplifiers is temporarily out of commission. The impedances of the input RC sections have been chosen so that for any of the switch positions the preamps are not loaded by less than 0.5 megohm at any frequency below 20,000 cps. Such a high impedance level is possible only because of the high effective input impedance of the cathode followers. This is the reason, for instance, that the amplifier stage in the bass section cannot be the input stage. The 0.5-megohm input impedance allows the crossovers to be used with any preamp, including those, such as the Dynakit, which will not drive an impedance smaller than 0.5 megohm without internal modification. Several crossover frequencies can be incorporated and selected with the switch SW_1 if one wants to build a more versatile crossover. However, if one goes to a lower frequency than 150 cps by, say, increasing R_3, the input impedance of the cathode follower will no longer be negligible; and if one goes to a higher frequency, the ω_o's for the RC sections should be staggered to provide uniform frequency response, since the acoustic signals will no longer add in phase, according to our earlier discussion. In my unit I have incorporated crossovers at 300 and 600 cps in case the power handling capacities of the speakers have to be used to the fullest; however, the occasion has never arisen.

Precision resistors and capacitors can be used for the elements of the RC networks, but this expense is not necessary. It would be sufficiently accurate to use 20 per cent elements and then adjust R_3 and R_4 and, if necessary, C_1 and C_2 until all three sections gave an e_o/e_i of ½ at the same frequency. If an oscilloscope with a horizontal input is available, a slightly more sensitive method would be to put the input signal from an oscillator on one axis and the output from the crossover on the other and to find the crossover point by finding the frequency at which the Lissajous figure can be made into a circle. The adjustment can be made simply by adding different resistors in series or parallel with R_3 or R_4 until the crossover point occurs at the proper frequency. The exact frequency does not matter so much, of course, as the matching of the crossover frequency in the three sections.

Two warnings should be given to the constructor: first, do not turn the switch SW_1 when the loudspeakers are on, since the discharging of the coupling capacitors will cause a loud pop. Second, be sure the B+ supply is sufficiently well filtered—the amplifier section is sensitive to hum. Although the circuit was designed to operate with a B+ of 300 v, at which the tube currents are 3 ma in the cathode follower sections and 6 ma in the amplifier section, I had to add several filter sections to my B+ source (the bass power amplifier), dropping the B+ to 200 v, before the hum was eliminated. Both the crossover unit and my Dyna preamps, however, operate well at 200 v.

Alignment of the Speakers

In order to phase the three speakers properly, the following procedure was adopted. First a signal at the crossover frequency of 150 cps was fed into the woofer (on the left) and the right speaker, and the phase of the latter was adjusted to the position in which the signal sounded louder. Then the left and right speakers were brought into phase by using the white noise signal on the Audio Fidelity Test Record. White noise, I find, is the most unambiguous method to check phasing. The gain of the bass channel was then adjusted to get smooth response with the low-frequency glide tone on the *Popular Science* Test Record No. 1.

The Speakers

The other components in the system and the reasons for their choice will now be described, starting at the back end—the speakers. The cabinet for the woofer is a Karlson, built from ¾-in. plywood, veneered with walnut, and finished with boiled linseed oil (*Fig.* 4). The joints were both glued and screwed (using a total of 130 screws), and weatherstripping was used to provide an air tight seal on the back. The inside surfaces were shellacked, with Fiberglas damping material on the two recommended surfaces. The Karlson has a reputation for making a cheap speaker sound good, and indeed it sounded fine with my old $20 15-in. woofer in it. In one splurge, however, I acquired an Altec 803B, the least expensive of the first-rate woofers, and now the speaker is too good for the cabinet. This speaker resonates at 25 cps in free air, but its output at the lowest frequencies is limited by the cabinet, and, to some extent, by the size of the room. The Karlson also has a peak from 70 to 90 cps; fortunately, this peak is rather broad, presumably because of the exponential slot. Some day I may get around to mounting the 803B in an exponential horn or an infinite baffle, although I would hate to part with my first veneering job, which turned out rather well.

The midrange-tweeter speakers are

Fig. 4. The woofer and left speaker.

Fig. 5. The right speaker.

both Wharfedale Super 8's, but mounted differently. The left unit is housed in a small matching cabinet on top of the Karlson, as shown in *Fig.* 4. The back of the cabinet is open, covered only with a grill cloth to keep out the dust.

This arrangement was my solution in the monophonic days to the problem of wanting to mount the midrange unit open-backed, but not having provision for it within the Karlson. The fact that the two cabinets are physically separate also allows for flexibility in arrangement of the furniture. The small rectangular opening in the open-back cabinet is not a port but a mounting hole for a University 4401 tweeter. This was originally added to the Super 8 with a crossover at 8000 cps, but it has since been disconnected since the Super 8 was found to need no help at all at the highest frequencies, and the tweeter merely served to unbalance the left and right channels.

The right speaker, as shown in *Fig.* 5, is mounted in an old corner bass reflex cabinet originally built for a cheap 12-in. speaker. Fortunately, the resonance of the Super 8 (a little below 70 cps), is close to that of the original 12-in. speaker, and no retuning of the port was necessary. Moreover, the Super 8 is not being used as a full-range speaker anyway. There was some worry that the difference in cabinetry for the left and right speakers, necessitated by considerations of decor, would unbalance the two channels. However, this did not turn out to be the case. The relative positions of the speakers are shown in *Fig.* 6. The chair between them is of course not the one used for listening, but the balance is so good that even from this chair one can hear a soloist apparently standing in the middle of the opposite wall and staying there.

You have no doubt noted the extreme separation of the speakers—some 13 feet—and the fact that they are "beamed" toward the center of the

Fig. 6. Placement of the speakers.

room. This means that there is only one really good listening position. However, this has not turned out to be a great disadvantage, since I have found that all serious listening has to be done alone anyway. When there is a crowd,

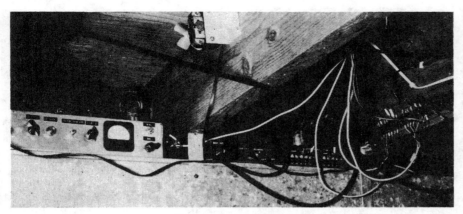

Fig. 7. The amplifiers.

there is always conversation. In spite of the speaker separation, there is absolutely no "hole-in-the-middle." On monophonic sources, the sound comes out of the middle of the wall between the two speakers. If a program has been recorded with too much separation, a turn of the blend switch fills in the hole to any degree desired.

The Super 8's seem to have presence peaks around 2000 and 5000 cps, which are accentuated by the sensitivity of the ear in this frequency range. A broadband RLC filter, constructed with a hand-wound choke, and centered around 4000 cps, was inserted in the speaker circuit to attenuate these peaks and provide a smoother apparent frequency response curve. However, most music did not sound as good with this filter in place as without, mainly because solo instruments would sound muffled and far away. With the filter in, the sound was more nearly like that from "colorless" speaker systems such as the acoustic Research series, but on A-B comparison I almost always prefer the Super 8's as they come. The filtering action of my wife's plants fortunately seems to have a negligible effect.

Amplifiers

The amplifiers, shown in *Fig. 7*, are located on top of a heating duct in the basement. Also visible are the connecting cables going through a hole in the floor to the living room, a patch panel for distributing speaker leads to different parts of the house, and a fan for cooling the output tubes of the bass amplifier. The emphasis, it should be quite apparent, has been on accessibility rather than neatness. The amplifiers are actually so close to the preamps that the standard length cables supplied with the preamps could be used.

The homemade bass amplifier, which also supplies all the preamp and crossover power, employs a modified Dynakit circuit. An ammeter has been added to check the current in the output tubes, the bias and balance being independently adjustable. Controls have also been added for independent adjustment of current and voltage feedback, and for changing from ultralinear to triode operation. These switches have since been found to be unnecessary, the damping being already optimal in the original design.

The treble amplifier is an Eico HF-86 dual 14-watter. This amount of power is entirely adequate for the efficient Super 8's, particularly since the amplifier does not need to supply any bass. With the wife and children safely out of the house, it is possible to turn the volume up to almost the threshold of pain without any sign of distortion.

The bass amplifier will deliver up to 50 watts. Since it is used only for the region below 150 cps, this is more power than is necessary for ordinary program material. However, if there are power peaks in the program sufficiently large to produce distortion, these peaks will occur in the bass, simply because such a large peak in the midrange would be painfully loud. Moreover, running the bass amplifier way below its power rating would practically eliminate harmonic distortion at the lowest frequencies.

Preamps and Control Circuits

The preamps are Dynakit PAM-1's and a DSC-1 stereo control. These were chosen for their versatility, desirable combination and arrangement of controls, and well thought out and sophisticated circuitry, as well as for the distortionless and humless reproduction they are known for. I have had only two complaints with the Dynakits: first, the master volume control had to be changed several times (at Dyna's expense) before one was found that tracked reasonably accurately; second, there is no provision for having both a tape head and a second RIAA input. I believe the latter has been fixed in the PAS-2, which came out after I had bought my preamps. The PAS-2 has several advantages over the PAM-1 plus DSC-1 combination, particularly in cost, but does not have quite the versatility. I still think the volume control should have been changed to a stepped one, even if only 20 per cent resistors are used.

The control panel is shown in *Fig. 8*. Under the Dynakits are two homemade chassis with etched brass panels and knobs to match the preamps. The unit on the left contains the speaker controls, phase reversal switches for the three speakers, a switch for inserting the presence filter mentioned before, and a switch for connecting the left amplifier to the remote outlets, to either an 8-ohm or a 4-ohm extension speaker or to both an 8-ohm speaker and the normal left speaker. The knob marked "VU meter" will be explained later. The unit on the right, under the end of the preamps which are insensitive to hum, contains three relays. Power to the entire system is turned on through a holding relay. This relay can be released, thus turning everything off, either manually or automatically at the end of a tape or record, as selected by the center switch. There is also provision for plugging in a timer to turn the system on and off, for recording radio programs *in absentia*. One of the a.c. switches on the preamps may be used to turn off the amplifiers alone; the other, for turning off the program sources alone.

The Program Sources

The preamps and program sources

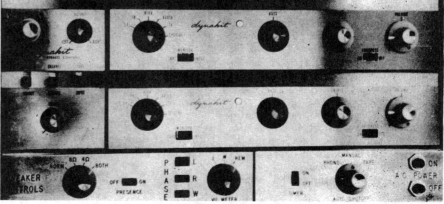

Fig. 8. The control units.

are shown in *Fig.* 9. The cabinet was a floor sample picked up for $10, to which lucite doors and my wife's artistic design have been added. The electronic crossover can be seen under the cabinet, together with the seemingly unavoidable mess of cables. Records are stored in the lower part of the cabinet, and tapes on the small shelf in the adjoining bookcase on the right. The tuner is a Bogen R660, the only component left from the first incarnation of this system and the first one due to be replaced, although it still works quite well.

The enclosures in the cabinet are only about 13-in. × 17-in. × 15-in. deep and originally housed only a Miracord XS-200 changer with a GE GC-7 cartridge. It was obvious that no ordinary turntable or tape recorder could fit in such a space, and that expanding to a larger cabinet would raise howls from you-know-who. Fortunately, there are two components of high quality which do not take up any more room than necessary. They are also very reasonably priced.

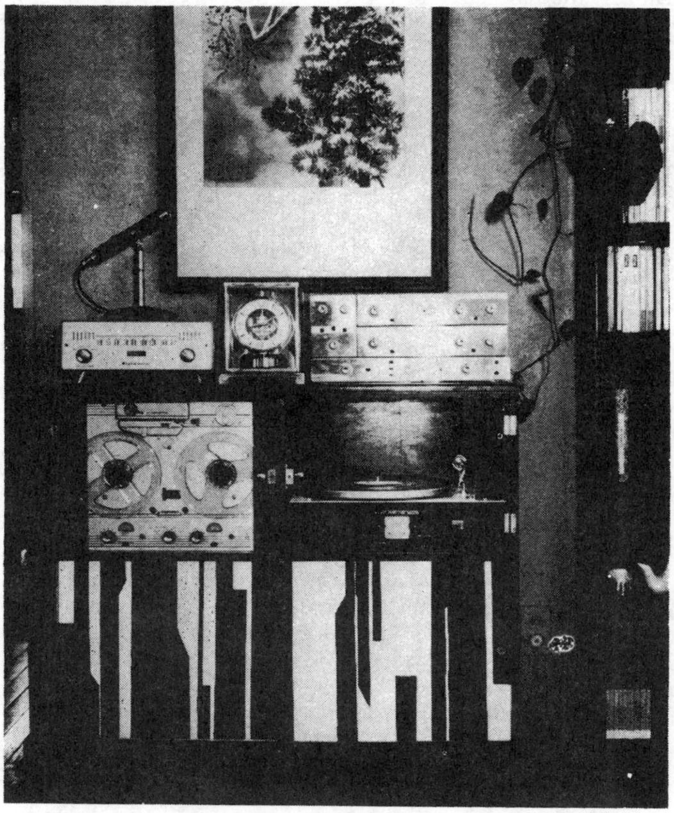

The turntable is a Weathers KL-1 kit, mounted on an aluminum plate suspended above a wooden base. The cartridge and arm chosen is the B & O TA-12 combination. In the turntable base are mounted an hour-counter to keep track of the stylus wear, and a Realistic dual VU meter. The latter is connected across the 32-ohm taps of the Eico dual amplifier to give a little more gain in monitoring the signal actually arriving at the speakers. By means of a switch on the speaker control panel, the left meter can also be put across the woofer or an external speaker. The turntable is lighted by a 6-watt fluorescent lamp at the top front of the cabinet. The light also serves as a strobe lamp. To reduce hum, the ballast and starter for the lamp are located in the basement. The lamp is switched by one of the unused loudness switches on the PAM-1's.

The tape deck is a Viking Stereo Compact RMQ quarter-track machine with two built-in recording amplifiers. This and the Tandberg 6 were the two high-calibre decks which would fit in my cabinet, and, unfortunately, the difference in price was a factor of 2. The Viking is an excellent deck, with a .00009-in.-gap playback head and a separate wide-gap record head, making slow-speed quarter-track recording a reality. I find that with sufficient treble boost, the loss in quality at 3¾ ips is quite acceptable for most music, except when there are high-pitched percussion instruments, and I now do most of the recording off the radio at 3¾ ips, getting 6 hours of music on a single 1800-ft. reel. The Viking is made for hobbyists like myself, and between the Viking and the Dyna preamps, versatility is virtually unlimited. At the moment I do not have playback preamps on the tape deck, and the playback heads are connected directly to the "special" input of the Dynakits, so that the tape-monitor switch cannot be used. However, because of the great flexibility, I can still monitor monophonic recordings by using one preamp for the program source and the other for tape playback, and using the channel reverse switch as a monitor switch. The VU meters on the recording amplifiers are very useful; however, they do not light up, and I had to add pilot lights to show when the amplifiers are on.

I now tape all of my new stereo records, using a slightly greater than normal stylus force, and play the records only on special occasions. Aside from reducing record wear, this practice also eliminates the necessity of meticulously dusting the record each time and of changing records every 20 minutes.

After writing this article, I got up the courage to add up the cost of this system. I came to the conclusion that, exclusive of the tape deck, it can be reproduced for less than $500, plus an awful lot of work. For $550 to $600, one can probably buy a "standard" system of similar quality, but without the versatility and luxury features of this system. Although I would hesitate to recommend this common-bass speaker system to the average music listener, I think it deserves consideration by audio hobbyists. The common-bass speaker system is particularly useful when the woofer can be mounted in a wall, using a closet, a garage, or another room as an infinite baffle. The location of the woofer in the room would be immaterial because of the low crossover frequency, and the Super 8's would require very little room. If the Super 8's were properly baffled as full-range units[4], the crossover could be reduced to below 100 cps; then our assumptions of in-phase addition of acoustic signals and of non-directionality of the bass would hold much more accurately. The common-bass system would also be useful to those who have "full-range" speaker systems which have insufficient output below 60 cps, and who wish to add a single woofer to supplement the extreme bass. Æ

Fig. 9. The home-decorated equipment cabinet.

PARTS LIST

$R_1, R_2, R_5, R_6, R_{13}$	1 megohm
R_3, R_4	2 megohms
R_7, R_8, R_{14}	2200 ohms
R_9, R_{10}, R_{15}	47,000 ohms, 1 watt
R_{11}, R_{12}	20,000 ohms
R_{16}	10,000 ohms
R_{17}	470,000 ohms
R_{18}	18,000 ohms, 1 watt
R_{19}	1000 ohms
R_{20}	8200 ohms, 1 watt
C_1, C_2	500 pf
$C_3, C_4, C_8, C_9, C_{10}$	0.1 µf
C_5, C_6	0.05 µf
C_7	0.002 µf
V_1, V_2	12AU7
SW_1	6-pole, 3-position

---

[4] J. L. Grauer, "#4, 80 Pounds, a Super 8, and the Shim Method," AUDIO, Jan. 1961.

New Design Chart for Bass-Reflex Enclosures

R. D. HERLOCKER

After a thorough study of bass-reflex enclosures and the effects of aspect ratio of the port, the author provides a design chart that takes into account all of the pertinent factors.

THE BASS-REFLEX is probably the most popular type of speaker enclosure that is available today. This is particularly true with those audio fans who build their own. There are several good reasons for this popularity: a properly designed and constructed bass-reflex enclosure will give a smoothly extended low end to a moderately priced speaker; the design is quite flexible, and can usually be modified as may be necessary to fit the available space; it is as economical of materials as any type of enclosure; and last, but certainly by no means least from the standpoint of the person who is building one, it can be built without cutting and fitting a lot of tricky corners, such as are needed with most horn designs.

The bass-reflex is a time-tested enclosure, which, however, seems to have been down-graded in the minds of many people, since the big publicity during the past few years has been for horn-type enclosures, some of which were actually horns by courtesy only. Since the reflex is basically a much simpler design than the horn, it is the writer's belief that, dollar for dollar, at least in the lower and middle price ranges, more performance can be obtained from a bass-reflex and a properly matched speaker than from a horn system of equivalent cost. How far down the bass response goes is, in either case, pretty much a function of the speaker used. While admittedly the electro-acoustic efficiency of the bass-reflex system is lower than that of a good horn, it is still more than adequate for normal home use.

New information on the effects of some of the variables involved in the design of a bass-reflex enclosure has been developed by the author. This information has been incorporated into a new design chart, the use of which will simplify the design calculations for such enclosures.

Some of the comparative disfavor felt toward the bass-reflex is undoubtedly due to the "boom-bass" resulting from gross mis-matching of the enclosure to its speaker. It is well known that, to obtain the best results from a bass-reflex enclosure with a typical speaker, its resonant frequency should match that of the speaker. (An exception to this rule is the case of the very highly compliant speaker, whose resonance will fall in the 15- to 25-cps range, and which is intended for use in an "acoustic stiffness" type of enclosure, which may or may not be ported). Matching, however, is not too critical, within 5 per cent being considered quite satisfactory, because of the low "Q" of the finished enclosure.

Then, too, the resonant frequency of a speaker can easily change that much with prolonged use, thus making more precise matching useless.

Matching may be done in more than one way. The crudest, and least satisfactory, is to cut the port somewhat larger than you expect to need; then, while listening to music containing tones in the desired low-frequency range (such as a good theater pipe organ record), block off increasingly larger areas of the port, until your ear tells you that the response is balanced. The block is now fastened to the inside of the port in this position. A somewhat more refined, but still tedious, method is to run impedance curves for the frequency range below 100 cps, first on the speaker alone, and then with the speaker mounted in the enclosure, adjusting the port size until the ratio of the resonant frequency of the speaker alone to that of the lower peak of the speaker in the enclosure equals the ratio of the upper peak frequency to that of the speaker alone.

If, while the enclosure is still in the planning stage, it can be matched to the speaker, even approximately, much of this cut-and-try can be eliminated. Use of the design chart will accomplish this end, which is especially important if a ducted port is to be used, such as in several designs by Jensen[1] and others[2,3], since the duct dimensions cannot easily be changed once the enclosure is completed.

Design Equations

The basic design equation for the bass reflex enclosure is:

$$F^2 = \frac{2150^2 A}{V(L + A^{0.5})} \quad (1)$$

where

F = resonant frequency, cps (taken as the speaker resonant frequency)

---

[1] D. J. Plach, & P. B. Williams, "The Bass Ultraflex enclosure." *Radio & Television News*, December, 1954.
[2] Monitor, "Pro-Plane Prismatic Speaker System," *Radio-Electronics*, February, 1956.
[3] F. Langford-Smith, *Radiotron Designer's Handbook*, 4th ed., Radio Corp. of America, 1953 (page 847).

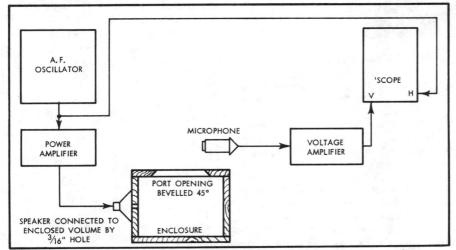

Fig. 1. Block diagram of equipment arrangement for determining resonant frequency of bass-reflex enclosures.

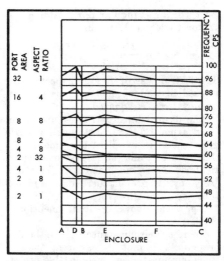

Fig. 2. Comparison of resonant frequencies of various shapes of enclosures.

A = port area, square inches
V = enclosed volume, cubic inches
L = duct length, inches (if no duct is used, L is the thickness of the enclosure wall).

It has been known for quite a while that the shape of the enclosure and of the port may affect the validity of this question. However, it has been generally assumed that enclosure shape has no significant effect, as long as the longest enclosed dimension is not over three times the shortest, to avoid an "organ pipe" type of resonance. Similarly, the effect of port shape was of little importance until recently, when slotted ports began to be widely used. Voigt[4] indicated that a definite relationship exists, though he did not fully develop it. On the other hand, Moir[5] claims that the shape of the port has no noticeable effect on the resonant frequency of an enclosure.

A set of experiments was designed by the author to determine the effect, if any, of changes in enclosure and port shapes on the resonant frequencies of bass-reflex enclosures.

Six enclosures were constructed, each of which contained 3036 cubic inches of air, within 0.5 per cent, but of varying dimensional ratios, as listed in Table I. All joints were carefully sealed, and an eight inch square port was cut into one side of each box. The edges of each port were beveled to 45 deg. to minimize the ducting effect of wall thickness. An "L" shaped piece of plywood, with its edges similarly beveled, was used to change the size and shape of the port as desired. By maintaining these restrictions, it was possible to eliminate volume and duct length from the variables, and to concentrate attention to the port itself, and to enclosure shape.

Resonant frequencies were determined for each of the six enclosures for a variety of port sizes and shapes. The test procedure was that given by Moir.[6] A hole, approximately 3/16 inches in diameter, was drilled in one wall of the enclosure, and a small TV-type speaker fastened tightly over the hole. The speaker was fed from an audio oscillator through a small power amplifier. While Moir states (and this writer has confirmed) that resonance can be quite accurately detected by ear, for these tests a microphone was placed adjacent to the port, within three to four inches, and its output fed through a voltage amplifier to the vertical plates of an oscilloscope, where the microphone output was compared to the direct output of the oscillator, which was simultaneously fed to the scope's horizontal plates. The equipment arrangement for these tests is shown in *Fig.* 1. The frequency of maximum pickup from the microphone was taken as the resonant frequency of the enclosure. Incidentally, the phase angle of the sound from the port changes very rapidly in he immediate vicinity of resonance, as can readily be seen on the oscilloscope screen.

There was no interference from the natural resonance of the speaker, as it was completely damped out by the small amount of air trapped between the speaker and the enclosure wall. Neither was there any noticeable interference from direct radiation from the speaker. Most measurements were made in the low-frequency range where the speaker's direct radiation efficiency is very low, and the use of a directional microphone helped.

A comparison of resonances of the six enclosures tested is given in *Fig.* 2. (See Table I for identification of the enclosures). They are compared here at a variety of port areas ranging from 2 to 32 square inches, and aspect ratios (port length divided by width) from 1 to 32. Enclosure "A" is a cube, and enclosure "C," (at the right of *Fig.* 2), with its 1:1:3 internal dimension ratio, represents the greatest departure from the cube of any of those tested. In general,

[4] P. G. A. H. Voigt, "All about the reflex enclosure" *Radio-Electronics*, April, 1959.
[5] James Moir, "Ported loudspeaker cabinets" AUDIO, October, 1956.
[6] James Moir, *"High Quality Sound Reproduction"* MacMillan: 1958, (page 446).

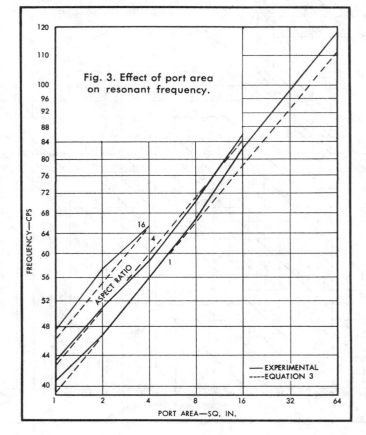

Fig. 3. Effect of port area on resonant frequency.

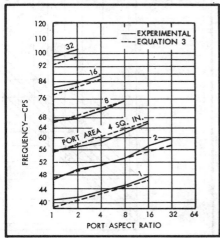

Fig. 4. Effect of aspect ratio on resonant frequency.

the enclosures whose shapes are furthest from the cube have slightly lower resonances than those whose shapes approach the cube. However, the difference is very slight, and is also rather erratic, so that, on the basis of these tests, it is believed best to ignore this factor in design, at least within the limit of the three-to-one maximum ratio of longest to shortest sides.

In *Fig.* 3 is shown the relationship between port size and resonant frequency. For this figure, and also for *Fig.* 4, frequency values for the six enclosures have been averaged, in the interest of clarity. This procedure is believed valid, in view of the lack of significant variation among them.

Since the enclosures were designed to eliminate (or at least minimize) the ducting effect of the wall thickness, we may set $L=0$ in Eq. (1), and we have

$$F^2 = \frac{2150^2 \, A^{0.5}}{V} \quad (2)$$

which is the dashed line in *Fig.* 3, corresponding to a port aspect ratio of 1 (square port). The solid line represents the test results.

The other two pairs of lines in this figure give similar test and calculated results for port aspect ratios of 4 and 16. The agreement between calculated and experimental data is quite good, verifying the validity of Eq. (2) for these test conditions, including the simplifying assumption of zero port thickness.

Change of Port Shape

The effect of changing the shape of the port is shown in *Fig.* 4. Each solid

TABLE I
ENCLOSURES TESTED

Enclosure	Enclosed Volume, Cubic Inches	Dimension Ratio
A	3050	1:1:1
B	3051	1:1:1.5
C	3022	1:1:3
D	3044	1:1.5:1.5
E	3048	1:2:2
F	3032	1:3:3

line in this figure represents the resonant frequencies obtained from a given port area by changing its shape. All ports involved here were rectangular, with aspect ratios varying from 1 to as high as 32. In all cases, there is a regular increase in resonant frequency as the aspect ratio is raised. This increase is proportional to the 0.06 power of the aspect ratio of the port, for a rectangular port. Adding this factor to Eq. (2), we have

$$F^2 = \frac{2150^2 \, A^{0.5} \, R^{0.12}}{V} \quad (3)$$

The 0.12 power of R (the aspect ratio) is used instead of 0.06 because the frequency term is squared.

The dashed lines of *Fig.* 4 are calculated from Eq. (3). The smaller ports give results in excellent agreement with this equation, but the largest ones (particularly 32 sq. in.) show a slight discrepancy. At present there is no explanation for this. However, it is within the 5 per cent working limit mentioned earlier.

There are circumstances where the discrepancies cannot be ignored, though. Such circumstances include the use of complex shaped ports, such as L-shaped, or multiple ports. Table II shows some test results from the use of L-shaped and twin rectangular slots as ports with enclosure B (see Table I). In all cases, the resonant frequency is raised well above that of an equivalent rectangular port, even taking the most favorable value of R for comparison. In those few cases where complex porting would be felt to be necessary or desirable, one must re-

(Continued on page 68)

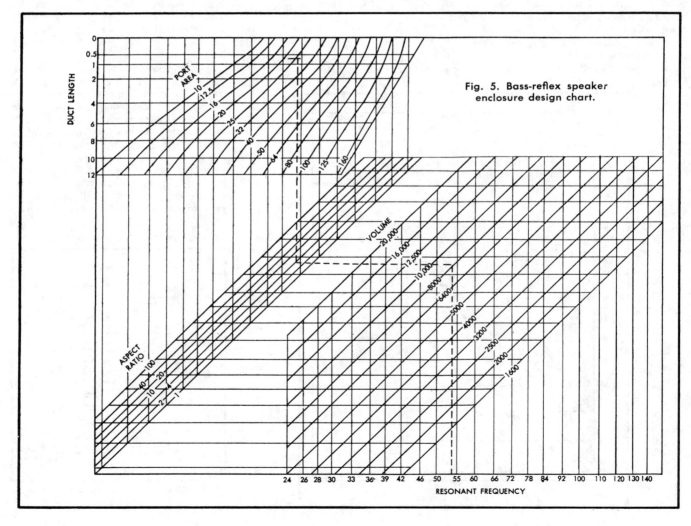

Fig. 5. Bass-reflex speaker enclosure design chart.

What Hath FCC Wrought?

DAVID SASLAW

After several years of study, the FCC has finally decided upon a system of FM stereo. The system approved has the capability of achieving fidelity as high as is now available with monophonic FM broadcasts. Here are some details—

ON APRIL 20TH OF THIS YEAR, the FCC made known its decision to permit FM stereo broadcasting commencing June 1. The decision, although actively sought for many years, seems to have caught many people by surprise. The surprise does not center about multiplexing as such, but rather the system chosen by the FCC. On the one hand there have been comments implying that the chosen system is "not as good" as one of the systems not chosen. On the other hand, there have been comments to the effect that many FM tuners were rendered "obsolete" because of the FCC decision. Actually these reactions are quite natural considering the enthusiasm with which particular systems were championed and the unusual lack of factual information about the system chosen. At this point it might be well to point out that the Zenith-GE system performed as well as the best of the systems not selected, and in addition is essentially more flexible (able to retain the existing commercial FM multiplex, SCA) than the runner-up system (Crosby). As for the fear that some types of tuners would be unable to be adapted to stereo operation, this is just not so. More about both of these points later.

Exactly What Is the New Stereo System?

First of all it should be noted that the FCC merely established what is to be transmitted, not how it is to be achieved. In essence, what was approved was a basic equation which defines the signal to be transmitted (for the exact equation see the Daniel von Recklinghausen article on page 8). In addition, the FCC defined the transmission standards for this basic signal in order to ensure high quality broadcasts (from a technical viewpoint). Following is the section of FCC Docket 13506 wherein the transmission standards are defined:

§ 3.322 *Stereophonic Transmission Standards.*

(a) The modulating signal for the main channel shall consist of the sum of the left and right signals.

(b) A pilot subcarrier at 19,000 cycles plus or minus 2 cycles shall be transmitted that shall frequency modulate the main carrier between the limits of 8 and 10 per cent.

(c) The stereophonic subcarrier shall be the second harmonic of the pilot subcarrier and shall cross the time axis with a positive slope simultaneously with each crossing of the time axis by the pilot subcarrier.

(d) Amplitude modulation of the stereophonic subcarrier shall be used.

(e) The stereophonic subcarrier shall be suppressed to a level less than one per cent modulation of the main carrier.

(f) The stereophonic subcarrier shall be capable of accepting audio frequencies from 50 to 15,000 cycles.

(g) The modulating signal for the stereophonic subcarrier shall be equal to the difference of the left and right signals.

(h) The pre-emphasis characteristics of the stereophonic subchannel shall be identical with those of the main channel with respect to phase and amplitude at all frequencies.

(i) The sum of the side bands resulting from amplitude modulation of the stereophonic subcarrier shall not cause a peak deviation of the main carrier in excess of 45 per cent of total modulation (excluding SCA subcarriers) when only a left (or right) signal exists; simultaneously in the main channel, the deviation when only a left (or right) signal exists shall not exceed 45 per cent of total modulation (excluding SCA subcarriers).

(j) Total modulation of the main carrier including pilot subcarrier and SCA subcarriers shall meet the requirements of Section 3.268 with maximum modulation of the main carrier by all SCA subcarriers limited to 10 per cent.

(k) At the instant when only a positive left signal is applied, the main channel modulation shall cause an upward deviation of the main carrier frequency; and the stereophonic subcarrier and its sidebands signal shall cross the time axis simultaneously and in the same direction.

(l) The ratio of peak main channel deviation to peak stereophonic subchannel deviation when only a steady state left (or right) signal exists shall be within plus or minus 3.5 per cent of unity for all levels of this signal and all frequencies from 50 to 15,000 cycles.

(m) The phase difference between the zero points of the main channel signal and the stereophonic subcarrier sidebands envelope, when only a steady state left (or right) signal exists, shall not exceed plus or minus 3 degrees for audio modulating frequencies from 50 to 15,000 cycles.

NOTE: If the stereophonic separation between left and right stereophonic channels is better than 29.7 decibels at audio modulating frequencies between 50 and 15,000 cycles, it will be assumed that paragraphs (1) and (m) of this section have been complied with.

(n) Crosstalk into the main channel caused by a signal in the stereophonic subchannel shall be attenuated at least 40 decibels below 90 per cent modulation.

(o) Crosstalk into the stereophonic subchannel caused by a signal in the main channel shall be attenuated at least 40 decibels below 90 per cent modulation.

(p) For required transmitter performance, all of the requirements of Section 3.254 shall apply with the exception that the maximum modulation to be employed is 90 per cent (excluding pilot subcarrier) rather than 100 per cent.

(q) For electrical performance standards of the transmitter and associated equipment, the requirements of Section 3.317 (a) (2), (3), (4) and (5) shall apply to the main channel and stereophonic subchannel alike, except that where 100 per cent modulation is referred to, this figure shall include the pilot subcarrier.

Broadcasting FM Stereo

Previously we indicated that the FCC had approved the form and the technical standards for the stereo signal, not how it would be achieved. The distinction is quite significant. For instance, the approved system is known as the GE-Zenith system. Both GE and Zenith have on file with the FCC diagrams of proposed methods for producing the signal in the desired form. Both diagrams happen to be essentially similar as to method of signal generation. (Compare *Fig.* 1 in this article with *Fig.* 1 in the article by Csicsatka and Linz, page 16.) Does this mean that this method must be used by all broadcasters?

Definitely not.

There are at least two fundamentally different methods for achieving the standard signal. One of them is the type illustrated by GE and Zenith and described in the Csicsatka and Linz article, the other system is described at length in the von Recklinghausen article.

To summarize the two methods, the GE-Zenith technique requires a matrix wherein the sum and difference signals are achieved. The sum signal is fed to the FM exciter after a suitable time delay (to keep it in step with the difference signal which follows a somewhat different path). The difference signal, on the other hand, goes through a 38,000-cps suppressed carrier AM modulator and then to the FM exciter. Of course, both the sum and difference signals are properly pre-emphasized (before the matrix in the GE diagram, after in the Zenith), and the difference signal is filtered to eliminate the harmonics of the carrier.

Figure 2 shows the block diagram of an AM modulator (Zenith proposal) which will generate the 15,000-cps-wide sidebands around the 38,000-cps carrier (which is suppressed) and, at the same time, provide the 19,000-cps pilot signal. The method used in this modulator is to mix the output of two crystal oscillators to provide the 38,000-cps carrier, or, by a 2/1 division, the 19,000 cps pilot. The pilot is then added to the sidebands which remain after the carrier is suppressed and the combined signal passed through a linear-phase-shift low-pass filter. If we refer back to the FCC specifications we note that the maximum difference permitted between the main and subchannel is 3 degrees. For this reason, care had to be exercised to avoid introduction of unwanted phase shift. This is especially true in a system such as this wherein the paths for the main and subchannel are not identical.

On the other hand, the second method for generation of the standard signal is far less critical as to phase shift. This system is a time-division multiplex switching system between left and right stereophonic program channels. In this system a switching device alternately takes the whole left or the whole right signal. In a way it is rather difficult to understand how switching rapidly between the two inputs will produce the standard equation, but the fact of the matter is that it does produce it. In actuality, the form of the standard equation is the key to its derivation; it is essentially a Fourier expansion of two variables. That is, if we were to take any two independent variables and expand them mathematically we would arrive at substantially the same equation. According to Carl Eilers of Zenith Radio Corp. (who worked on it), the idea for the now accepted system had its inception through mathematical analysis of the time-division multiplex signal. Strangely enough, in its own presentation, Zenith did not propose this method of signal generation. Instead they proposed the system shown in *Fig.* 1. As far as we know at present, H. H. Scott, Inc. is the only proponent of the time-division method of signal generation on a practical basis; they have announced the availability of equipment utilizing this principle. Although on the surface this method is much more sophisticated from an engineering viewpoint, it is quite possible that the matrixing method is much easier to integrate with existing equipment. Most likely this is the reason most of the emphasis has been placed on the matrixing method. It is also possible that the time-division method is not as well known as it should be.

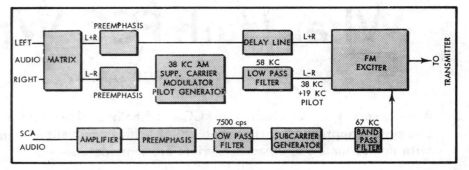

Fig. 1. Block diagram of Zenith proposal for achieving the stereophonic signal.

Receiving FM Stereo

Just as there are two fundamentally different methods for transmitting the stereo signal, there are an equal number of methods for receiving it. It might be said that these methods are "mirror images" of the transmission systems; they essentially reverse the procedure of the broadcast station.

For example, let us consider the matrixing method. The sum signal is derived from a matrix and, except for some normal processing, is transmitted in that form. Referring to the block diagram of the GE adapter (*Fig.* 2 in the Csicsatka and Linz article) we can see that the receiver just reverses this process. The difference signal, however must first be recovered from the sidebands in which it was transmitted. This involves reinserting the 38,000-cps carrier which was suppressed at the transmitter and then separating the audio from the carrier. In order to reinsert the carrier precisely where it should be, the 19,000 cps pilot is used; either to synchronize a local 38,000-cps oscillator or directly in a doubler circuit. We know that this pilot will give us the precise time location we need since it was transmitted with the signal. Then we can demodulate. The difference signal is now ready for matrixing to recover the original left and right signals which started the whole process.

The switching (time-division) method, used in the receiver, is a "mirror image" of the time-division transmission method although there are several ways of effecting the time division. *Figure* 3 is the block diagram of the method used by Zenith and H. H. Scott uses another method as shown in *Fig.* 4 of the von Recklinghausen article. In both cases the switching is synchronized by a 38,000-cps signal derived from the 19,000-cps pilot. The H. H. Scott adapter is explained in detail by Mr. von Recklinghausen. A schematic of the Zenith adapter is given in *Fig.* 4. In reality very little explanation is necessary for this system once the time-division method of generation is clear; the receiver is only required to switch between signals at the precise rate used in broadcasting to reverse the process and extend the original left and right signals. The switching rate (as explained by von Recklinghausen) is 38,000 cps, the second harmonic of the pilot signal.

The natural question now is whether the matrix receiver will operate with a signal generated by the time-division method and vice versa. The answer to this, of course, is that the signal generated by both the matrixing and time-division methods is exactly the same in the air; the receiver sees the same signal no matter which method is used to generate it. I must admit, however, that to date I have never actually "heard" the time-division method (I did attend a demonstration of the matrixing method), but then how many people have?

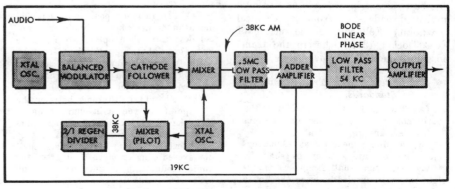

Fig. 2. AM modulator and 19,000-cps pilot generator.

Existing Equipment

Now to return to the apprehensions expressed by some people about the forced obsolescence of some existing tuners. Apparently it was felt that tuners that use Foster-Seely type discriminators would be unable to be adapted. The reality of the situation is that there is no valid technical reason for this fear. It has been stated that discriminators of the Foster-Seely type are inherently too narrow in bandwidth to handle FM stereo transmission, a situation made more difficult by the fact that in stereophonic transmission more energy is concentrated near the edges of the passband than heretofore. In fact, a well-designed FM tuner, whether it uses a ratio detector or Foster-Seely discriminator, can have a sufficiently wide bandwidth to handle stereophonic transmission under the rules adopted. The key words here are "well-designed." Certainly a poorly designed tuner with a discriminator will probably yield unacceptable distortion with stereophonic signals. But the fact of the matter is that such a tuner will provide distorted monophonic signals too. A poorly designed FM tuner, even with a ratio detector, will provide unacceptable distortion. This whole matter can be summed up by noting that a high-quality tuner will provide a high-quality signal, whereas a poor-quality unit will distort—no matter what system of FM detection is used. In other words, the existing high-quality FM tuner is not obsolete.

Is the Monophonic Signal Degraded?

In our enthusiasm for the marvels of FM stereo, we tend to overlook the fact that many people will want to continue receiving monophonic FM programs for some time. Will they "pay the piper" for those who wish to have stereo now. In other words, is the monophonic signal degraded? The answer is no—and in fact this was one of the important reasons for selecting the GE-Zenith system. The following excerpt from FCC Docket 13506 indicates that the degradation is only experienced in the stereo channel (system 4-4A is the Zenith-GE system, 1 is the Crosby system):

15. In comparing FM stereophonic systems, it is customary to use as the standard of comparison the signal-to-noise ratio obtained with monophonic transmission and reception for a given amount of transmitted power and other specified conditions, including height of antenna, transmission path and receiver sensitivity. When stereophonic transmission is substituted under the same set of conditions, the main carrier output and subcarrier output at the receiver will have reduced signal-to-noise ratios. The amount of reduction depends upon a number of transmission parameters, including the subcarrier frequency, the frequency swing of the main and subcarriers and the deviation of the main carrier caused by the subcarrier or subcarriers. The calculated loss of signal-to-noise ratio, compared to monophonic transmission and reception for each System is:

Monophonic	System 1	System 4-4A
receiver output	6 db	less than 1 db
Subcarrier output	15 db	23 db
Left signal output	13 db	20 db
Right signal output	13 db	20 db

16. It will be observed that System 1 has the greater loss in signal to noise ratio for monophonic reception and the lesser loss for stereo; conversely, System 4-4A has a smaller loss for monophonic reception and a greater loss for stereo. Both the monophonic and stereo losses for System 4-4A would be greater if SCA subcarrier frequencies were also used.

Clearly the public has been served by this FCC decision—it not only preserves the existing monophonic transmission, but in addition provides the new dimension of stereo.

Our title paraphrases the statement made at the inauguration of the telephone which ushered in a new era. I trust this decision by the FCC may be as momentous for FM broadcasting. Æ

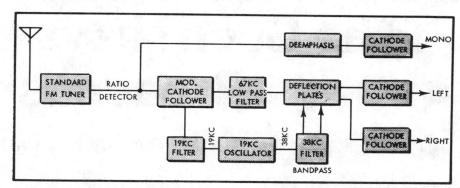

Fig. 3. Block diagram of Zenith time-division adapter.

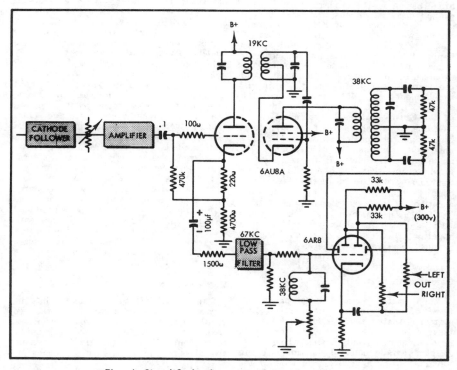

Fig. 4. Simplified schematic of Zenith adapter.

The Good News – Stereo Broadcast

EDWARD TATNALL CANBY

Well, the first glossy photo of a real, honest-to-goodness FM-multiplex stereo adapter has just come in the mail (cost $99.50), the F.C.C. decision on stereo broadcast is still a breathless bit of news—and it seems I made another of my left-handed prophesies. (A left-handed prophesy is one that was written before but published after the fact.) I said last month that maybe broadcast stereo would be this season's coming hi fi sensation. It will be, as witness this issue of AUDIO, given over to the good news that took so long in coming.

You will of course look elsewhere in AUDIO for the technical description of the wining system, if you haven't had it lying around in your files for a year or so already. (My General Electric press release on this system came to me in the spring of 1960.) It is my business to take the wide view and observe what I may as to the implications.

Well frankly, I am delighted. For it seems to me that a situation which three years ago looked like a real "mess," insoluble unless either an existing industry were to be wiped out (the music stations broadcasting multiplexed restaurant background material on a subscription basis), or unless the quality of FM stereo broadcasting were to be seriously compromised, has instead been resolved with all parties happy. We have found a way to have our cakes and eat them too, without compromise, or so I gather at this point. And the method is ingenious to the point of wonderment—why didn't somebody think of it earlier, back in the acrimonious days of the Crosby-Halstead arguments?

Genius, to paraphrase the familiar phrase, is an infinite capacity to be simple. If I am right, the men at G.E.-Zenith (I'll make no distinctions) are to be congratulated for just that sort of simplicity. They figured a way to avoid all the clashes and conflicts and compromises and counter-accusations of the rival camps that first tried to launch stereo broadcasts three years back, and, it seems, their system is possibly superior even to the best of the original propositions. Hard to believe.

Components Plus . . .

You'll probably read that the prime feature of the now-accepted G.E.-Zenith system is that it requires only a relatively inexpensive conversion unit, the multiplex decoder, utilizing only a single tube in its simplest arrangement. True, this is a big advantage. It's a vital one in view of the huge importance in radio of the mass market, both the small mass-produced radio sets and the millions of large, mass-produced consoles. The fact is, of course, that component-style radio is all very fine, but radio itself, i.e., broadcasting, can't get along only on components. If we are to have broadcasting of any sort, it must be based on components plus mass-produced equipment. Period.

The broadcast system that will be successful obviously must provide a dual basis: (a) the means for top-quality sound, in the broadcast and in the reception, and (b) simultaneously, the means for simple, low-cost, mass-production receiving equipment.

AM radio has never, except in special cases and special areas, provided the full top-quality potential in actual practice, but has fulfilled the mass-production aspect admirably. That's why AM still horses along, in spite of TV and FM combined. FM broadcast, mono, has done an admirable job from the beginning on the first proposition—top quality sound for top-quality reception. Unfortunately, FM's mass-production capabilities were underplayed at the beginning, and the medium almost went under for good. (I was in on that near-debacle myself; I lost a good job right after the war when the FM boom failed to materialize as predicted.)

Now, happily, in the last few years FM has found itself, partly boosted up by increasing component sales and the general impact of the idea of hi fi, but also buoyed by an increasingly solid mass-production basis in the receiver area. Now—who'd have guessed it even ten years ago?—we have transistor FM-AM pocket radios, not to mention all other sizes. Sound quality is not exactly the strong point in these models. Their mere existence proves my point, that *quality broadcast depends on non-quality mass-production reception.*

FM, then, is on solid ground at last, in the mono medium. And meanwhile stereo, in spite of its not-very-handsome progress, has slowly taken over more and more ground, everywhere but in radio. The persistence of the clumsy AM-FM stereo broadcasts and a few FM-FM arrangements shows that the missing link of broadcast stereo has been a very real one. After all, it is the last link in the chain of linked elements that includes stereo discs, stereo tape, tape recorders and players, the home hi fi component system and mass-produced stereo "hi fi" (quotes are traditional with me). (You'll note that we now have another new noun in the business. A few years back, "hi fi" became "*the* hi fi"; now one goes out and buys a "stereo.")

Compromises

Thus, to go back to my earlier point, the new stereo broadcast system faced the two vital requirements that would somehow have to be met, both of them—top quality throughout, plus adaptability for inexpensive mass-production receiver equipment.

If both a stereo hi fi band and one or more background music bands were to be carried piggyback on a single FM transmission via multiplex, there just didn't seem to be any way in which they could keep out of each other's hair and yet remain uncompromised in sound quality. Even the reasonable compromise in the assigned bandwidths prepared by Editor McProud was still based on the premise that we had to have both stereo and background music, on the same broadcast signal, imposed in the same manner and thus still potentially subject at least to accidental interference.

What nobody saw, at that time, was the brilliant possibility—so simple—that if we could multiplex these two different and competing signals—one hi fi music, the other mood music—by *different systems* upon the same FM broadcast carrier, there would be no interference. The two would not recognize each other, nor, so to speak, be aware of each other's existence, though part of the same basic signal wave. This was the blindingly happy G.E.-Zenith idea. The question was merely—how. And it must have taken a while to work out the details, since this system did not appear in public and before the F.C.C. until a good while after the earlier proposals.

Again, I refer you to more detailed explanations elsewhere, but just marvel with me at the neatness of the very concept itself: Put your second stereo signal, (the difference signal), onto the main carrier via an *amplitude modulated* subcarrier, suppress the carrier, and use the remaining AM sidebands to *frequency modulate* the main FM carrier. Then, higher up (you have to 75 kc), put the background music signal on the same carrier via the present FM multiplexing. Thus the two will be mutually exclusive; the detector that brings one of these signals to life can't even "hear" the other signal at all. It's wonderful!

It Looks Like This

As the Boss explained the inner details to me (he's the best teacher I've ever run into), the FM stereo transmission under the new system will look like this—I'll give it to you in my lay words, just as a side-prop to the more proper engineering accounts elsewhere.

From 50 to 15,000 cps, on the FM carrier is the main or sum channel, receivable as a standard FM mono transmission. At 19,000 cps there's a control "synch" signal, to grab the local oscillator (producing the subcarrier) in the home multiplex receiver, similar to the TV arrangement. That oscillator runs at 38,000 cps, the second harmonic of the control signal.

Then from 23,000 to 53,000 cps, still on the FM main transmission, is the vital AM multiplexed signal, 15,000 cps wide on each side of the 38,000 cps AM carrier frequency, suppressed at the transmitter—only the side bands go out on the air.

What? said I, the sides without the middle? Sounds to me like a bottle of milk without the bottle. Yet so it is.

The oscillator in your home receiver puts back the 38,000 cps, synchronized by that 19,000 cps control tone. And the simple fixed AM detector just tunes in on it, removing an AM sound-signal.

And 'way, 'way up in the FM stratosphere, still on that same main FM carrier, going on up to the limit, 75,000 cps, is the piggyback FM-multiplexed background music signal, different music, for different receivers. It's far away from the lower FM signal, whose top is 15,000 cps. And it can't even hear the AM signal that fits in between. Neat, neat, neat!

No interference, top stereo quality (50–15,000 cps on both the sum and difference signals) and everybody happy. Cheap receivers too, or so I hear, in spite of that $99.50 job I mentioned. That is—cheap equipment is inherently possible and satisfactory within its own sphere of operations. Mighty important, I tell you, even if the top product does cost $100, self-powered. Imagine if it cost $99.50 to buy *any* stereo adapter. Or even $39.50. It won't, if all goes well.

Now there's only one pay-off to mention here. I wonder just how many hopefully thinking audiofans have jumped to a conclusion that is as likely as a short circuit in a wet distributor. Namely, that AM for the second (difference) stereo channel is bad, because it'll be subject to static and noise (whereas the main signal, FM, is immune to the noise)?

This, of course, was one of the most violent objections most of us had to the old FM-AM stereo broadcasts, one stereo channel via the FM transmitter and the other via AM. The noise, in one channel but not in the other, was far more distracting than the difference in tonal quality between the two channels.

Do we have that problem again here, with FM *and* AM both in use?

Of course not! I almost fell into this trap myself, but managed to save myself and my reputation just before I sprawled. This AM is multiplexed upon a carrier FM signal; it doesn't exist in "free air" so to speak. The receiver picks up only the FM signal, carrying with it three different messages. The AM part is purely internal, if you see what I mean. So—no static.

Sure, sure, elementary; but I'll bet a lot of folks get tangled up on this little point, just the same.

* * * *

Sometimes I wonder what it must be like to be a "lay" member of that august body, the F.C.C. The explanations that must go on, with millions of dollars hanging on every word! Lay or no, I'd say that the new F.C.C. has done a good job this time, seeing a good thing when it was presented, coming to a relatively quick decision in view of the "recency" (as Mr. Harding might say, he of that famous word, normalcy)—the recency of the new set-up.

We'll all be talking more, much more, about the forms of equipment that are now likely for stereo broadcast and the forms of stereo entertainment that may crop up. *That*, of course, is very much within my own province and if anybody around here wants a guy who knows the technical ropes of stereo *and* the parallel ropes of its main signal—the art of music—I'm all set to produce stereo music broadcasts galore. I can even explain sum and differences to the folks "out thar," if I have to. That takes explainin', let me tell you, especially to Aunt Mamie, who thinks "stereo" is something she saw in the movies, with a giant screen.

Have recorder, can edit. End of plug.

Æ

CUSTOM CONSOLE

(from page 50)

of fast changing developments in the audio component field, and their effects on obsolescence, we designed the control panel and the attached supports, such that the unit is entirely removable. Two screws are used to fasten the panel assembly. For instance, should it become necessary due to a future acquisition of a multiplex unit, the sum of less than $5 (coupled with a morning's effort) will find a new panel and support assembly framing the revised "front ends".

In the same vein, future purchase of the almost inevitable tape machine, will find a generously spaced home in the right hand lower compartment now occupied by a patented, slide-out, record storage rack (*Fig.* 6). The space available for the tape instrument is designed to accept rack mounting type recorders as well as the more diminutive portable instruments. As the reader may well reason, audio custom design is not all black and white! This addition of a possible tape instrument is obviously confined deep in stage one!

Lastly, several features were incorporated which are well nigh invisible, but do much towards adding convenience and mobility to the entire system. Previously mentioned, there is the inclusion of the sliding record rack. In connection with the turntable compartment, looking at *Fig.* 7, one notes the use of a lamp. This lamp is automatically switched by a push-pull pressure switch operated by the door closure. This feature was added at very reasonable cost when I discovered a General Electric Co. automatic closet lamp fixture for under $2.00! In conclusion, the entire console (in operating condition weighing some 200 pounds) is mounted on four 2-inch rubber tired casters so that the efficient housewife is not hampered in her tidying up. The console rolls quite easily at such times!

Many excellent treatises are available on the manner and methods of wood finishing. The only comment necessary here is that the basic ingredient needed to acquire that professional hand-rubbed look so desired by all, is simply hand sanding and polishing! No small contribution to the sheen of the final finish, were the long hours spent sanding and polishing (with 4-0 steel wool) the unstained raw wood. This was an extra smoothing operation in addition to the polishing done after staining and between each of the three successive coats of lacquer. The whole ensemble is stained a platinum walnut. The color of this stain is beautifully enhanced by this lengthy but rewarding process.

Fig. 6. Record storage now—tape recorder later.

It is rather evident that there has been a deliberate attempt on the part of the author to minimize the description and attendant elaboration on detail. This elaboration was purposely avoided. Any audiofan proceeding seriously from stage one to stage two and including in the planning stage the requirement for custom cabinetry, must of necessity tailor his design to his individual specifications, both in choice of instrumentation and decor. Very briefly: The author's console scales 50 inches long, 30 inches high, and 23 inches in depth, while the speaker enclosures were modified (in decor only) from factory plans included with each speaker component kit.

The solutions presented herein, particularly that of laminated case construction, are sufficiently adaptable to permit their inclusion in the majority of instances where the audiofan is desirous of having the type and quality of furniture housing that will be commensurate with the quality of his carefully chosen components. It is hoped that the methods of achieving this custom look as presented herein may satisfy his needs and desires, while consuming less of his exchequer than were he to purchase the equivalent in the trade marts.

Æ

Fig. 7. The turntable—with an automatic lamp to light the way to the spindle.

FM Stereo—The General Electric System

ANTAL CSICSATKA and ROBERT M. LINZ

Here, in abbreviated form, is an explanation of the General Electric FM stereo system by two of the engineers responsible for it. In addition, a description of a one-tube stereo adapter is given.

THE KEY CHARACTERISTICS of the FM stereophonic broadcasting system adopted by the Federal Communications Commission are that it satisfies all the requirements the Commission set forth and can operate with a one-tube adapter to produce stereophonic sound from a conventional tuner and stereophonic amplifier.

An examination of system specifications (Table I) quickly enlightens the serious listener to the capabilities of the adopted system. In particular, reference is made to the fully separated stereo (30 db from 50–15,000 cps), while maintaining existing monophonic distortion requirements.[1]

The transmitter[2] is most easily explained with reference to the block diagram of *Fig.* 1. The left, L, and right, R, signals are developed conventionally and then pre-emphasized separately before being fed to the matrix where the sum (L+R) and difference (L−R) are produced. The L+R signal is fed directly to the FM modulator in the usual fashion to frequency modulate the main carrier thus providing one portion of the stereophonic signal while simultaneously serving as the compatible monophonic signal. The L−R signal is fed into a balanced modulator where proportional sidebands are generated above and below the subcarrier frequency of 38,000 cps. The subcarrier is automatically suppressed, but the L−R sidebands frequency modulate the main carrier.

It should be noted that the carrier input to the balanced modulator comes from frequency doubling the input of a 19,000-cps oscillator. A parallel output from the same 19,000-cps oscillator goes into the FM modulator to act as the pilot carrier.

---

[1] FCC "Report and Order," April 20, 1961.

[2] Comments by the General Electric Company to FCC Docket 13506, October 28, 1960.

TABLE I
SYSTEM SPECIFICATIONS

Main Channel
- L + R audio (FM modulating main carrier)
- 50–15,000 cps audio band
- 90 per cent maximum main carrier deviation
- Standard 75 μsec. pre-emphasis

Subchannel
- L − R 38,000 cps suppressed carrier AM subcarrier (FM modulating main carrier)
- 50–15,000 cps audio band
- 90 per cent maximum main carrier deviation
- 19,000 cps pilot carrier (FM modulating main carrier 8–10 per cent)
- Standard 75 μsec. pre-emphasis

Separation between Left and Right Signals—30db between 50 and 15,000 cps
Distortion—Maintain existing FCC requirements

The receiver operates generally as depicted in the block diagram of *Fig.* 2, and is conventional to the discriminator output which is, however, taken ahead of any de-emphasizing networks. The L+R signal in an existing monophonic receiver would produce a compatible program, but in the stereophonic receiver it is fed directly to the matrix. The L−R sidebands and the pilot signal which are near or above the range of normal hearing would not be heard in the monophonic receiver. However, in the stereophonic receiver they must be decoded to produce the L−R audio signal. This takes place when the 19,000-cps pilot signal is filtered and doubled to recover the 38,000-cps subcarrier which is in turn added with the filtered sidebands to form normal amplitude modulation. This is detected to produce L−R audio for the matrix.

The matrix outputs, after passing through separate de-emphasis networks, are then the original left and right stereophonic signals.

A study of the spectrum of signals appearing in a discriminator output will help in understanding the system. Such a spectrum is shown in *Fig.* 3. Shown is a monophonic or what would normally be the 50–15,000-cps audio program, and the SCA signal (storecasting) with the SCA subcarrier at 67,000 cps and maximum deviation of 6700 cps by the SCA program.

Also shown is the stereophonic signal which is made up of a lower sideband

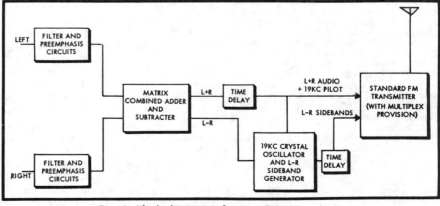

Fig. 1. Block diagram of stereo FM transmitter.

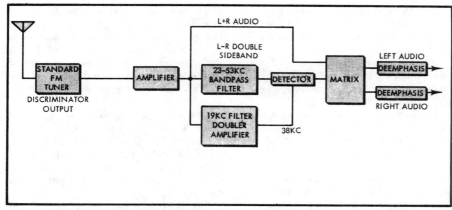

Fig. 2. Block diagram of stereo FM receiver

from 23,000 to 38,000 cps and an upper sideband of 38,000 to 53,000 cps, plus the 19,000-cps pilot.

Interleaving

There are important advantages for suppressing the carrier and transmitting a subharmonic pilot. One of these results in an interleaving effect which permits a 90 per cent maximum deviation on the main channel as well as 90 per cent on the subchannel, with the other 10 per cent in each case being reserved for the pilot carrier.

Interleaving, or nesting, of the L+R main channel signal and the L−R generated sidebands is one of the most interesting and important aspects of the newly adopted system.

Because of this effect, 90 per cent of normal deviation can be used on the main channel, and also the subchannel, because one is producing peak main carrier deviation while the other is zero, and vice versa. Thus, the monophonic listener experiences a signal-to-noise loss of less than 1 db.

This interleaving effect arises from the fact that the sum of two variables (L+R) is high when their difference (L−R) is low and *vice versa*. Since the amplitude of the sideband envelope produced by the L−R signal is directly proportional to L−R, this relationship between a sum of two variables and their difference is maintained and the main channel and subchannel will interleave.

Perhaps a reference to *Fig.* 4 will help in developing an understanding of this phenomenon. In *Fig.* 4 (A) represents the L signal input; (B) shows an imag-

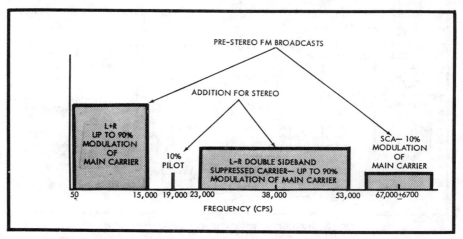

Fig. 3. Spectrum of signals appearing at output of discriminator.

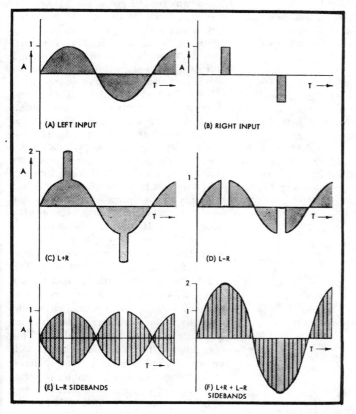

Fig. 4. Interleaving of L+R and L−R sidebands.

inary square wave pulse on R, used for illustrative purposes; (C) shows L+R (sine wave plus pulses); (D) shows the L−R (sine wave minus pulses); (E) shows the L−R subcarrier sidebands and (F) the composite signal (minus the pilot for illustrative purposes) consisting of L+R and the L−R sidebands that would be the signal fed to the FM modulator. Note that its peak amplitude is not greater than the peak amplitude of L+R or the L−R sidebands. Also observe that there is a depression (caused by −R, the pulse) in the L−R sidebands, while there is a simultaneous peak (caused by +R, the pulse) on the L+R signal. When they add to form the composite, the L+R peak fills the L−R sideband depression.

If the subcarrier carrier is suppressed, the main and subchannels can have peak FM deviations limited only by the necessity to provide for the pilot carrier.

Another advantage of the 19,000-cps pilot can best be explained at this time. Note that the 19,000-cps pilot falls in a clear channel portion of the discriminator output, with the L+R audio 4000 cps below and the L−R lower sideband 4000 cps above. It will be recognized that this affords the use of relatively simple filter circuits in the receiver for isolating the pilot signal so that is may be doubled to recover the subcarrier. This is an important feature of the system and is one of the primary reasons that a simple one-tube adaptor can be employed.

Circuit Description

Figure 5 is a schematic[3] of a doubler circuit which employs a tuned doubler amplifier to recover the subcarrier from the pilot.

The discriminator output from a tuner is applied to the control grid of the 12AT7 amplifier. The plate of this tube provides an amplified output signal which is applied, via an amplitude adjusting potentiometer and a time delay network, to the L + R signal input line of the matrix.

The output signal of the 12AT7 is also attenuated by a resistive voltage divider and is applied via a bandpass filter[4] to the input of the detector. The filter is tuned to provide a bandpass from 23,000 to 53,000 cps with the series arm displaying an "anti-resonance" at the SCA frequency which, for the NSRC field tests, was 67,000 cps. If the "anti-resonance" is not designed into the filter, an annoying whistle may be heard because of mixing between the 67,000-cps SCA subcarrier and harmonics of the stereo subcarrier.

The output signal of the 12AT7 is also fed, via a resistor or capacitor, to the pilot filter, which is shown tuned to the pilot signal frequency of 19,000 cps. The output of the filter is coupled to the grid of the second triode of the 12AT7. This tube is operated as a doubler-amplifier.

The plate is connected to a circuit which is tuned to double the frequency of the pilot (38,000 cps). This frequency-doubled signal is applied, via a secondary on the coil of the tuned circuit or a capacitor, to the input of the detector.

The detector consists of a pair of rectifiers connected to the input with opposite polarities, as shown. Filter capacitors and resistors are respectively connected between the output electrodes of the detector rectifiers and ground.

The matrix circuit contains the resistor bridge shown. It is important to note that stereo separation is dependent on the degree of balance of this bridge.

The de-emphasis network comprises the usual resistor and capacitor providing the standard 75 μsec de-emphasis. The output signals, L and R, are taken after the de-emphasis.

It should be noted that the output impedance of this one-tube device is quite high, and, also, that the insertion loss will range from 6 to 10 db. Æ

[3] *Reply Comments by the General Electric Company to FCC, November 7, 1961.*

[4] *"Reference Data for Radio Engineers,"* Federal Tel. and Radio Corp., 1948, p. 176.

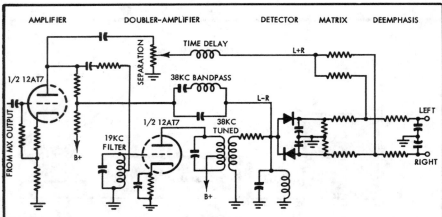

Fig. 5. Simplified schematic of one-tube adapter.

DESIGN CHART FOR ENCLOSURES *(from page 60)*

vert to trial and error to determine the enclosure's resonant frequency, since here the available mathematics does not tell enough of the story.

Referring back to Eqs. (1) and (3), they may be combined to give a general design equation:

$$F^2 = \frac{2150^2 \, A \, R^{0.12}}{V(L + A^{0.5})} \quad (4)$$

which takes into account variations in port aspect ratio (or shape), within the limitation that it is applicable only to single, simple ports. This equation, with its exponential factors, is somewhat difficult to use. Therefore, it has been put into chart form, *Fig.* 5, from which can be read off the several factors involved in the design.

Use of Chart

An example will make clear the method of using the chart: Supposing it is desired to match a speaker with 54-cps resonance to a reflex enclosure of 8000 cu. in. It is further desired to hold the port aspect at 4, and to use no duct other than the ¾-in. thickness of the enclosure wall. To solve this using *Fig.* 5,

follow up from 52 cps to the 8000-cu. in. diagonal. Then to the left to meet the "4" aspect-ratio diagonal, and then up again to ¾-in. duct length. The port area required is read off as 24 sq. in., and its dimensions will be 2.45 by 9.8 in.

Since these enclosures were built strictly for experimental use, with no thought of using any of them for mounting a speaker for actual listening, no attempt was made to damp out any resonances, either by internal padding, or by bracing, though some of the panels were certainly large enough to need bracing. While not really pertinent to the present investigation, a quick check was made for the presence of resonances above the fundamental Helmholtz resonance. Enclosure "B" used was with a 64 sq. in. port. Resonances were detected at 347, 362, 390, 462, 730, 800, 810, 830, 900, and 980 cps, along with at least twenty others above 1000 cps. Several were pronounced enough to be quite objectionable, should they have been present in an enclosure intended for normal use. This is mentioned to emphasize what is well known, but apt to be forgotten at times, that adequate damping, both by padding and by solid construction, with bracing of large panels, can be just as important as the basic design of the enclosure. Æ

TABLE II
EFFECT OF COMPLEX AND MULTIPLE PORTS ON RESONANCE OF REFLEX ENCLOSURE B

Port Shape	Dimensions[1] Inches	Area Sq. In.	Aspect Ratio[2]	Resonant Frequency Measured	Resonant Frequency Calculated (Eq. 3)
Symmetrical L	¼ × 15¾	3-15/16	63	78	62.1
"	½ × 15⅝	7-15/16	31	84	72.4
"	1 × 15⅛	15⅛	15	99.5	83.3
"	2 × 14	28	7	102	94.9
"	3 × 13	39	4-1/3	108	101.6
"	4 × 12	48	3	110	105.9
Parallel Twin Slots, 7½" c.l. spacing	½ × 8⅛ (each)	8⅛	32	86.5	72.9
" 7¼" spacing	¾ × 8⅛ (each)	12-3/16	21-2/3	88	79.8
" 6¾" spacing	1¼ × 8⅛ (each)	20-5/16	13	110	90.8

Notes
[1] Length of L-shaped slots is the total length of both legs.
[2] Aspect ratio for L slots is that for the equivalent rectangle having length of both legs of the L. For twin slots, is that of the rectangle formed by placing the two slots end to end.

FM-Stereo: Time-Division Approach

CARL G. EILERS

Here is an explanation of time-division FM-stereo by one of the innovators of the approach. He explains the approach in mathematical terms as well as by means of a specific FM-stereo demodulator.

THE SYSTEM of FM stereophonic broadcasting which has been adopted by the Federal Communications Commission had its inception in a time-division multiplex switching system between left and right stereophonic program sources.

In a time-division multiplex switching system a switching rate is chosen to be at least twice the highest frequency of modulation to be transmitted. If one analyzes the signal mathematically it becomes apparent that the signal basically consists of sum (L+R) and difference (L−R) components. If the switching waveform is a square wave, as shown in Fig. 1, then the sum of the left and right stereophonic channels appears as audio modulation on the main radiated carrier and the difference between the left and right stereophonic channels appears as a suppressed carrier amplitude modulation of a series of odd harmonics of the switching rate.

Equation (1) describes the resulting modulation function of the radiated carrier:

$$\frac{(L+R)}{2} + \frac{2(L-R)}{\pi}\cos\omega_{sc}t - \frac{2(L-R)}{3\pi}\cos 3\omega_{sc}t + \frac{2(L-R)}{5\pi}\cos 5\omega_{sc}t - \ldots = M(t)$$

$$Eq.\ (1)$$

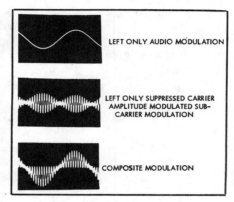

Fig. 2. Illustration of interleaving effect of sum and difference signals (left only).

where: $M(t)$ is the composite modulation
L is the left channel audio
R is the right channel audio
ω_{sc} is the subcarrier angular frequency

Since the fundamental subcarrier term contains all the necessary stereophonic information in the form of (L−R) modulation and, in order to prevent radiation outside the 200,000 cps channel and, further, in order to allow for the possible addition of an SCA background music channel, it is desirable to limit the spectrum of the modulation to the necessary stereophonic components. We can then describe a new composite modulating function:

$$(L+R) + (L-R)\cos\omega_{cs}t = M'(t)\quad Eq.\ (2)$$

where: $M'(t)$ is the new modulating function

It will be noted that the maximum peak-to-peak amplitudes of the sum (L+R) audio and the difference (L−R) modulated subcarrier are equal. It is also true but not obvious that the composite modulation function maximum peak-to-peak amplitude is equal to the maximum of either of the components alone. Thus, the FM transmitter may be fully modulated with (L+R) audio and then fully modulated with the (L−R) subcarrier without having to reduce the modulation percentage for either component as applied to the radiated carrier. This interleaving property of the sum (L+R) and difference (L−R) signals is directly related to the original concept of time-division multiplexing between left and right sterophonic signals.

The photographs of *Fig.* 2 illustrate the concept of interleaving. The upper photograph displays the sum (L+R) audio component of the composite modulating signal. The radiated carrier deviation is ±75,000 cps. The center photograph displays the difference (L−R) subcarrier component. In the lower portion of the photograph is shown the result of the addition of these two components which make up the composite modulating signal. It is evident that the maximum peak-to-peak amplitudes are identical for all three photographs.

The photographs of *Fig.* 3 show the makeup of the modulating signal when a left only signal is applied to the FM transmitter. For this illustration the audio modulating frequency was higher than that of *Fig.* 2, so that the actual subcarrier cycles may be viewed. It may be noted that the subcarrier reverses phase each time the zero axis is crossed. This shows that the suppressed carrier type of moduation is taking place.

For clarity, the pilot subcarrier is not

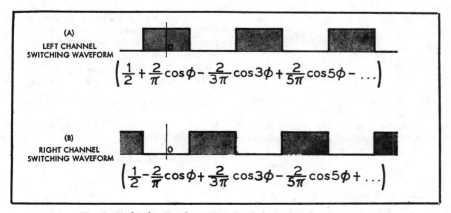

Fig. 1. Multiplexing functions for left and right channels.

included in the photographs of *Fig.* 2 and *Fig.* 3.

Stereophonic Subcarrier Demodulation Methods

Since the stereophonic system that has been described is a sum and difference system, the left and right signals may be derived by demodulating the stereophonic subcarrier by the use of a synchronous detector for recovery of the (L−R) modulation and then matrixing at audio frequencies with the main channel (L+R) modulation.

Left and right signals may also be derived directly in one operation of the composite modulation. In order to illustrate this method of demodulation refer to *Fig.* 4.

(A) of *Fig.* 4 shows the composite waveform of a time-division multiplexed signal. The envelopes of both the left and right signals are clearly discernable. (B) shows the difference (L−R) component of the time-division multiplexed signal with a superimposed suppressed carrier amplitude modulated subcarrier being modulated by difference (L−R) audio. The similarity between the two waveforms is evident. (C) shows the sum (L+R) component of the time-division multiplexed signal. The addition of the suppressed carrier component of (C) results in the waveform shown in (D). Once again, the envelopes of both the left and right signals are clearly discernable.

If the composite waveform shown in (D) of *Fig.* 4 and repeated in (A) and (C) of *Fig.* 5 were sampled with two interleaved unit impulse functions, as shown in (B) and (D) of *Fig.* 5, synchronized to 38,000 cps, the carrier tips of the left signal would be followed by one of the interleaved set of unit impulse functions and the carrier tips of the right signal would be followed by the other set of unit impulse functions. Thus, it is possible to recover the left and right stereophonic signals from the composite waveform using a direct method of demodulation.

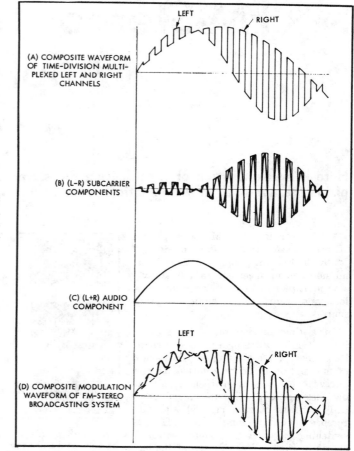

Fig. 4. Relation of time-division multiplexing to FM-stereo broadcasting.

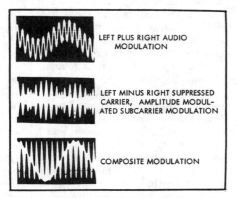

Fig. 3. Illustration of interleaving effect of sum and difference signals (left plus right).

We will now illustrate a method of left and right stereophonic signal derivation using partial demodulation and partial matrixing. This method is directly related to the original concept of time-division multiplex demodulation. If one multiplies the composite modulating function $M'(t)$, as shown in *Eq.* (2) with: $1+2\cos\omega_{sc}t$, the result would be demodulation for recovery of the left channel:

$$[(L+R)+(L-R)\cos\omega_{sc}t][1+2\cos\omega_{sc}t]$$
$$=(L+R)+2(L-R)(\cos2\omega_{sc}t)+\ldots$$
$$=(L+R)+2(L-R)(\tfrac{1}{2}+\tfrac{1}{2}\cos2\omega_{sc}t)=2L$$
$$Eq.\ (3)$$

If one multiplies the composite modulating function $M'(t)$ with: $1-2\cos\omega_{sc}t$, the result would be demodulation for recovery of the right channel:

$$[(L+R)+(L-R)\cos\omega_{sc}t][1-2\cos\omega_{sc}t]$$
$$=(L+R)-2(L-R)(\cos2\omega_{sc}t)+\ldots$$
$$=(L+R)-2(L-R)(\tfrac{1}{2}+\tfrac{1}{2}\cos2\omega_{sc}t)=2R$$
$$Eq.\ (4)$$

There is no electronic waveform which corresponds to the multiplier: $1+2\cos\omega_{sc}t$. However, a square wave may be used which corresponds to the multiplier: $1+\tfrac{4}{\pi}\cos\omega_{sc}t$. A half sinewave may also be used which corresponds to the multiplier: $1+\tfrac{\pi}{2}\cos\omega_{sc}t$. Since the square-wave function is easily derived electronically, this would seem an appropriate multiplier to explore further.

If one multiplies the composite modulating function $M'(t)$ with $1+\tfrac{4}{\pi}\cos\omega_{sc}t$, the result would be partial demodulation for the left channel:

$$[(L+R)+(L-R)\cos\omega_{sc}t][1+\tfrac{4}{\pi}\cos\omega_{sc}t]$$
$$=(L+R)+\tfrac{4}{\pi}(L-R)\cos2\omega_{sc}t+\ldots$$
$$=(L+R)+\tfrac{4}{\pi}(L-R)(\tfrac{1}{2}+\tfrac{1}{2}\cos2\omega_{sc}t)$$
$$=(L+R)+\tfrac{2}{\pi}(L-R) \qquad Eq.\ (5)$$

It is obvious by examining this result that the sum (L+R) signal is somewhat larger than the difference (L−R) signal, thus resulting in mostly left but some right signal in the left channel.

If a sum (L+R) signal having an amplitude: $-1+\tfrac{2}{\pi}$ is added to the above partially matrixed left channel, the result would be:

$$(L+R)+\tfrac{2}{\pi}(L-R)+(L+R)(-1+\tfrac{2}{\pi})=$$
$$\tfrac{4}{\pi}L.$$

This $(L+R)(-1+\tfrac{2}{\pi})$ signal is readily

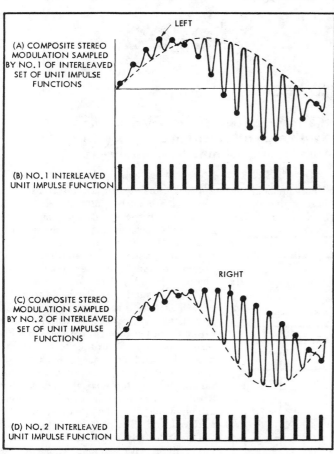

Fig. 5. Direct method of deriving left and right signals from composite stereophonic modulation.

available because it is an inverted main channel signal.

Likewise, if one multiplies the composite modulating function $M'(t)$ with: $1 - \frac{4}{\pi}\cos\omega_{sc}t$, the result would be partial demodulation for the right channel:

$[(L+R) + (L-R)\cos\omega_{sc}t][1 - \frac{4}{\pi}\cos\omega_{sc}t]$

$= (L+R) - \frac{4}{\pi}(L-R)\cos^2\omega_{sc}t + \ldots$

$= (L+R) - \frac{4}{\pi}(L-R)(\frac{1}{2} + \frac{1}{2}\cos 2\omega_{sc}t)$

$= (L+R) - \frac{2}{\pi}(L-R)$ Eq. (6)

Again, the sum $(L+R)$ signal is somewhat larger than the difference $(L-R)$ signal, thus resulting in mostly right but some left signal in the right channel.

Once again, if a sum $(L+R)$ signal having an amplitude: $-1 + \frac{2}{\pi}$ is added to the above partially demodulated right channel, the result would be:

$(L+R) - \frac{2}{\pi}(L-R) + (L+R)(-1 + \frac{2}{\pi}) = \frac{4}{\pi}R$.

Multiplication by a square wave may be accomplished using an electronic switch operated in synchronism with the modulated subcarrier signal, thus becoming a synchronous demodulator. Such a demodulator may take on many forms—one of which is shown in Fig. 6. In this case the synchronous demodulator is a beam deflection tube, 6AR8A, which has: one electron stream, two anodes, two deflection plates, and one control grid. The composite stereophonic signal is applied to the control grid after any SCA background channel which may be present has been removed with a low-pass filter type of trap.

The deflection plates of the beam deflection tube are driven by a push-pull sinewave source generated by an oscillator-doubler circuit having an output frequency of 38,000 cps. A sufficient amount of sinewave drive is applied so that switching of the beam results in essentially a square wave of anode current. In this way the products of the two polarities of square wave and the composite stereophonic signal are generated.

At one anode the product of one polarity of square wave function and the composite stereophonic signal appearing at the control grid is formed for the left channel, as shown in Eq. (5). Similarly, at the other anode the product of the opposite polarity of square wave function and the composite stereophonic signal appearing at the control grid is formed for the right channel, as shown in Eq. (6). The addition of a sum $(L+R)$ signal having an amplitude: $-1 + \frac{2}{\pi}$ is accomplished by the 500 ohm variable matrix adjustment in the cathode circuit of the beam deflection tube demodulator, as shown in Fig. 6. Thus, the combination demodulator and matrix network accomplishes the complete derivation of left and right signals. The necessary deemphasis in left and right channels is achieved by simply adding a 1500 pf capacitator at the matrix outputs.

The push-pull sinewave drive which is applied to the deflection plates of the beam deflection demodulator tube is generated by an oscillator frequency doubler combination, as shown in the carrier regenerator portion of Fig. 6. The oscillator operates at a frequency of 19,000 cps. The anode circuit of this carrier oscillator is tuned to the second harmonic, or 38,000 cps. The tube acts as an electron-coupled oscillator and frequency doubler.

Magnetically coupled to the oscillator tank circuit is another parallel resonant tank which is in the anode circuit of the modified cathode follower, shown at the left in Fig. 6. This parallel resonant circuit is tuned to the 19,000 cps pilot subcarrier frequency, thus forming a tuned amplifier for extracting the pilot subcarrier from the composite stereophonic signal appearing at the grid of the cathode follower.

(Continued on page 84)

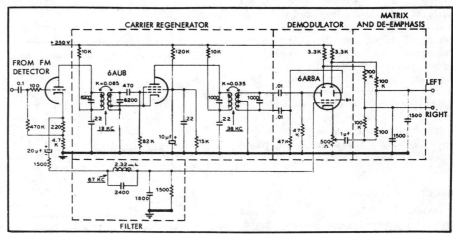

Fig. 6. Schematic diagram of FM-stereo demodulator.

Filters For FM-Stereo

NORMAN H. CROWHURST

When the regenerated subcarrier and the main carrier are rejoined in the FM-stereo decoder they must be in phase with each other within 3 degrees. In order to effect this, several of the filters and circuits must be phase-linear.

THE CHOICE OF SYSTEM for FM-stereo caught many people in the industry by surprise and, whichever variety of receiver circuit individual designers may fancy, they have been finding themselves in filter design problems, somewhat different from problems encountered before in audio circuitry. Of course, the basic design of a stereo adapter is audio, althought the tuner end is r.f. But, given a wide-band tuner, that can demodulate the FM carrier linearly when it carries modulation frequencies up to 53,000 cps (without SCA subcarrier, or 75,000 cps with a subcarrier), the handling of that demodulated signal to reconstruct undistorted stereo is a problem in audio engineering.

Other articles have explained basic approaches used in receiver adapter circuits. What we are concerned with here is solving the various problems posed by the filters that will be needed. Audio engineers have long been familiar with crossover filters, and to some extent with band-pass, band-reject, and m-derived filters; but time delays, the requirement of phase-linearity and the phase adjustment of regenerated subcarriers is a new subject to them.

Time Delays

Time delays filters are needed to equalize for delays, either along lines, or in other filters. Delay filters to compensate for difference in the transmission time along lines, are a transmission man's headache, and need not be considered here. At the transmission end, the problem is simpler in one respect: the filter can be of standard form, using as many components, with whatever cost and precision, is necessary for the job. At the receiver, the cost of this kind of filter would price the receiver or adapter out of the market, so other methods are sought. Fortunately, as we shall see, phase compensation is relatively simple, once the basic requirements of phase linearity have been met.

Basic Design: Low-Pass Filter

Considering the low-pass filter first will clarify some of the issues involved, such as what is meant by phase linear and how to achieve it. Let's consider first a simple "half-section" arrangement, as shown at *Fig. 1*. The half-section designation derives from classic filter design

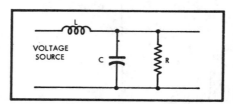

Fig. 1. Basic low-pass filter section, values for which are discussed in the text.

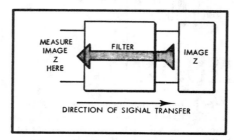

Fig. 2. Conventional filter design is based on a hypothetical image impedance, which is not a simple value; this means the results obtained are only approximate.

approach, using the *image impedance* concept. For such a filter to realize its theoretical response, it must be terminated at a specified end, say the output, with a theoretical impedance value; when this is done, the impedance reflected to the input terminals is an identical one (*Fig. 2*).

What is usually overlooked in this approach is that this image impedance is a resistance of *almost* constant value through the pass range, passing through zero or infinity (according to configuration) at cut-off frequency, to become a reactance in the rejection or attenuation range. Only by terminating such a filter with an impedance that fulfills this condition, would its calculated response be achieved.

This is usually overcome by terminating a filter with sections whose image impedance shows minimum deviation from constant up as close as possible to cutoff frequency, so that termination with a constant resistance will minimize deviation from calculated response. In effect, we finish up with a double approximation situation.

The approach is made much simpler

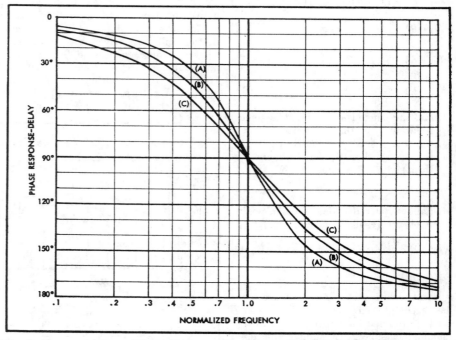

Fig. 3. Phase response of three low-pass filters, using the configuration of Fig. 1 with different values. At the reference frequency (cutoff) the reactance of L and C each is: Curve (A), equal to R; Curve (B), equal to 1.414 times R; Curve (C), equal to 2 times R.

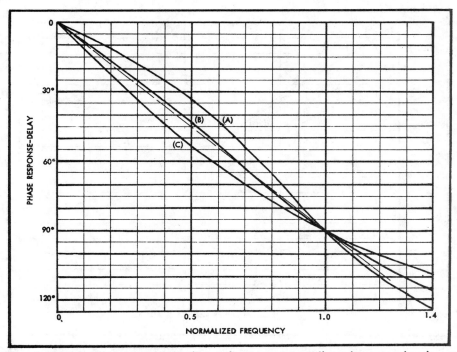

Fig. 4. Plotting phase response to linear frequency, as well as degree scale, shows how the values for curve (B) optimize phase-linearity up to cutoff frequency.

sponses is linear. But we are concerned with phase linearity from the viewpoint of the relations between various side-bands of a subcarrier, which are determined by their sum and difference from the subcarrier frequency. So we must use a *linear* frequency scale, if we want to represent corresponding pairs of side-bands as equidistant on either side of the subcarrier.

Replotting phase responses this way, we take the same three sets of values and the results are shown at *Fig.* 4. The corresponding amplitude or attenuation responses, using a db against log frequency scale, are shown at *Fig.* 5. Notice that the middle curve (B), using the so-called "constant-resistance" values, maintains maximum flatness up to the 3-db rolloff point (or cutoff frequency) and deviates from phase linearity by considerably

by being more direct. Instead of starting with an assumed terminating impedance which can never be realized, only approximated, we start with an assumed terminating resistance, which we can eventually put a value to, and get out of the resistor box.

First let's clarify what we mean by phase-linear. The usual response is plotted against the usual frequency scale, which is logarithmic. To these scales, any low-pass filter will have a curved "tangent-law" response, of the type shown at *Fig.* 3. The curves are for three sets of values, each with the same 90 deg. phase reference frequency. At this frequency, the value of each reactance is identical and the relation to the terminating resistance at this frequency can identify differences in response. Curve (A) is for reactances each equal to terminating resistance at cutoff curve (B) is for the constant resistance case, where each reactance is $\sqrt{2}$ times the terminating resistance; curve (C) is for reactances each twice the terminating resistance.

Plotted this way, none of the phase re-

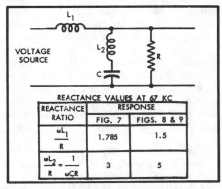

Fig. 6. Circuit of so-called m-derived filter configuration, used in this article to develop a phase-linear filter with a rejection frequency.

less than 2 deg. over the range from zero to cutoff frequency.

The phase linearity, as well as the attenuation response, deteriorates when this condition is deviated from. However, the fact that only three basic circuit elements are used, one of which is the terminating resistance, means that reasonable deviation, say by using 5 per cent values, will result in a filter that will easily stay within 3 deg. deviation from phase-linear.

Where the filter operates from a cathode follower into a resistance load many times the source resistance of the cathode follower these basic values can be used. Where the filter is interposed between two impedances, both of which must be regarded as finite for design purposes, these basic values need changing, in accordance with design data we have given elsewhere.

Low-Pass Filter, M-Derived

Some may criticize this designation of the next type of filter. The term "m-derived" has been applied to a good many ways of deriving a filter, which do not conform to the classic method, so we trust one more will be permitted. Our

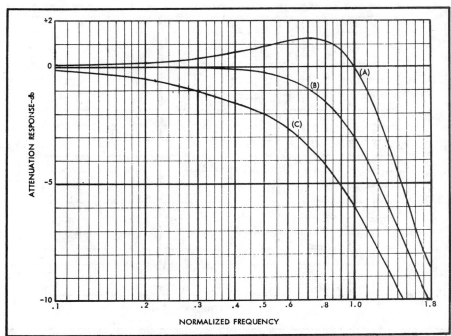

Fig. 5. Attenuation responses of filters using same values as for the phase responses of **Figs.** 3 and 4.

reason for using the term is that the configuration is identical with the classic m-derived form (*Fig.* 6), *but the values are very different.*

The first step may be regarded as somewhat similar to m-derivation in concept. We use a value of series L identical with that for the constant resistance type filter and with the same cutoff frequency, in this case 53,000 cps, we select values for L_2 and C that will (a) resonate at the rejection frequency, in this case 67,000 cps, while (b) providing the correct capacitive reactance at cutoff frequency.

Plotting out the phase response of these values (*Fig.* 7), we find it to be reasonably phase linear up to about 45,000 cps, which is not quite good enough. Obviously what is needed is a narrowing down of the margin between the basic cutoff frequency and the maximum rejection frequency.

To change our approach slightly, since the definitive point of this configuration is really the maximum rejection frequency, we use it as the reference frequency for design. On this basis, 53,000 cps is 0.79 times the maximum rejection frequency. Using this technique, we have only two variables to explore:

(1) the relationship between each 'rejection' reactance and the terminating resistance at rejection frequency;

(2) the value of the additional series inductance.

Exploring various possibilities in this way, we find the combination shown in *Fig.* 8 gives a good approach to phase linearity, being well within 2 deg. up to the required cutoff frequency. The corresponding attenuation response is shown at *Fig.* 9. Notice that the values (*Fig.* 6) are such that exact adjustment is not necessary for at least acceptable performance. It is recommended that the rejection pair be carefully tuned to 67,000 cps, after which the series element can be within 5 or 10 per cent, without serious change in performance. What changes is phase slope, within this range, to a much greater extent than phase linearity. Phase slope can be compensated for by slight readjustment of the subcarrier reinsertion phase.

As with the low-pass circuit, this design assumes zero source impedance with finite load impedance of known (design) value. If both values are finite, which approach may well save a stage in some circuit configurations, the filter design is a little more involved, but follows the general method outlined here.

Band-Pass Filters

Now all we have to do is apply this to band-pass filters. Unfortunately, it's not that easy. Remember that a high-pass

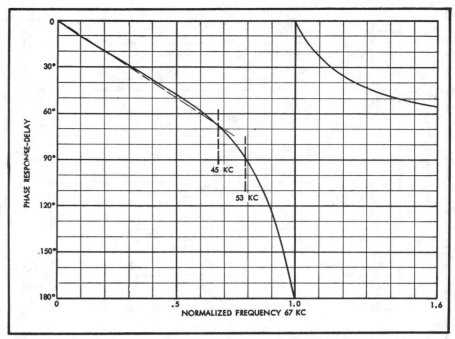

Fig. 7. Using a form of m-derivation from the constant-resistance values of Figs. 2 through 5, this is the phase response achieved.

filter always presents a phase *advance*, that is complementary to an equivalent low-pass filter, *plotted to a logarithmic frequency scale*, as at *Fig.* 2.

The combination of values that achieves such good phase linearity in a low-pass filter requires frequency to be plotted to an *inverse* scale for the phase response to look linear. To a linear scale, the rate of change is always greater nearer cutoff and falls away to zero at higher frequencies, approaching infinity.

A band-pass filter is essentially a combination of high-pass and low-pass action. In a narrow band-pass filter, the range of absolute frequency is such that presentation on logarithmic, linear, or inverse law scale makes relatively little difference. Slight asymmetry in design could make the phase more nearly linear, if the departure is great enough to make the effort worth while.

But when the "carrier" is 38,000 cps, with sidebands from 23,000 to 53,000 cps, we are no longer narrow band. Only a linear frequency scale can provide 23,000 and 53,000 cps at equal distances from the 38,000 cps carrier. Some fudging might conceivably get phase linearity

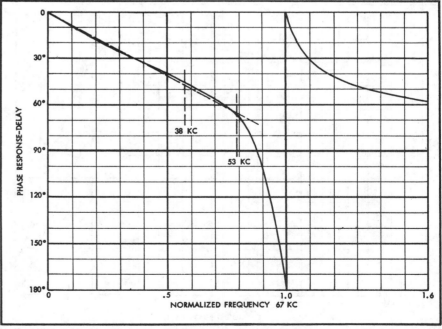

Fig. 8. Optimizing values for phase linearity over the desired range, this is the response achieved.

down to two "bites" of equal slope, but it could never achieve full linearity from 23,000 to 53,000 cps.

The band-pass filter has been "needed" to separate the subcarrier sidebands from the main or sum channel. If this matrixing method is used at the transmitter, no such filter is required. Maybe low-pass filters are used to ensure that the audio contains nothing above 15,000 cps, to cause interference between main channel and subcarrier sidebands. But the actual matrix needs no selective filters. So why should the second (reception) matrix need them? A little further thought reveals that it doesn't.

If we consider the composite of frequencies fed to the subcarrier detector, after reinserting the subcarrier, we have: 50–15,000 cps, main channel information; 23,000–38,000 cps, subcarrier lower sidebands; 38,000 cps subcarrier, which should be the biggest single component at this point; and 38,000–53,000 cps, subcarrier upper sidebands. (*Fig.* 10). If this is demodulated after the manner of an ordinary AM carrier, only the lower and upper sidebands will come out in the audio range; the 50–15,000 cps will be equivalent to asymmetric sidebands for modulation frequencies of 23,000–38,000 cps (inverted).

It's just as if an adjacent carrier of zero frequency carried this sideband. The separation of 38,000 cps between carriers ensures the sidebands of one are ultrasonic in the demodulation of the other.

There's just one snag: using ground reference, the detector gets the full 50–15,000 cps audio, unless there's a band-pass filter, without being able to tell it's supposed to be related to a 38,000 cps carrier. In short, the presence of the 50–15,000 cps will act as variable bias on the detectors in handling the 38,000 cps and its sidebands.

But this is not difficult to overcome: the detector load needs unhooking from ground, so the proper bias can be inserted. In other words, the whole detector circuit "floats" at the main audio waveform (*Fig.* 11). There are undoubtedly other ways of doing this, but this illustrates the general method. Now we don't

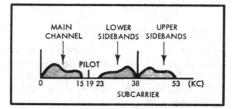

Fig. 10. Audio spectrum of demodulated stereo program; the main channel, as sideband of the subcarrier, is ultrasonic.

Fig. 11. A method of connecting the difference detector that avoids the need for a band-pass filter and makes matrixing more efficient.

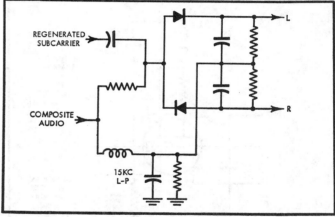

need the conflicting phase slope that is inevitable with a wide-band band-pass filter.

Relative Phase on Second Matrix

If phase linearity is maintained, the relative phase, between sum and difference, is simple to correct. The carrier needs inserting at the correct phase, along the phase slope of the sidebands (*Fig.* 8), so that a flat output, free from distortion, is obtained. Now the bigger time delay will be in the sum channel, which is equivalent to 90 deg. at 15,000 cps, or 16⅔ microseconds (due to the 15,000 cps, phase-linear low-pass filter, *Fig.* 3).

If the m-derived filter is in the sidebands channel after the take-off point for the 15,000 cps low-pass, this will introduce a delay of almost exactly 45 deg. at 38,000 cps (*Fig.* 8) or 3.3 microseconds. So, depending on the location of the m-derived filter, the demodulator needs to introduce a delay of 16⅔ or 13.3 microseconds, which can easily be arranged by means of the "r.f." (in this case 38,000 cps and harmonics) filtering and its time constant.

As there is a nice wide gap between the residual subcarrier components to be eliminated and the highest audio—15,000 cps—simple R-C combinations can care for this time delay and filtering (*Fig.* 12).

Subcarrier Phase

This part is more a practical item than a basic design item, but it's related to the rest of this problem, so here it is. Each way of regenerating the subcarrier will have slightly different methods of adjusting the phase to meet the requirements of the rest of the circuit.

Where the pilot frequency is just isolated and frequency doubled, extremely good isolation (with a high-Q circuit) is essential, so the tuning of the pilot frequency circuit should not be changed appreciably. In this case, probably the best compromise is to effect part of the phase compensation at 19,000 cps and

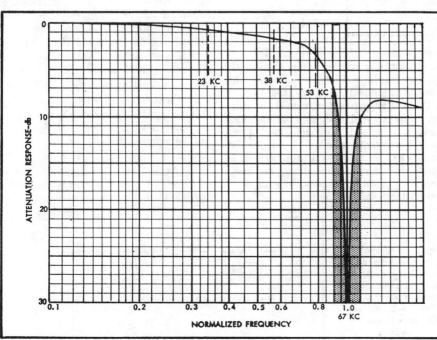

Fig. 9. Attenuation response of the filter whose phase response is shown in **Fig. 8**. Shaded area represents the maximum excursion an SCA subcarrier may make.

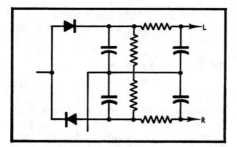

Fig. 12. Additional filtering of the subcarrier (as shown here), can, by correct choice of values (see text), correct for delay difference between sum and difference channels.

part after doubling, at 38,000 cps, because serious detuning of the 38,000 cps subcarrier can result in either asymmetry or distortion of the regenerated subcarrier.

Where the circuit uses a pilot-frequency oscillator, locked to the incoming pilot, slight adjustment of the free-running pilot oscillator frequency gives good control of phase. The doubler is then left tuned to exactly 38,000 cps. Detuning it would only have slight effect on phase anyway.

Where the circuit uses a subcarrier oscillator, synchronized by the pilot frequency, appreciable shift of the 38,000 cps free-running frequency is not desirable, because it will result in inequality of alternate cycles of subcarrier, even if the pilot holds it. Slight detuning of the 19,000 cps pilot will do the best job, by changing the point at which the (true) 38,000 cps is locked.

Frequency Doubling

Some spurious components can be eliminated from the output by removing all of the 19,000 cps from the regenerated subcarirer. A simple tuned circuit will never do a perfect job. A relatively simple circuit that will is shown at *Fig.* 13. The 19,000 cps oscillator plate current is drawn through the 38,000 cps tuned circuit in such phase as to neutralize the amplifier 19,000 cps component due to the doubler stage (pentode).

Careful adjustment of values (operating voltages and coupling values) can make this circuit produce almost pure second harmonic with no tuned circuits at all (*Fig.* 14). When a tuned circuit is used as common plate load, virtually complete rejection is possible with relatively uncritical setting.

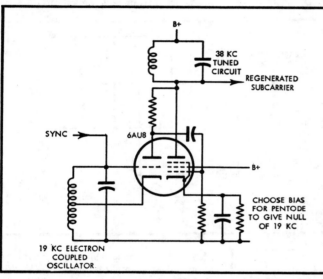

Fig. 13. A refined frequency-doubling circuit, that allows the 19,000-cps pilot to be completely nulled out.

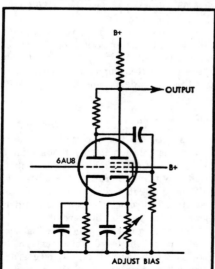

Fig. 14. With resistance coupling, the circuit of **Fig.** 13 will frequency-double any signal, and null out the input fundamental.

Output Filtering

By now, many will have discovered that the output, even with deemphasis, contains appreciable signal at 38,000 cps and its sidebands, which can be troublesome, if you want to tape record, particularly. A broad resonant circuit does not produce particularly good rejection. A narrow one, or a conventional twin-T filter, only takes the 38,000 cps and leaves most of its sidebands.

A very neat solution is a special twin-T that incorporates the deemphasis as well, provides a wide, deep null over the 38,000 cps region and requires values that are not nearly as critical as the normal twin-T circuit (*Fig.* 15).

Correct deemphasis is achieved by adjusting the total series source resistance (including matrix or diode load elements in their series-parallel arrangement) to an impedance (resistive) of 25,000 ohms for the values shown in the twin-T. Other values can be used if different circuit values have to be matched. In the original twin-T network design, the ideal values would be $a/c = 7.5$, $b/c = 1.15$. Using values $a/c = .6$, $b/c = 1.2$ allows for some series source resistance.

The great advantage of this circuit is that all values are relatively non-critical. Changing any pair of values by 5 per cent retains a rejection of better than 40 db. In fact when this filter is used, it is found that there are second harmonics of 38,000 cps present, that were completely masked by the much larger 38,000 cps component previously.

If the 76,000 cps components should prove troublesome, these can be removed by additional R and C, using modified values so this can be part of the deemphasis too, and thus obtain some 30 db rejection of 76,000 cps. If more or better rejection is wanted, a similar twin-T design can be used, centered on 76,000 cps, with even greater latitude in tolerances.

This article has given the basic design data or approach, rather than completing any individual circuit design. This can be applied to any of the adapter circuits where such filtering is needed, by using the normalized design factors we have given, in whatever arrangement appeals to the individual designer. Æ

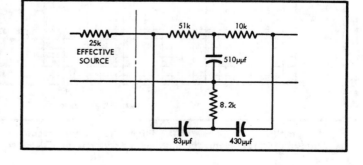

Fig. 15. A useful asymmetrical twin-T filter that achieves deemphasis and 38,000-cps band rejection in the same circuit.

An FM Multiplex Stereo Adaptor

DANIEL R. VON RECKLINGHAUSEN

This adaptor utilizes time-division to achiieve stereophonic FM reception; a method whereby switching rather than matrixing is used to recover the left and right signals.

AFTER YEARS OF STUDY and considerable field testing, the Federal Communications Commission has set the specifications for the compatible stereophonic signals to be transmitted from a single FM station. The system chosen transmits the sum of the left, A, and right, B, input channels on the frequency modulated main carrier. The difference between A and B amplitude modulates a 38,000 cps subcarrier with the subcarrier itself suppressed. The subcarrier signals in turn frequency modulate the main carrier. A 19,000 cps signal is also transmitted for stereo demodulator synchronization.

These and other pertinent specifications can be obtained from the FCC rules and regulations. Written in legal and engineering language, they are clear to the engineer engaged in this type of work but not necssarily enlightening to persons unfamiliar with the proceedings of the National Stereophonic Radio Committee (which performed the majority of the work of analyzing and testing the various stereo broadcasting schemes proposed).

The recording and reproduction of stereophonic signals had its inception in the 1920's and 1930's, then described as binaural signals. It was only natural that the attention of scientists and engineers was also focused on means of transmitting these signals to remote points. Carrier current telephony over cables and also radio links were investigated intensively.

The stereophonic system chosen by the FCC may be accomplished by the time multiplex system shown in *Fig.* 1. Here, the input of a cable or radio link is switched rapidly between the two inputs A and B. The output of the cable or radio link is also switched rapidly to the two output terminals. Switch synchronization has to be provided so that the channel A input signals will not accidentally appear in the channel B output.

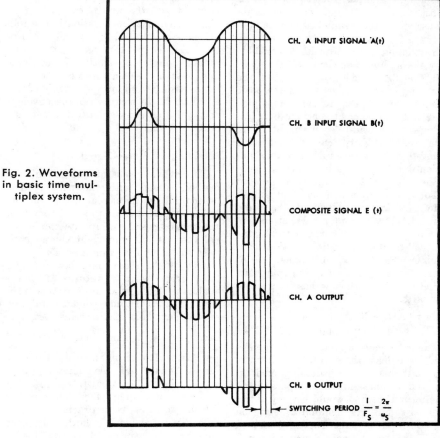

Fig. 2. Waveforms in basic time multiplex system.

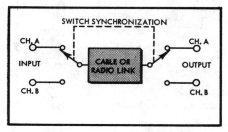

Fig. 1. Basic time multiplex stereo system.

The signal waveforms of such a system are shown in *Fig.* 2. Here, the input signals, $A(t)$, and $B(t)$, are switched at a rate f_s to the link. The composite signal $E_o(t)$ now shows portions of the two input signals in quick succession and a good representation of the two input signals is evident. If the lead containing the composite signal is then switched in synchronism and at the proper time to the two output leads, the channel A and B output waveforms result. These waveforms can contain all the information present in the original two signals. The highest input frequency which can be transmitted by such a method is exactly equal to one half the switching rate, f_s.

Mathematically, the composite signal, $E_o(t)$, by switching $A(t)$ and $B(t)$ at a rate $f_s = \frac{\omega_s}{2\pi}$ becomes:

$$E_o(t) = \left[\frac{A(t) + B(t)}{2}\right] + \frac{2}{\pi}[A(t) - B(t)]$$
$$\{\cos \omega_s t - \tfrac{1}{3}\cos 3\omega_s t + \tfrac{1}{5}\cos 5\omega_s t \ldots\}$$
Eq. (1)

where $A(t)$ and $B(t)$ are the instantaneous input signals, A and B, as a function of time. It can be seen from this that the sum of the input signals is transmitted directly. This is also the compatible monophonic signal which can be utilized without any further demodulation. The difference between the input signals appear as amplitude modulation of a series of odd harmonics of the switching rate f_s.

The composite signal might be utilized for modulation of an FM broadcast station. However, the transmission of the higher harmonics of the switch-

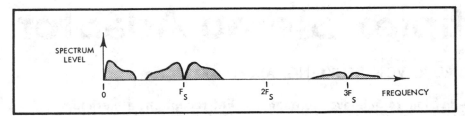

Fig. 3. Spectrum of basic time multiplex system.

ing frequency would result in radiation of signal components from the station outside its allotted 200,000 cps bandwidth, assuming 15,000 cps audio response with inputs switched at a 38,000-cps rate.

Restriction of bandwidth to include only the audio frequencies of A and B, and the first order sidebands of the switching frequency, results in the basic specification of the present multiplex system. This also has the benefit of a less stringent bandwidth requirement along with only slightly changed effective signal-to-noise ratio or change in separation due to phasing errors of the subcarrier employed for detection.

To be able to utilize the full amplitude handling capability of the radio channel (i.e. maximum deviation capability of an FM transmitter), the relative amplitude ratio of the main channel, $A+B$ and subchannel, $A-B$, has to be changed to give the composite signal $E(t)$ which now has been made standard:

$$E(t) = [A(t)+B(t)] + [A(t)-B(t)] \cos \omega_s t \quad Eq.\ (2)$$

The composite signal can be generated by at least three different methods. The first one uses a switching modulator and a phase-linear low-pass filter. The second method would employ a two channel square law modulator acting on the two input signals. The third method would require the use of an audio matrix network (with transformers or resistors) and a suppressed carrier modulator. Here, two separate outputs could be obtained: the main channel output for direct modulation of the FM transmitter and the stereo subchannel for modulation of the FM transmitter at a later stage of multiplication where higher frequency modulation is possible.

The third method is most likely to be incorporated in FM transmitters employing phase modulation in conjunction with a frequency-to-phase correcting network and proper audio delay equalization to correct for the envelope delay of the early stages of the transmitter.

Receiving the Stereo Signal

The above discussion of modulation methods along with the mathematical description of the signal waveform and its development leads to several methods for separating the composite stereo signal into its left and right components.

If the composite signal of Eq. (2) is passed through a square law demodulator driven by a waveform $[1+2 \cos \omega_s t]$, the left channel output will be equal to $2A(t)$ plus fundamental and second harmonic of the reinserted subcarrier, $f_s = \frac{\omega_s}{2\pi}$. The right channel output $2B(t)$ can be obtained by driving a second square law demodulator with a waveform $[1-2 \cos \omega_s t]$.

Since good square law detectors are difficult to come by in practice, a linear detector driven with a large reinserted subcarrier, f_s, can also be used. This, in effect, uses a square wave for demodulation which gives the detector a multiplying function $[1 \pm \frac{4}{\pi} \cos \omega_s t]$. This in turn requires a gain adjustment of the subcarrier signal by $\frac{\pi}{2}$ prior to detection or the same gain adjustment of the difference signal after detection.

A third method of detection would employ the use of a bandpass filter for the selection of the subcarrier sidebands in addition to the use of a suppressed carrier AM detector and a resistive sum and difference audio matrix network with proper main channel audio delay.

All of these stereo detection schemes have one common advantage and one common disadvantage. The advantage is that they are relatively economical in parts cost, employing only two diodes or a beam deflection tube (such as 6AR8) for demodulation. The disadvantage is that all of these detectors produce a very large output at the subcarrier frequency, f_s (38,000 cps), and its harmonics (76,000 cps, 114,000 cps, etc.). These are removed only with difficulty by filtering without disturbing audio frequency response. This, in itself, is not harmful, since these frequencies are above the range of human hearing. However, these signals, if not filtered out, will tend to overload amplifiers and tweeters because of reduced power handling capability at high frequencies. The result is considerably higher distortion. More serious is that the bias frequencies of tape recorders fall in the frequency range of the subcarrier frequency and its harmonics. It is made poorer by the increased amplification of tape recorder circuits to compensate for the required tape pre-emphasis and high-frequency losses of the tape. Whistle tones known as "birdies" are the result of subcarrier signals causing tape overload. Therefore, it is absolutely necessary that subcarrier frequencies are prevented from appearing at the audio outputs of the multiplex adaptor.

In the H. H. Scott adaptor, the subcarrier is balanced out by means of balanced bridge demodulators and a 15,000 cps sharp cutoff filter is used at the

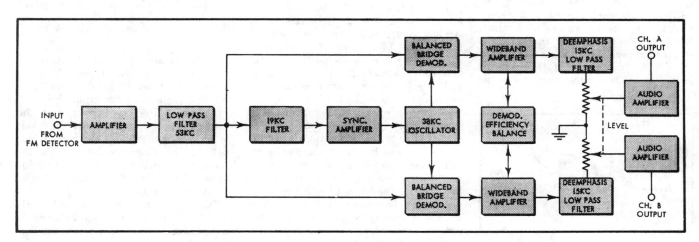

Fig. 4. Block diagram of H. H. Scott type 335 multiplex adaptor.

Fig. 5. Front view of H. H. Scott adaptor.

audio outputs. This effectively eliminates any of the subcarrier frequencies at this point.

The forerunners of the multiplex stereo system chosen by the FCC have been described by a magazine writer as "the Radio Manufacturers' Dream". This writer was correct in his estimate since this system allows radio manufacturers to design sets of relatively modest cost which produce stereo of some sort. However, to produce high quality stereo worthy of the name "High Fidelity" a great deal of engineering and complex circuitry has to be expended in both tuner and multiplex adaptor design. For example, to achieve 30 db of separation, the phase response of the tuner-adaptor combination may not differ more than ± 3 degrees between main and subchannel at all modulating frequencies. Similarly, the amplitude response may not vary more than ± 0.3 db, and the phasing of the subcarrier with respect to the pilot carrier has to be constant at all r.f. input levels. For this reason, it is extremely important that the tuner and adaptor match each other and have a wide and phase-linear response including the required connecting cable. Any level controls connected between the tuner's multiplex output and the stereo demodulator circuitry will have a severe effect on phase and frequency response. Similarly, the frequency and phase response of the audio circuitry of existing tuners is not controlled closely enough to use the tuners regular audio output and maintain high separation between left and right outputs while deriving the subchannel information separately.

For these and other reasons, almost all adaptors derive both main and subchannel information directly from the tuner's multiplex output circumventing all audio stages of the tuner. The wideband ratio detectors used in most high fidelity tuners have a sufficiently wide bandwidth, wide frequency response, and low internal impedance to permit the use of up to a 3-foot connecting cable with negligible effect on amplitude and phase response. To maintain good low-audio-frequency separation, it was found necessary to maintain the low frequency input impedance of the adaptor in excess of 50 megohms. Sufficient amplification had to be provided in the adaptor to produce a 2.5-volt minimum output at 100 per cent modulation of either left or right channel from the relatively low output of the wideband detector (0.3 volt typical at 75,000 cps deviation).

The Adaptor

Figure 4 shows the block diagram of the H. H. Scott type 335 Multiplex Adaptor. The signal from the FM detector (multiplex output) is first amplified in a high-input-impedance stage and then passed through a phase-linear filter attenuating frequencies above 53,000 cps. This removes any background music signals from the stereo demodulator inputs. A narrow band and noise immune 19,000 cps filter selects the pilot carrier. After further amplification it synchronizes the 38,000 cps subcarrier oscillator. All tuned circuits are temperature compensated so that the oscillator exhibits a warmup drift of only .01 per cent in the absence of a pilot carrier, the 38,000 cps subcarrier oscillator remains phase locked to the pilot signal so that maximum separation is maintained at all r.f. signal levels. Measurements with a wave-analyzer have shown that separation of left and right audio signals is maintained even with such low r.f. signals that the signal-to-noise ratio makes listening impossible.

The output of the 53,000 cps low-pass filter and the 38,000 cps oscillator drive the two balanced bridge stereo demodulators. Two wideband amplifiers following the demodulators have a common efficiency-balance circuit (as required by the difference of *Eq.* (*1*) and (*2*) above). This assures best separation. The de-emphasis and 15,000 cps cutoff circuits are in the separate audio channels rather than ahead of the demodulator or matrix networks. By this method, any component tolerance will not affect channel separation as it otherwise would. A stereo level control with low impedance output amplifiers complete the actual adaptor circuit.

A number of circuit refinements have been incorporated in this self-powered adaptor. A front panel switch permits listening to either multiplex stereo or AM-FM stereo broadcasts if an AM-FM stereo tuner is the signal source. Other switches engage noise filter circuits permitting stereo listening of weak signals with reduced noise and full frequency response or full separation. Both stereo amplifier and stereo tape recorder outputs are provided.

The FCC-approved system can be received with extremely simple adaptors that will provide adequate results with inexpensive FM radios and tuners. Fortunately, for the more demanding music listener it is possible to design a multiplex equipment with the same high engineering standards found in our wideband tuners. Æ

Fig. 6. Top-rear view of H. H. Scott adaptor.

Signal Sampling For FM Stereo

R. SHOTTENFELD and S. ABILOCK

Using a new technique in the FM-stereo field, diodes and a synchronized oscillator provide a sampling circuit which eliminates all vacuum-tube circuitry from the signal path, avoids the need for user-operated controls, and provides an indication of the presence of a stereo signal.

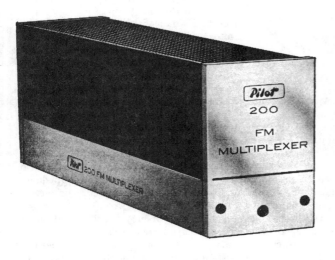

External appearance of Pilot Model 200 stereo adapter.

WITH FEW EXCEPTIONS, present day multiplex adaptors and FM-stereo tuners use circuits that operate on the principles demonstrated to the Federal Communications Commission at the Uniontown, Pa., field tests. The successful system had two proponents, and two prototype adaptors were shown. The circuits of these two units, while outwardly different, were similar in their use of frequency separation filters.

In one prototype (G. E.) the composite signal is first separated by filtering into two components: the sum signal, $(L+R)$, and the sidebands of the difference signal, $(L-R)$. After carrier reinsertion (the carrier having been suppressed in the transmitter) the difference signal, $(L-R)$, is recovered from the second component by means of AM detection. Finally the left and right channel audio signals, L and R respectively, are obtained by matrixing the sum and the difference signals.

In the other prototype (Zenith), direct but imperfect demodulation of the composite signal is accomplished by means of a synchronous switching circuit. Separation of L and R is then completed by partially matrixing the switch output with the sum signal, $(L+R)$, obtained from the composite signal by filtering.

A serious disadvantage in the use of frequency separation filters and matrixing soon makes itself evident when an attempt is made to use these circuits in the design of a high-performance multiplex adapter. In order to obtain essentially complete separation it is necessary to match both the signal levels at the matrix and the transmission time delay through the filters to an extremely high degree of precision for every frequency in the audio spectrum. A circuit to match the signal levels is not very difficult to construct especially if a gain control is used. However, the time delay problem is not so easily solved. It would take a large number of filter sections to match the frequency response and the time delay of a low-pass filter and its corresponding band-pass structure to the precision necessary for complete separation of L and R over the entire audio spectrum. It is practical to use only a few filter sections in consumer equipment. Therefore, separation can be made high only over a limited frequency range, and it must decrease at the upper and lower ends of the audio spectrum where the frequency response and the time delay of the filters fail to match.

When we began the development of a commercial multiplex adaptor, our first approach was to work with filters and matrices. As soon as we encountered the difficulties outlined, we set about searching for a different way to do the job. The outcome of our subsequent investigation and the resultant radically different circuits form the substance of this article.

What are the most desirable properties of a good commercial FM-stereo adaptor? First of all, the unit should perform well with the widest possible variety of tuners—those made in past years as well as current models. It should have sufficient inherent stability to be able to maintain its high level of performance despite varying input signal levels, environmental changes, and reasonable aging of tubes and components. Installing and connecting the unit should be simple and require no mechanical skill. To fit easily into existing custom installations or consoles, the unit should be compact and require a minimum of panel space. And, *there should be a minimum of controls to operate.* The Pilot Model 200 multiplexer has these desirable properties plus two very unusual circuits, one of which is the subject of a patent application.

The Adapter Block Diagram

The block diagram, *Fig.* 1, shows the major circuit functions and the signal paths for the Pilot Model 200. Three paths extend from the input terminal. The upper path is for a monophonic signal and it contains only a de-emphasis network. The central path contains all the circuits which have to do with demodulation of the composite signal. The lower path contains the circuits which work with the pilot and the stereo subcarriers. At the right end of the diagram all the signal paths terminate in the Stereo-Mono Switch circuit block. From this, in turn, two paths extend to the outputs.

In the demodulator path the first circuit block is the tuner compensating network. The purpose of this network is to provide phase and amplitude correction to composite signals from tuners with frequency response rolloffs at 53,000 cps. The ideal tuner, of course, does not have any rolloff, and for such a tuner the compensating network is not necessary and should be removed. However, the great majority of tuners do require this network. Its use provides a very substantial improvement in separation. The network shown at (A) on the schematic diagram, *Fig.* 2, consists of a 680,000-ohm resistor, a 39-pf capacitor shunting it, and a 1-megohm resistor to ground. The configuration is that of a

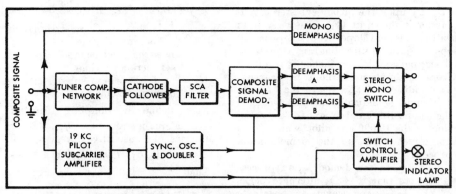

Fig. 1. Simplified block diagram of Pilot Model 200 multiplex adapter.

phase-lead high-frequency boost network. The values were selected to make it the inverse of that circuit in an average tuner, which produced the undesirable rolloff.

That a modern tuner should require a corrective network does not indicate that it is a poor tuner. The wide frequency response needed for stereo reception is not necessary for monophonic reception, and most tuners designed before the advent of stereo broadcasting do not have it. (It is worthwhile noting that the capacitance of the connecting audio cable contributes to the rolloff, and therefore it should be as short as possible.)

Another factor contributing to poor separation of some tuners made prior to the FCC approval is that the multiplex output jack is coupled to the FM detector through either a small capacitance or a high resistor. Such a tuner cannot furnish a correct composite signal to a multiplex adapter that has a conventional value of input impedance. The remedy is to increase the coupling capacitor to at least 0.1 µf, or to short out the resistor.

After passing through the tuner compensating network, the composite signal, with all its frequency components in correct time relationship and proper relative amplitudes, passes through a low-distortion cathode follower to provide a low-impedance driving source for the circuits that follow.

Next in order is the SCA filter. This is a band-rejection network, not a simple resonant circuit that provides only an attenuation notch. This network provides uniform attenuation over the range of frequencies from 67,000 to 76,000 cps and affects the response at 53,000 cps by much less than 1 db. The advantage of such a filter is that there is attenuation for the SCA sidebands as well as the carrier, and that the upper sidebands which produce low-frequency beats with 76,000 cps are uniformly attenuated.

At this point the composite signal enters a circuit that represents a radical departure from any presently used for stereo demodulation. With this method the left- and right-channel audio signals are extracted directly from the compos-

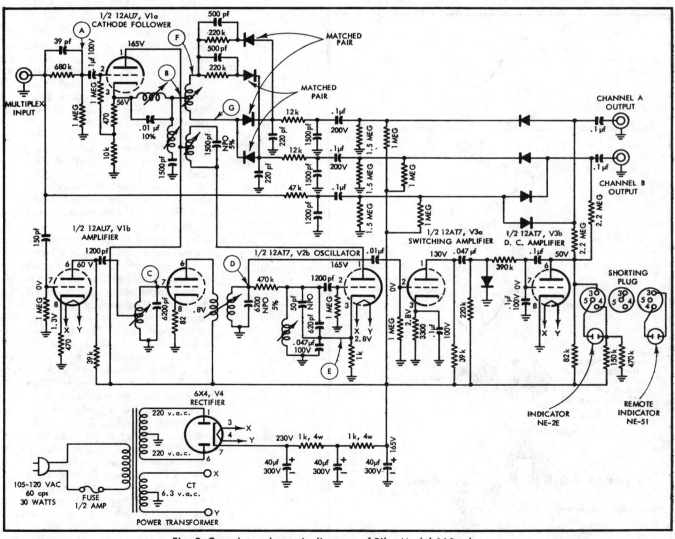

Fig. 2. Complete schematic diagram of Pilot Model 200 adapter.

ite signal without using either filtering or matrixing. The circuit is capable of perfect separation over the entire audio spectrum, and its output is substantially free of stereo subcarrier. It consists essentially of a synchronous switch that samples the composite signal on an impulse forming basis, and a "data hold."

Synchronous Switching

In order to explain how the circuit works we will begin by first showing how the composite signal is formed. When the waveform is clearly in mind it will become much easier to understand how the demodulation takes place.

The various waveforms that contribute to the formation of the composite signal are shown in *Fig. 3*. As the L and R signals are entirely independent, each may be chosen arbitrarily. For convenience the L signal will be represented by a sine wave, and the R signal by a rectangular waveform as shown at (A) and (B). The sum signal $(L+R)$ is shown at (C), and the difference signal $(L-R)$ at (D). Double-sideband suppressed-carrier amplitude modulation of $(L-R)$ upon the 38,000-cps stereo subcarrier is shown at (E). The wavy lines within the envelopes represent 38,000 -cps, and the entire diagram represents the upper and lower sidebands resulting from modulation, without the carrier. The addition of $(L+R)$ to the $(L-R)$ sidebands is indicated at (F). This is the composite signal except for the omission of the 19,000-cps pilot, which plays no direct part in the demodulation process.

The key to the demodulation process is to note that in (F), alternate peaks of the 38,000-cps waveform terminate on envelopes which have the same waveshapes as the original L and R audio signals. Thus, in the composite signal, the L and the R signal information is maintained separately on the two envelopes which form the boundaries of the 38,000-cps waveform. And, the L and R signals can be recovered separately and independently by circuits which allow the appropriate output to "view" the composite signal at suitably timed instants.

The mathematical expression for the composite waveform at (F) in *Fig.* 3 is:

$$E_t = [L+R] + [L-R][\cos 2\pi f_{sc} t]$$

Where: L is the left channel signal as a function of time
R is the right channel signal as a function of time
f_{sc} is the stereo subcarrier frequency, 38,000 cps.

Now define a sampling function $\overline{E}_t = E_t$ at the instant $t = n/2f_{sc}$, n being the series of consecutive integers 1, 2, 3, ..., and: $\overline{E}_t = 0$ at all other times.
Then:

$$\overline{E}_t = [L+R] + [L-R][\cos n\pi]$$

Notice that for even values of n:

$$[\overline{E}_t]_{n,\,even} = [L+R] + [L-R][+1]$$
$$= 2L = pure\ left\ signal$$

And for odd values of n:

$$[\overline{E}_t]_{n,\,odd} = [L+R] + [L-R][-1]$$
$$= 2R = pure\ right\ signal$$

The result of applying the mathematical sampling function to the expression for the composite signal is the same as physically viewing the waveform for very brief instants, a series of narrow spikes emanating from the base line and terminating at the values of E_t at those instants.

From sampling theory, it is known that the original continuous function can be completely and accurately recovered from the sampled function by low-pass filtering, provided that the sampling rate is more than twice the highest-frequency component of the original signal. This can be done for the signals involved in FM stereo, because 38,000 cps is more than twice 15,000 cps, the upper limit of the audio band.

When the sampling consists of very narrow pulses, the energy content is quite low and the recovered continuous function should have a very low amplitude compared to the input signal. The impulse signals also contain a large component at sampling frequency that require an extensive filter to remove.

The desired output can be greatly increased and the sampling frequency component decreased by making use of a circuit which has the ability to hold the output constant at the level of the preceding sample until the next one comes along. Such circuits are called "data holds," or "memory circuits." The output waveform now becomes a "staircase" or "step" function approximation to the continuous function. There is a good deal more energy in this and therefore the output signal level is more nearly the same as the input signal.

Other circuits used for stereo demodulation have used switching signals of long duration approaching one-half cycle at 38,000 cps in order to build up the energy of the output signal. Unfortunately this does not provide com-

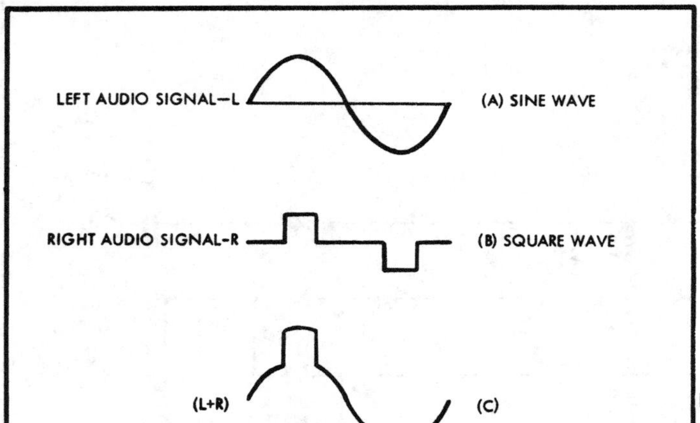

Fig. 3. Composition of composite stereo signal. For simplicity, the 19,000-cps pilot signal is omitted.

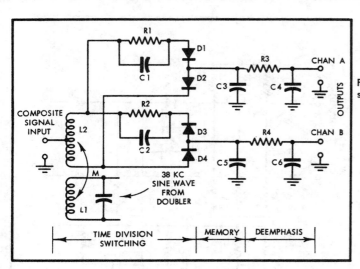

Fig. 4. Composite stereo signal demodulator.

plete separation and so subsidiary matrixing must be resorted to.

In the Model 200 circuit, just the opposite is done. Narrow sampling impulses are used because complete separation can be obtained that way. When the "data hold" is used to build up the energy of the recovered signal, the output can be within 1 db of the input, and the component at sampling frequency is much reduced.

The actual circuit is shown in *Fig. 4*, which is a part of the complete schematic *Fig. 2*. L_1 and L_2 are the primary and secondary of a transformer tuned to 38,000 cps. L_1 receives a 38,000-cps signal from the frequency doubler. L_2 is an accurately center-tapped coil. The composite signal is fed into the center tap of this coil. The voltage developed across L_2 causes the diodes D_1 and D_2 to conduct for a small fraction of a cycle each time that terminal 1 of L_2 is positive with respect to terminal 2. The network R_1 and C_1 controls the conduction time interval by biasing the diodes into non-conduction. As long as the diodes do not conduct, their common terminal is disconnected from the input. When they do conduct, there is a low-impedance path between their common junction and the input. By properly phasing the 38,000-cps signal, the conduction of the diodes may be made to coincide in time with the peaks of the composite signal waveform, and by making the conduction period very short the high separation associated with impulse sampling is achieved.

Diodes D_3 and D_4 are similarly connected to L_2, but their polarity is reversed so that they conduct on the peaks of the other half cycle. Their common junction is therefore connected to the input only when the other set of diodes is not. In other words we have a momentary contact SPDT switch that alternately connects the input to the Channel A and the Channel B outputs.

The "data holds" are simply the capacitors C_3 and C_5. The value of each is made small enough so that it can charge rapidly to the composite signal voltage in the small time that the diodes conduct, and the leakage to ground of the entire output circuit, which tends to discharge the capacitors, is made so small that the voltage across them remains essentially constant during the time interval between conduction instants. This, then, is the mechanism of step function formation. $R_3 - C_4$ and $R_4 - C_6$ are standard de-emphasis networks.

Notice that the two halves of L_2 and the diodes D_1 and D_2 form a bridge circuit. The 38,000-cps energy injected into L_2 does not appear between the input terminal and ground; neither does it appear between either output terminal and ground. In practice the bridge need not be balanced to a high degree of precision. A fair balance will give quite good suppression of 38,000-cps and the de-emphasis network also helps. The diodes must have a quite rapid turn-on and turn-off characteristic, otherwise suppression of stereo subcarrier will depend too strongly on matching this characteristic between the two diodes.

Needless to say, the diodes must conduct synchronously with the peaks of the 38,000-cps waveform in the composite signal. Failure to do so will result in loss of separation. The separate L and R signals are next fed into the stereo-mono switch.

Returning now to the block diagram and schematic, consider the lower signal path. A two-stage tuned 19,000-cps amplifier selects and amplifies only the pilot subcarrier. It is important to remove all vestiges of modulation from the subcarrier before using it to synchronize the oscillator, otherwise distortion may result. The oscillator section, V_{2b}, strongly locked to the pilot subcarrier, generates the 38,000 cps that actuates the demodulator. The oscillator coil, 79–113, is loosely coupled to the output of the 19,000-cps pilot subcarrier amplifier by the 0.47-megohm resistor only.

This kind of coupling ensures that both tuned circuits operate at resonance.

If inductive or capacitive coupling were used there would be a detuning effect that would shift the phase of the oscillator when the amplitude of the synchronizing signal varied. This undesirable affect is avoided by using pure resistance coupling, building up a large synchronizing signal and loosely coupling it to the oscillator so that strong synchronizing action is obtained even with the weak coupling that is necessary to avoid interaction between the two tuned circuits.

All the tuned circuits must be accurately temperature compensated to avoid phase shift as the unit heats. It is most important to compensate each circuit by itself. Unless this is done there will be a variation in the ability to maintain proper phase with different signal levels as the unit heats up.

The Stereo-Mono Switch

Next consider the stereo-mono switch. A schematic of this circuit, broken out of the complete schematic, is shown at *Fig. 5*. Some FM stations broadcast SCA programs in the stereo subchannel frequency range in addition to the regular monophonic programs. This will cause interference with a monophonic program if listened to through a multiplex adapter. For this reason it is advisable to use the multiplex circuits only when actually listening to a stereo program. Many multiplex adapters provide a manual switch for this purpose. Such adapters cannot be tucked away unless the associated equipment has switching that can bypass the adapter.

Automatic switching, actuated by the pilot subcarrier, is used to bypass the multiplex circuits when they are not needed. Therefore, this multiplexer, which has no controls for the user to operate, can be installed in any out of the way spot.

The circuit consists of two major parts: the switch-control amplifier, and the switch element itself. The switch-control amplifier consists of a stage of amplification for 19,000-cps pilot V_{3a}, a forward biased diode D_1, a slow-acting d.c. amplifier V_{3b}, and a neon lamp indicator. V_{3a} receives a 19,000-cps signal from the pilot-subcarrier amplifier V_{2a}. The output of V_{3a} appears across the diode D_1 which is forward biased by the current through R_2. As the input voltage increases, starting from zero, at first nothing happens at D_1 because the current through it causes it to be a short circuit across the load resistor R_1. As soon as the 19,000-cps current exceeds the bias current in D_1, the junction of R_2 and R_3 starts to go negative with respect to the cathode of V_{3b} and this tends to lower the plate current of the tube.

The action so far is this: With no input signal, the *grid* of V_{3b} is at the same potential as its cathode. The tube conducts strongly, and its plate voltage is low—about 50 volts. When the 19,000-cps input signal to V_{3a} exceeds the threshold value, V_{3b} becomes biased and its plate voltage rises. Because of the high gain, the transition of V_{3b} plate current from saturation to cut off is very rapid and the plate voltage swings from 50 volts at saturation to almost 160 volts at cutoff. A filter consisting of R_3 and C_2 allows only the d.c. component of the rectified output of D_1 to reach the grid of V_{3b}, and the capacitor C_3 slows the time of response of V_{3b} to prevent transients from reaching the switch itself.

The stereo indicator lights up when the voltage across it is 60 volts or more. One terminal of the neon indicator lamp is connected to the junction of R_5 and R_6 which maintains that terminal at about 80 volts positive with respect to chassis. When the plate of V_{3b} is at 50 volts to ground, the indication lamp is across only 30 volts, and therefore does not light. When V_{3b} is cut off its plate voltage of 160 exceeds 80 volts by more than 60 volts, and therefore the lamp lights up.

The switching elements themselves are silicon diodes. They also are controlled by the plate potential of V_{3b}. For the mono signal, D_3 and D_4 have a common anode potential, maintained at about 80 volts by the voltage divider consisting of R_8 and R_{13} whose impedance is very high compared to the signal circuits. Their cathodes are separate, one going to Channel A output and the other to Channel B output. The cathode potentials of D_3 and D_4 can be varied by the plate swing of V_{3b} acting through R_{10} and R_{11}. When the plate of V_{3b} is at 50 volts, the cathode of D_3 and D_4 are

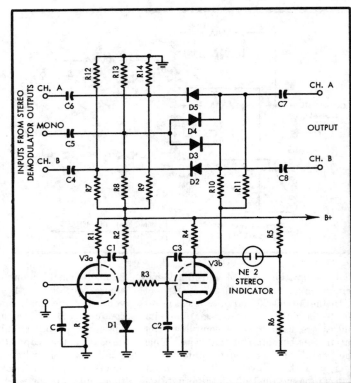

Fig. 5. Composite of pilot-controlled stereo-mono switch.

negative with respect to their anodes and the diodes conduct. The signal path through them is closed and the mono signal is connected to both the outputs.

Conversely when V_{3b} is cut off, the cathodes of D_3 and D_4 are positive with respect to their anodes and, therefore the diodes are cut off. The signal path through them is open and the mono signal is disconnected from the output.

The stereo switching diodes, D_2 and D_5, are connected in reverse polarity, therefore they conduct when the mono diodes are open and vice versa. Thus when no 19,000-cps pilot subcarrier is present at the input to the multiplexer, the diodes D_3 and D_4 conduct, setting up a signal connection between the mono-de-emphasis network and the output. At the same time, D_2 and D_5 are made non-conducting, thus opening the signal connection between stereo de-emphasis networks and output. Similarly, when the 19,000 cps is present at the input, the stereo connection is set up and the mono connection is inhibited.

Conclusion

In conclusion, this multiplex adapter circuit has been in production for some time now. The product has proven itself to be an excellent performer, its stability has been very good, no servicing problems have arisen. Æ

FM—STEREO
(from page 71)

A sufficient amplitude of pilot subcarrier is injected into the oscillator tank circuit to lock the oscillator not only on frequency but within a few degrees of phase relative to the incoming pilot subcarrier appearing in the composite stereophonic modulation.

For monophonic reception it is desirable to remove the contribution of the stereophonic subcarrier channel from the output of the left and right channels by killing the oscillator so that the beam deflection demodulator tube electron beam rests between the two anodes, thus equally dividing the beam current between the two anodes.

Adjustment

The circuitry as shown in *Fig. 6* is quite simple to adjust for optimum demodulation. The sequence of tuning is as follows:

The 67,000-cps trap shown in the filter portion of the schematic is tuned until a null at 67,000 cps appears at the output of the filter. The anode resonant circuit of the triode portion of the 6AU8 is tuned for maximum pilot subcarrier amplification. The oscillator tank circuit is then adjusted so that the oscillator is locked to the incoming pilot subcarrier frequency (an aural test is adequate here). The doubling tanks, both primary and secondary, in the anode circuit of the electron-coupled oscillator are tuned for maximum output at 38,000 cps, as observed at the deflection plates of the beam deflection demodulator tube. The signal generator is modulated with a composite stereophonic signal having the modulating condition L = −R. Each of the carrier tank circuits, namely: anode circuit of the cathode follower, oscillator tank circuit, electron coupled frequency doubling tank circuit, and the driving tank circuit for the beam deflection tube is then retouched for maximum recovered audio at both left and right outputs.

The modulating condition of the signal generator is then changed to left channel modulation only. The 500-ohm matrix control in the cathode circuit of the beam deflection demodulator tube is adjusted for a null at the right channel output. If the matrix resistors are well matched and the gain characteristic of each anode circuit of the beam deflection demodulator tube are well matched, then a null condition will also exist at the left output when a right only signal is applied to the signal generator. Æ

Distortion in Tape Recording

HERMAN BURSTEIN

Types and causes of distortion should be understood by the recordist if he is to obtain the best results. Various compromises are shown to be effective under different conditions.

FROM TIME TO TIME the writer visits some friends who have in their living room a pre-war radio console, which cost over $500 and was considered one of the finest units of its day. It receives good care and enough service to maintain it in "as good as new" condition. To one whose ears have become attuned to modern high fidelity equipment, this console falls noticeably short of the mark in terms of frequency response and noise characteristics. But its most obvious deficiency, the greatest deterrent to pleasurable listening, concerns distortion. The instrument just does not have the smoothness and ease of reproduction afforded by modern equipment.

The foregoing illustrates the point that one of the most noteworthy developments in the audio art in the high fidelity era, at least to the ears of this writer, has been the reduction of distortion. While the power amplifier has come in for a great share of attention, it is also true that designers of control amplifiers, tuners, cartridges, speakers, and other components have concentrated on reducing distortion to imperceptible amounts.

In these days when most audio equipment is built to exacting standards with respect to clean reproduction—i.e. low distortion—one looks for comparable refinement in the tape recorder. The meticulous recordist will wish to preserve the original quality of the sound so far as possible. While satisfactorily low distortion can be achieved in tape machines, this is far from a simple matter. Overcoming distortion remains considerably more of a challenge in tape machines than, say, amplifiers. When used with today's better amplifiers, tuners, and speakers, a tape recorder must indeed be of high quality, and must be properly used, in order not to add noticeable distortion.

In tape recording, distortion is inextricably linked with several other aspects of the process—signal-to-noise ratio, frequency response, equalization, bias current, and tape speed. Therefore in the following discussion we shall discuss distortion in terms of its relationship to these factors. First, however, it would appear profitable to devote some space to a review of what is meant by distortion. Such an understanding can prove useful in various ways; for example, it enables one to appreciate why a given recording level results in no noticeable

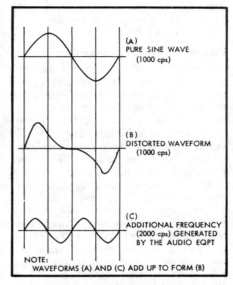

Fig. 1. Example of harmonic distortion.

distortion for some kinds of sound and quite perceptible distortion for other kinds.

Meaning of Distortion

Reproduced sound is never totally devoid of distortion. But in the present state of the art it can be kept so small in most audio components as to be unnoticeable, permitting the reproduced sound to retain the ease and naturalness of the original. In somewhat larger quantity, it may still not be immediately discernible but instead may produce a consciousness of aural fatigue after one has been listening for a moderate period of time. In successively larger quantities, distortion causes the sound to become grainy, gritty, coarse, and finally so broken up as to be partially or completely unintelligible.

Distortion consists of a change in the original waveform, due to improper functioning of one or more audio components. Such improper functioning is called non-linearity; that is, the waveform turned out by the component is not an exact replica of the incoming signal.

It can be demonstrated, mathematically and by suitable test equipment, that the change in the waveform actually consists of the addition of new frequencies to those that were originally produced by the sound source. This is illustrated in *Fig.* 1. At (A) we see the original waveform, a pure sine wave; (B) shows a distorted version of the original. The distortion consists of the waveform shown at (C). If the distortion frequency in (C) is added to the original frequency (A), the result is the distorted waveform of (B).

The new and undesired frequencies, which are termed distortion products, are produced by the audio equipment. Unlike noise and hum, which are also undesired frequencies produced by (some) audio components, distortion products appear only in the presence of an audio signal.

The principal kinds of distortion, those most offensive to the ear, are harmonic and intermodulation distortion. Harmonic distortion denotes the generation of frequencies that are multiples of the original frequency. To illustrate, in the course of reproducing a 1000 cps tone the audio equipment may, as the result of its non-linear behavior, also generate frequencies of 2000, 3000, 4000, etc. cps.

To an extent, the ear is not unduly offended by extraneous frequencies if they are harmonically related to—exact multiples of—the original note. As a rough rule, harmonic distortion products are compatible with pleasant listening when, in total, they constitute no more than about 1 to 2 per cent of the total sound; generally, 3 per cent is considered too great.

At the same time, the amount of harmonic distortion which is tolerable depends upon whether the distortion products are even or odd multiples of the original frequency. Even multiples tend to be less offensive. Furthermore, the "order" of the harmonic products is a determining factor. High-order products are many times the original frequency; low-order products are a few times the original frequency. High-order products tend to be more offensive. Thus if the original frequency is 1000 cps, distortion products of 8000 and 9000 cps would be more disagreeable than 2000 and 3000 cps. (It is appropriate to intersperse here that a tape recorder which cuts off sharply above 9000 or 10,000 cps may offer cleaner sound than one which goes out to 15,000 cps because the former eliminates high-order distortion products to a greater extent.)

Intermodulation distortion—IM for short—occurs only when two or more frequencies are simultaneously reproduced by the audio equipment. Deformation in the waveform of one frequency results in deformation of a second frequency, although it could well be that the second frequency, if reproduced alone, would not have been distorted by the equipment in question. Thus IM distortion refers to interaction among frequencies, with new frequencies being born out of this interaction. When a substantial number of frequencies are reproduced at once, as is often the case with music, the interaction, namely IM distortion, becomes very complex.

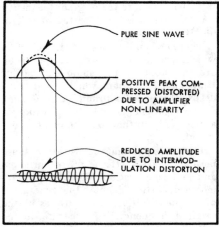

Fig. 2. Example of intermodulation distortion.

Figure 2 illustrates the process of intermodulation distortion. For simplicity, it is assumed that only two tones, 100 and 1000 cps, are present and that they are fed through an amplifier. Let us assume that the 100-cps signal is of substantially greater magnitude than the other, so that it causes the amplifier to operate in non-linear fashion at the positive peaks of the waveform. During these moments of non-linear operation of the amplifier the 1000 cps signal is also being treated in non-linear manner, despite the fact that this signal in itself is of too small magnitude to cause the amplifier to behave in non-linear fashion. At (A) we see the effect of amplifier non-linearity upon the 100 cps waveform. (B) shows the resulting effect upon the 1000-cps waveform due to the fact that the amplifier is periodically operating in non-linear manner. The 1000-cps waveform is compressed 100 times per second by the 100 cps signal. In other words, the 100-cps frequency is now present in the 1000-cps one.

Unfortunately the new frequencies created by IM distortion are not multiples of the original frequencies. The distortion products consist of various multiples of one frequency plus or minus multiples of the other frequency. For example, 100 and 1000 cps will form IM products of 1100 cps (sum of the original signals) and 900 cps (difference between the original signals). They will form 1200 cps (twice 100 cps plus 1000 cps) and 2100 cps (twice 1000 cps plus 100 cps). They will form 1900 cps (twice 1000 cps minus 100 cps) and 800 cps (1000 cps minus twice 100 cps). And so on and so forth. If there were more than two original frequencies involved, the distortion products would be still more complex.

IM distortion not exceeding 1 to 2 per cent is often considered compatible with high fidelity. On the other hand, it has been found that the ability to reduce IM to as low as 0.1 per cent in voltage amplifiers and power amplifiers has produced noticeable improvement.

Distortion Ratings for Tape Recorders

Extremely seldom does one find the specifications for a tape machine having anything to say about IM distortion. The reason will appear later, when we compare harmonic and IM distortion produced by tape recorders. The nigh-universal practice instead is to rate tape machines in terms of harmonic distortion at a stated signal-to-noise ratio, for example 50 db in moderate-quality machines or 55 db in high-quality machines. The record-level indicator is adjusted to provide an indication of maximum permissible recording level when the level is such as to produce anywhere from as low as 1 per cent to as high as 5 per cent harmonic distortion (at a frequency of 400 cps or so). The low-priced machines typically use 5 per cent harmonic distortion as maximum permissible recording level, while the top quality ones use 1 or 2 per cent. Many machines, of varying quality, use 3 per cent harmonic distortion as the reference. The official standard, applicable to 15 ips recording, considers 2 per cent harmonic distortion to be the maximum permissible quantity.

Distortion and Signal-to-Noise Ratio

In the process of recording and playing back a tape, there are two principal sources of noise to contend with: tape noise and amplifier noise. Tape noise is of two kinds. One, known as tape hiss, is due to incomplete cancellation of magnetic fields when the tape is erased. These magnetic fields are of random character and therefore produce random frequencies with a characteristic "hissy" quality. The other kind of tape noise is known as modulation noise, which appears only in the presence of an audio signal on the tape. Modulation noise is due to imperfections in the base and/or magnetic coating of the tape. When an audio signal is recorded, corresponding imperfections appear in the recorded signal and are manifest as noise. As the result of the improvements that have taken place in tape manufacture, modulation noise is less serious a problem than tape hiss.

Tape-amplifier noise occurs both in recording and playback. However, the signal fed to the tape amplifier is generally of much smaller magnitude in playback—the tape delivers but a fraction of a millivolt at many frequencies—so that it is principally noise of the tape playback amplifier which presents a problem.

In sum, the principal obstacles to a good signal-to-noise ratio are tape hiss and the noise (including hum) produced by the tape playback amplifier.

To achieve an adequate signal-to-noise ratio it therefore becomes vital to record as much signal as *practical* upon the tape. But the practical amount of signal that can be impressed on the tape is determined by the distortion characteristics of the tape, the tape head, and the record amplifier. Ordinarily, the tape sets the bounds to how much signal can be recorded. That is, the tape overloads, or should do so, before the tape head and the tape amplifier go into serious distortion.

However, there have been instances where a poorly designed head has produced significant distortion in recording, particularly at low frequencies, although the signal level was not such as to produce appreciable distortion on the tape. Laminated heads, which contain a greater amount of magnetic material, are generally apt to have superior distortion characteristics compared with those of non-laminar construction.

There have also been instances where an improperly designed recording amplifier has gone into serious distortion at too low a recording level. For example, one instance of this kind involved a machine of professional calibre. Although the amplifier did not produce appreciable distortion when conventional tape was employed, it went into excessive distortion when the recording level was increased to a point consistent with the use of high-output tape, which can accept several db more signal for the same amount of distortion.

For the most part, however, we can assume that it is the tape which sets the limit to the recording level by overloading before any of the other components do.

Figure 3 indicates the variation of harmonic distortion and of IM distortion with changes in input signal. The measurements were taken on a professional-quality tape machine operating at 15 ips. While the results doubtless would be different with other machines, tapes, and speeds, nevertheless these curves can be viewed as representative.

It may be seen in *Fig.* 3 that distortion, either harmonic or IM, increases quite slowly for a while as signal level is increased, but that the rise in distortion becomes precipitous after a point. Severe IM distortion occurs much earlier than harmonic distortion. Hence at recording levels which breed innocuous amounts of harmonic distortion the IM distortion will have risen to unacceptable levels. It is understandable, therefore, why a recording may sound grating if made un-

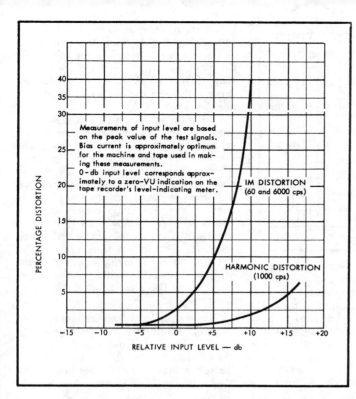

Fig. 3. Variation of tape distortion with changes in input level.

der conditions where the record level indicator permits 5 per cent maximum harmonic distortion.

On the other hand, a recording that permits IM distortion to reach 20 per cent or more is not always unacceptable. Sounds recorded at such distortion levels are tolerable if their duration is sufficiently brief. Characteristically, many sounds have peak levels 10 db, 20 db, or even more above their average level. While the peaks may be severely distorted, the major part of the sound may be at a level that escapes significant distortion. Whether the distortion in the peaks is tolerable depends upon their duration and how frequently they come along. If the peaks are occasional and very brief, large amounts of IM distortion in the reproduction of these peaks may escape attention.

The extent to which distortion is acceptable also depends upon the nature of the sound being recorded. Certain kinds of music must be recorded at lower levels than other kinds in order to maintain clean reproduction. Generally, higher levels of distortion are acceptable in reproducing speech than music. In recording a solo voice or a solo instrument, IM distortion is less apt to be serious than when recording a group of voices or instruments, because there will be fewer intermodulation products when there are fewer frequencies reproduced at one time.

In deciding how high a recording level one may employ for different source material, there is no substitute for experience. The neophyte recordist does well to invest a certain amount of time in experimenting with various recording levels for various kinds of material. In any event, he should remember that the desire for a slight improvement in signal-to-noise ratio—i.e., by raising the recording level just a few db—may bring with it a great increase in distortion if one happens to be at the point where distortion rises rapidly with a slight increase in recording level.

All in all, the recordist has three choices. First, he may be willing to accept occasional noticeable distortion, principally on signal peaks, for the sake of a relatively high recording level and therefore a superior signal-to-noise ratio. Second, he may be unwilling to accept any noticeable distortion whatsoever, but at the cost of a significant reduction in recording level and therefore in signal-to-noise ratio. Third, it is possible in a sense to eat one's cake and have it too by "riding gain." That is, one can record at a moderately high level, well below the point of noticeable distortion, during normal and quiet passages, then reduce the recording level just before loud passages come along. The last alternative requires one to be prepared with a score or other means of knowing when loud passages are about to occur. Also it implies that one is willing to compress the dynamic range (difference between the softest and loudest passages) in exchange for an improvement with respect to distortion.

It must be taken into account that the need to exchange signal-to-noise ratio, or possibly dynamic range, for a reduction in distortion depends upon the tape machine one is using. If the playback amplifier has superior characteristics in terms of low noise and hum, and if the head is specifically designed for playback and therefore has higher output than one intended for both recording and playback, the recordist's task of achieving a satisfactory compromise between the conflicting considerations of noise and distortion is lightened. On the other hand, if amplifier noise is relatively high and head output low, the recordist might conceivably decide he is willing to accept a fair amount of distortion in order to keep noise down relative to the audio signal.

Tapes differ somewhat in their distortion characteristics. This is illustrated by a test that was made of four brands of conventional tape. At a relatively high recording level, the input signal was adjusted in each case so as to produce the same output level in playback; after all, it is the playback level in which we are ultimately interested. At the same time, bias current was adjusted so as to produce minimum IM distortion. The results appear in Table 1. While the differences in distortion are not profound, still the difference between minimum and maximum IM, 3.4 per cent, is not insignificant. On the other hand, Tape A would not necessarily be one's choice, assuming that one goes by laboratory tests. It would further be necessary to investigate the tape's characteristics with respect to frequency, noise, and other factors.

A high-output tape was tested in the same manner as the four conventional tapes just discussed. In this instance, minimum IM distortion was only 3.5 per cent, a substantial improvement.

The ability of the tape to accept a

TABLE I
Minimum Level of IM Distortion Obtainable With Four Brands of Conventional Tape at a High Recording Level

Tape	Minimum IM Distortion %
A	7.6%
B	9.0
C	11.0
D	10.0

high recording level without serious distortion varies somewhat with frequency. At a tape speed of 7.5 ips, it appears that there is a rise in the amount of signal which can safely be presented to the tape. The rise starts at about 1000 cps and attains a maximum of some 4 or 5 db. The nature of this rise may vary among brands and kinds of tape. Conversely, it is indicated that the acceptable signal decreases at the low end of the audio range. *Figure* 4 suggests in approximate and relative terms the permissible recording signal that may be presented to the tape at various frequencies for the same amount of distortion. In view of what happens at the low end, it

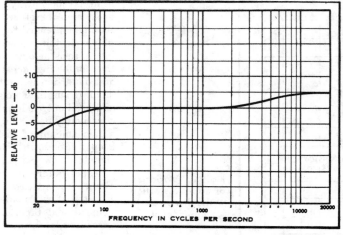

Fig. 4. Relative permissible recording level (approximate) at various frequencies at 7.5 ips.

may be advisable to record at a somewhat lower level than usual when dealing with a sound source dominated by low notes.

Distortion and the Record-Level Indicator

To record at a level high enough for a good signal-to-noise ratio yet low enough for tolerable distortion depends a great deal upon the record-level indicator. This may be either of the electronic-eye type, which indicates peak recording level, or of the meter type, which tends to indicate average level. In either case, it is of paramount importance that the meter be properly calibrated in the sense of indicating accurately when maximum recording level is reached. Thus an electronic eye that is supposed to close at a level producing 3 per cent harmonic distortion but actually does so at 6 per cent can account for an unsatisfactory recording in terms of clean reproduction. On the other hand, an eye that closes at a level resulting in only 0.5 per cent harmonic distortion would lead to very clean recordings but probably with an unnecessarily low signal-to-noise ratio.

Accordingly, the individual who is meticulous about the maintenance of his tape recorder will see to it that on occasion the calibration of the record level indicator is checked and adjusted if necessary.

The VU meter presents a special problem. The electronic-eye indicator has an advantage in that it indicates the peak level of the signal (although the meter has other advantages discussed in an earlier article.) In the case of the VU meter, which tends to indicate average level, it is necessary to allow for the inability of the pointer, a mechanical rather than electronic device, to follow rapid signal changes. Hence the VU meter understates maximum signal level. Accordingly, it is important that an off-setting adjustment be made in the calibration of the VU meter. This means that the meter should be adjusted to indicate maximum permissible recording level when the average signal (or a steady sine wave) is actually about 6 db to 10 db below the level that would cause maximum permissible distortion. Thus the meter "reads ahead," providing a safety margin to compensate for the fact that signal peaks tend to be much higher than the average signal level. Even with this safety margin, the recordist must employ experience and judgment in setting his recording level.

Distortion and Frequency Response

With rising frequency there are increasingly severe losses that take place in the recording process. These losses have to be made up by treble boost in the record amplifier. In many tape machines this treble boost goes beyond 20 db by the time the upper end of the audio range is reached. Such amounts of boost carry with them the danger of overloading the tape.

To a substantial extent the danger is mitigated by the fact that in most musical material the amplitude of the high frequencies is considerably less than that of the middle frequencies. *Figure* 5 shows for a typical orchestral selection the relative peak amplitude of frequencies over the audio range; while it should be kept in mind that this figure applies only to one particular orchestral selection, nevertheless it is typical. To the extent that the high-frequency peaks are lower than the peaks of the other frequencies, there is an offset to treble boost used in recording.

However, in many musical sources the relative amplitude of the high frequencies may be considerably greater than shown in *Fig.* 5, so that excessive distortion may occur in recording unless the recording level is appropriately reduced.

The problem of excessive treble signal is often raised by the fact that in recording a phono disc (ultimately reaching the audiofan via a broadcast and then transferred by him to tape) the engineers may deliberately emphasize the treble range or a portion of it in order to impart a false brilliance that is frequently mistaken for high fidelity. It may be possible for the tape recordist to reduce this false treble boost, or some part of it, by means of the treble control in his control amplifier before the signal reaches the tape recorder. This can be done in those control amplifiers where the tape output jack is located after rather than prior to the tone controls.

A substantial part of the treble losses in recording are due to bias current. To reduce these losses and cut down the need for treble boost, it is expedient to

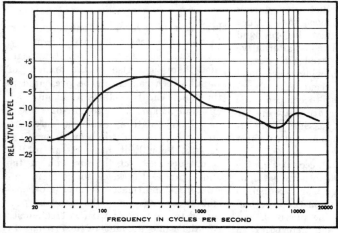

Fig. 5. Relative peak amplitude of various frequencies for a typical orchestral selection.

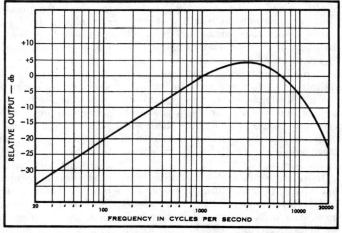

Fig. 6. Typical unequalized record-playback response of a tape machine operating at 7.5 ips.

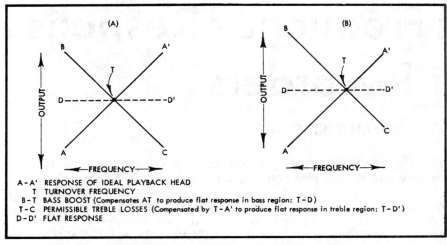

Fig. 7. Patterns of equalization in a tape machine.

reduce bias current. Unfortunately, reduced bias causes an increase in distortion. Were it not for the necessity of preserving treble response well out to the upper limits of the audio range, it would be feasible to increase bias and minimize distortion.

Distortion may also be traced occasionally to the desire to preserve response at the bass end. To maintain flat response to 50 cps and below, a slight amount of bass boost is often employed in recording. This boost reaches 3 db at 50 cps and increases as frequency declines. But, as mentioned previously, the tape is more susceptible to overloading at low frequencies than in the mid-range. If the sound source contains an abundance of very low frequencies at high amplitude, distortion may be appreciable unless, of course, care is taken to reduce the recording level.

Distortion and Equalization

Tape recorders require bass boost and treble boost, as indicated by *Fig.* 6, which shows the record-playback response of a tape machine at 7.5 ips in the total absence of equalization. The manner in which equalization is supplied affects distortion. For minimum distortion, bass boost should take place entirely or mainly in playback. Bass boost in recording imposes an excessive magnetic field on the tape. However, a number of tape machines employ half-and-half equalization, which consists of equal and ample amounts of bass boost in record and playback; and similarly for treble boost. The NAB standard, which applies to 15 ips recording, stipulates that bass boost shall take place essentially in playback.

For minimum distortion, it would be desirable to provide all or most of the necessary treble boost in playback. But this conflicts with signal-to-noise considerations. Playback treble boost emphasizes the noise of the playback amplifier, reducing the signal to noise ratio. Accordingly, it is the practice of quality tape recorders, in conformity with the NAB standard, to supply treble boost essentially in the recording process. Correspondingly, it becomes desirable to employ a pattern of equalization which minimizes the treble boost required in recording, thereby minimizing distortion.

The pattern of equalization revolves about the choice of a turnover frequency.

It is possible to employ varying patterns of equalization, each with a different turnover frequency, entailing different amounts of treble boost and therefore different amounts of distortion.

This may be explained with the aid of *Fig.* 7. Output voltage of an ideal playback head rises steadily with increasing frequency, as shown by curve A-A'. This upward sloping line may be viewed as either treble boost or bass cut, depending upon our standard of reference, namely the turnover frequency, which is labeled T. The portion of the line above T may be said to represent treble boost, while the portion below T represents bass cut.

If the turnover frequency is low, as at (A), this signifies that bass boost begins at a low frequency. Thus the bass boost in playback, B-T, compensates the bass droop A-T of the playback head. And the rising response T-A' of the playback head compensates the net recording loss (after treble boost) T-C. The large rise of T-A' in (A)—resulting from the choice of turnover frequency—signifies that substantial treble losses are permissible in recording. This in turn means that less treble boost need be used in recording, which leads to less distortion.

(B) in *Fig.* 7 represents a different pattern of equalization, one with a high turnover frequency. Consequently the playback head supplies less treble boost compared with the scheme of things in (A). This signifies that less treble loss is permissible in recording, thereby necessitating more treble boost in recording and greater distortion. (On the other hand, the equalization pattern of (B) permits a better signal-to-noise ratio because more signal is recorded on the tape and because there is greater de-emphasis of the treble frequencies—the noise region—in playback.)

The pattern of equalization is not a matter for the tape recordist to decide. It is an industry decision. At the time of writing the question of suitable equalization patterns (turnover frequencies) for the tape speeds principally in home use, namely 7.5 and 3.75 ips, was still unsettled and undergoing discussion by industry committees. But it does not ap-

(*Continued on page 96*)

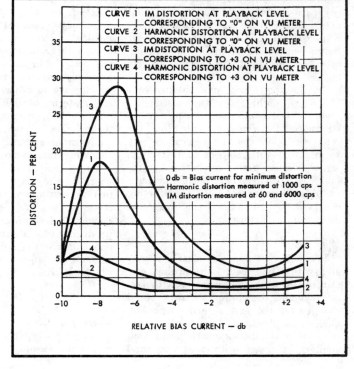

Fig. 8. Variation of tape distortion with changes in bias current.

Maintaining Frequency Response in Recorders

HERMAN BURSTEIN

If you are not satisfied with the frequency response you are getting from your tape recorder, this article may tell you why and it may also tell you what you can do to correct it.

As is usual in discussions of frequency response, we are concerned with response that is smooth, full at the low end, and maintained substantially to the upper limit of the audio range—at least to 12,000 cps and preferably to 15,000. In the case of tape recording and playback, the greatest problem is to preserve the high frequencies, and so it is this particular aspect of the subject of frequency response that will figure largest in the following discussion. Preservation of the upper audio range is relatively more difficult in tape recorders than with other elements of an audio system. This is particularly true when tape speed is 7.5 ips or less.

While a tape machine may perform well initially, its frequency response may deteriorate with age, use, or mishap. Maintaining frequency response as it should and can be requires care on the owner's part. On the other hand, frequency response of a tape machine may not initially be all that it should be. Alert to this possibility, the purchaser is in a position to reject a particular model or a particular unit of a given model that cannot deliver a desired standard of performance. Through comprehension of the various factors that enter into a tape machine's frequency response, the prospective purchaser or the present owner maximizes his chances of obtaining suitable frequency response.

At the same time, one obtains very little for nothing in the electronics realm. When maximizing frequency response in the sense of extending the range to 15,000 cps or so, sacrifices may be required with respect to distortion, signal-to-noise ratio, or both. Accordingly, the problem may be of finding a suitable compromise among conflicting considerations, namely treble response, distortion, and noise.

Tape Speed

Frequency response is closely associated with tape speed. In recording, certain losses occur that increase with frequency, and the slower the tape speed the greater the loss at any given frequency. In playback, there are losses associated with the playback head that similarly increase with frequency and become more acute as speed is reduced.

For some time it has been possible at 7.5 ips to achieve results consistent with the concept of high fidelity—namely, response extending to 15,000 cps or at least to 12,000 cps. Quite recently, it has appeared feasible to reach out to 12,000 cps or better at 3.75 ips as well, and even 1.875 speed has been gaining a place in home use. While it may be adequate for moderate quality reproduction of the voice and some forms of background music, as yet this last speed is incapable of high fidelity performance. Nevertheless, considering the constant progress that takes place in the tape art, it is conceivable that not too many years from now it will be possible to have high fidelity at 1.875 ips. At such a time the 15/16 ips might then play the role of a secondary speed where results of only moderate quality are required. Returning to the present, it may be said that nothing less than 3.75 ips is compatible with a first-rate home music system, and that to be really sure of good results it is still necessary to operate at 7.5 ips.

Head Losses

Head losses are of two kinds: (1) frequency-dependent and (2) speed-dependent. Frequency-dependent losses have nothing to do with tape speed and are electrical in nature. Specifically, they are eddy current and hysteresis losses, which have to do with the construction and material of the head, and they increase with frequency. In modern heads, these losses are very small within the audio range and may be left out of the following discussion.

The principal head loss is due to gap width of the playback head and varies inversely with tape speed. The narrower the gap, the higher is the maximum frequency that the head is capable of reproducing. As a rough approximation, one can use the following formula to estimate the upper response limit of a playback head:

$$f = \frac{S}{2G},$$

where f is the approximate upper frequency limit in cps, S is tape speed in inches per second, and G is the physical gap of the head, in inches, as specified by the manufacturer.

To illustrate, assume that tape speed is 7.5 ips and the gap of the playback head is .00025 in. according to its manufacturer. Then $f = 7.5/(2 \times .00025) = 15,000$ cps. At a tape speed of 3.75 ips, however, the upper response limit for this head would be only about 7500 cps. It is therefore apparent why gaps considerably narrower than .00025 in.—heretofore widely used in machines of good quality—are required if extended response is to be achieved at 3.75 ips. The newer heads have gaps in the vicinity of .0001 in. Inserting this value into the above formula, with speed at 3.75 ips, the upper response limit appears to be 18,750 cps. This is the feasible response *in playback*. In recording there are very serious losses that make it difficult to maintain this kind of treble response at 3.75 ips.

It should be noted that the physical gap is not the same thing as the magnetic gap. The above formula takes into account that in a well-made head the magnetic gap tends to be about 10 per cent wider than the physical gap. However, in a poorly constructed head, where the gap is not extremely straight and sharply defined, the magnetic gap may be considerably more than 10 per cent in excess of the physical gap, so that the upper response limit is correspondingly lower than indicated by the

Fig. 1. Erosion of the gap of a tape head due to abrasive action of the tape.

formula. As a result, it is quite possible that a head with an advertised gap of .0001 in. may afford better treble response than another head with an advertised gap of .00009 in., or 90 microinches.

While a head may initially have a gap sufficiently narrow and linear for good treble response at the speed in use, the gap may widen due to head wear and thereby cause a noticeable fall-off in reproduction of high frequencies. The rapidity and extent of head wear depend upon the following factors:

1. *Head Construction.* Laminated heads generally wear better than non-laminated ones.

2. *Smoothness of the Tape.* Depending upon the brand and quality of the tape and therefore upon the extent to which the tape has been lubricated and polished, head wear will vary. *Figure* 1 *suggests* the nature of head wear due to abrasive action of the tape; for visual clarity, the effect of abrasive action has been exaggerated in the drawing.

3. *Pressure of the Tape Against the Head.* For good treble response it is important that the tape and the heads maintain intimate contact. However, the pressure required for close contact results in friction as the tape moves past the heads. Thus the pressure should be just enough to maintain good contact and no more. The reels *tend* to pull the tape in opposite directions, so that the tape is held more or less taut against the heads, as illustrated in *Fig.* 2. This is the scheme generally employed in semi-professional and professional tape recorders to achieve close contact between the tape and the heads. Excessive pressure can result from excessive back tension exerted on the tape by the supply and takeup reels. There is usually provision for adjusting back tension.

Most home machines rely on pressure pads to obtain firm contact between the tape and the heads, because the path followed by the tape does not assure such contact. If the pressure pad holder is improperly adjusted, head wear may take place at an excessive rate.

On the other hand, it sometimes happens that a brand new head will offer improved treble response after a moderate period of wear. What happens is that the head wears down to the point where the gap is narrowest. But eventually the gap will begin to widen with increased wear and high-frequency response will deteriorate.

4. *Manner in Which the Tape is Wound and Rewound.* When the tape is wound rapidly in the forward or reverse direction, some machines "lift" the tape slightly away from the heads. Many, possibly most, home machines fail to space the tape away from the heads during rapid wind and rewind, thereby causing appreciable head wear, perhaps more wear than occurs during normal record and playback. To avoid this unnecessary head wear, it is generally possible to wind the tape directly from one reel to the other without going past the heads, as illustrated in *Fig.* 3. It is merely necessary to lift the tape out of its normal path past the heads—usually a guide slot—and allow it to take the shortest path between reels. The possible disadvantage of this procedure is that the tape may not be wound as smoothly as if it were following its normal path.

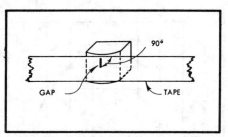

Fig. 4. Meaning of azimuth alignment.

5. *Care of the Heads.* Head wear can be minimized through suitable care, which includes regular cleaning of the heads to remove accumulated tape oxide, and the application of lubricants to minimize friction between the heads and the tape. Once the gap of the playback head has widened appreciably, nothing can be done except to replace the head, which is a good deal more costly than preventive maintenance. The gap does not have to widen very much before the head becomes unable to reproduce high frequencies. To illustrate, a gap of .0001 in. permits response to 18,750 cps at the 3.75 ips speed. If the gap widens by just one ten-thousandth of an inch, the upper response limit is reduced to 9375 cps at 3.75 ips, which is too low for high-fidelity purposes.

Azimuth Alignment

Improper azimuth alignment is one of the most common reasons for inadequate treble response. The gap of a correctly aligned head forms an angle of exactly 90 deg. with respect to the length of the tape, as shown in *Fig.* 4. If the angle differs from 90 deg., however slightly, there are losses that increase with frequency. For a given degree of misalignment, the loss at any given frequency goes up as tape speed is reduced. On the other hand, the narrower the track—for example a half-track recording compared with a full-track one, or a four-track stereo tape compared with a two-track one—the proportionately smaller are the azimuth losses.

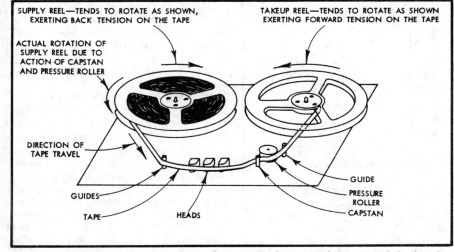

Fig. 2. Use of tape tension to achieve firm contact between the tape and the heads.

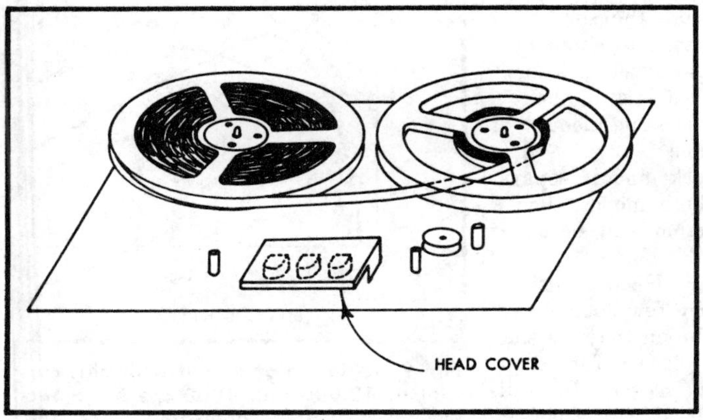

Fig. 3. Winding the tape directly from reel to reel can reduce head wear during rewinding.

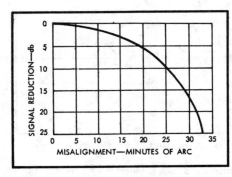

Fig. 5. Effect of azimuth misalignment upon response at 7500 and 750 cps at 7.5 ips.

The foregoing assumes that different heads are used for recording and playback. If the same head is employed for both modes of operation, the azimuth error cancels. However, when a misaligned record-playback head is used to play a commercial recorded tape, then treble response of course suffers.

Figure 5 shows how severe the losses due to azimuth misalignment can be. The curve shows the drop in response at 7500 cps for various degrees of misalignment when a half-track head is employed at 7.5 ips. It may be seen that misalignment of one half of 1 degree reduces output by more than 17 db under the stated conditions. If the frequency were greater than 7500 cps, if the tape speed were reduced, or if the track width were increased, the losses would be greater.

In the case of stereo heads, it is important to realize that it is possible for the two gaps to be out of alignment with each other, as shown in *Fig.* 6. Ideally, the two gaps should be in a perfectly straight line. If they are not, as sometimes happens, then it is not possible to obtain correct azimuth alignment on both tracks. Aligning one gap automatically throws the other gap out of alignment. Else one has to find a compromise position where both tracks are equally affected. The best solution is to replace the head, unless the degree of misalignment is so slight as not to cut response more than two or three db at the upper end of the audio range.

As stated before, the effect of azimuth misalignment decreases as track width is reduced. Hence four-track stereo heads have an advantage over two-track heads, because for a given degree of misalignment between gaps the effect upon treble response will be less with four-track heads.

Tape-to-Head Contact

Intimate contact between the tape and the heads is vital to preservation of high-frequency response. Failure of the tape to hug the heads may be due to various factors: inadequate pressure when pressure pads are used; inadequate tension when pads are not used; accumulation of tape oxide on the heads.

Losses due to separation of the tape and the heads can occur in recording as well as playback, although they are generally more severe in playback. *Figure* 7 shows the losses at 7500 cps and 750 cps for various amounts of separation in recording at 7.5 ips. The curve for 7500 cps shows that separation of about 0.4 thousandths of an inch, which can occur due to accumulation of tape oxide on the record head, will reduce the recorded level to about one-tenth of the level in the case of perfect tape-to-head contact—a loss of 20 db. The curve for 750 cps exhibits considerably smaller, but nevertheless substantial, losses. At frequencies above 7500 cps, the losses would be much greater than indicated in *Fig.* 7.

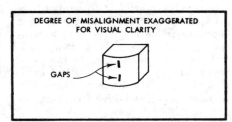

Fig. 6. Relative misalignment of the gaps of a stereo head.

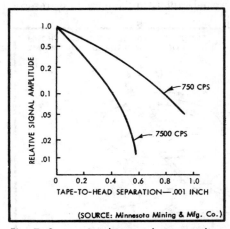

Fig. 7. Separation losses when recording frequencies of 7500 and 750 cps at 7.5 ips.

Figure 8 shows the separation losses for a playback head. Here it may be seen that at 7500 cps at a speed of 7.5 ips a separation of 0.1 thousandth of an inch reduces the signal to about one-fourth of its potential level, a loss of 12 db, compared with a loss of about 6 db in recording for the same amount of separation. If the same head is used in recording and playback, and if separation is .0001 in. at both modes, then a loss of 18 db altogether can occur at 7500 cps.

The moral is clear. Heads must be regularly cleaned every few hours to remove tape oxide. To be on the safe side, cleaning should take place before every recording or playback session. In addition, at suitable intervals, the machine

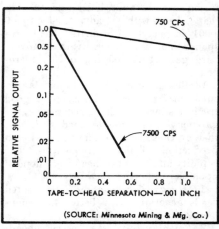

Fig. 8. Separation losses when reproducing frequencies of 7500 and 750 cps at 7.5 ips.

should be checked to ascertain that tape tension is correct or that pressure pad holders are properly adjusted.

The high-frequency bias current fed to the record head, if applied in sufficient quantity, causes the head to behave in the manner of an erase head. This is shown in *Fig.* 9, which represents the variation in output of a record-playback head operated at 7.5 ips at 1000 and at 10,000 cps as bias current is varied in magnitude. At first the recorded level goes up with an increase in bias, but eventually the level goes down as bias is increased further. By comparing the curves for 1000 and 10,000 cps, it may be seen that the level goes down faster at higher frequencies.

The effect of bias current upon frequency response can be observed more directly in *Fig.* 10, which shows the unequalized response of a record-playback head at two values of bias current. At the larger bias current, the drop in treble response is considerably greater.

The reason that the higher frequencies are more susceptible to an increase in bias current is that such frequencies, when recorded on the tape, do not penetrate the tape as deeply as do the lower frequencies. Therefore the upper frequencies are more easily erased by the

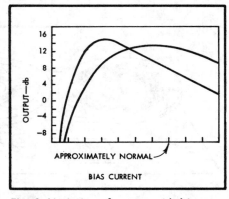

Fig. 9. Variation of output with bias current at 10,000 and 1000 cps for a record-playback head operating at 7.5 ips.

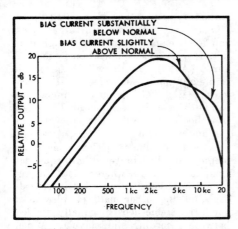

Fig. 10. Unequalized response of a record-playback head at 7.5 ips for two different values of bias current.

alternating magnetic field due to the bias current in the record head.

One of the simplest measures that can be taken to improve high-frequency response is to decrease bias current or, from a different point of view, to prevent bias current from exceeding the value specified by the manufacturer of the tape machine or of the record head. In fact, in the attempt to maintain full-range response at 3.75 ips, appreciably smaller amounts of bias current are often used than at 7.5 ips.

The better tape machines often contain a control (variable resistor or variable capacitor) that permits one to adjust readily the amount of bias current fed to the record head. In other machines, however, it is necessary to change the value of a component—resistor or capacitor—in the circuit supplying the bias. In either case, measuring bias current and adjusting it is a procedure requiring technical competence and suitable instruments. While the layman is ordinarily not equipped to do this, it is well for him to be aware that deficient treble may be simply due to excessive bias rather than to something which is much more expensive to remedy, such as a playback head with a gap that is too wide. It can happen that one goes to the effort and expense of replacing a playback head, only to find that the fault lay in too much bias.

One must guard against excessive reduction of bias in order to achieve the desired treble response. The penalty for too little bias is excessive distortion. And the increase in distortion is quite sharp as one reduces bias. It can easily happen that in the effort to extend treble response by a relatively moderate amount, say from 12,000 to 15,000 cps, one decreases bias to an extent which results in a severe increase in distortion. More about this in the next article.

One can partly or completely avoid an increase in distortion by reducing the recording level, but then one has less recorded signal on the tape, and this means a reduced signal-to-noise ratio in playback. In sum, the effort to extend frequency response by decreasing bias involves an increase in distortion or a reduction in signal-to-noise ratio or a combination of the two. On the other hand, it is quite possible that for accidental reasons bias current is above the level consistent with reasonably low distortion and a satisfactory signal to noise ratio. In such cases, particularly at speeds of 7.5 ips and less, it is important that bias be reduced to its proper value.

Record Equalization

In recording there are two kinds of magnetic losses which become increasingly severe as frequency rises. One of these has already been described—the loss due to bias current in the record head. The other, known as demagnetization loss, refers to the fact that as frequency goes up the equivalent bar magnets recorded on the tape grow shorter, with the consequence that the opposite poles of each magnet are closer and therefore tend to cancel each other to a greater extent.

Altogether, the recording losses require a great deal of treble boost in order to make it possible to achieve response out of 12,000 cps or beyond. *Figure* 11 indicates how much treble boost is necessary by showing the typical treble equalization for a machine operating at 7.5 ips and designed to yield a relatively flat response when playback equalization conforms to the NAB (formerly NARTB) curve. With such a large amount of boost required, it may be realized that anything which prevents correct operation of the treble boost circuit—a faulty resistor, capacitor, inductor, or other component—can deal a severe blow to treble response.

In some tape machines, especially the ones of semi-professional and professional quality, the amount of record treble boost can be controlled, within limits. Accordingly, as one adjusts bias current or as one changes to a different kind of tape with different high-frequency characteristics, one can make a compensating change in treble boost. In most home machines, however, there is no such adjustment. Unless precision components have been employed in the treble boost circuit, it may be necessary to replace a component in order to achieve treble response as flat as possible. It sometimes happens that there is a peak in treble response—usually in the region of about 6000 to 10,000 cps—which may be great enough to warrant removal through a change in the equalization circuit. Conversely, treble response may be deficient due to a component that is too far from design value. All in all, variable treble equalization in recording is a desirable feature in a machine to be operated by the audiofan who is meticulous about flat frequency response.

As tape speed is changed, the required amount of treble boost and the frequency (turnover) at which boost commences also change. In many tape machines, treble boost is automatically changed as speed is changed. In others, however, particularly those with external tape amplifiers, the equalization change must be made manually. It is quite easy for the recordist to forget to make this change when shifting speeds. If he has been recording at 7.5 ips and then goes to 3.75 ips without changing equalization, the result is deficient treble in the signal recorded on the tape. If he goes from 3.75 ips to 7.5 ips without changing equalization, then excessive treble is the result.

Some tape machines contain no provision for changing record equalization when shifting from one speed to another. Accordingly, inadequate frequency response is achieved at one speed or the other, unless a compromise equalization is used, which produces less than the best results at both speeds.

Playback Equalization

Figure 12 shows the NAB equalization that is standard for playback at 15 ips and virtually, though unofficially, standard for 7.5 ips as well. *Figure* 13 shows the two playback equalization curves that have been most commonly employed at 3.75 ips, although some tape machines have used the NAB curve at this speed. At the time of writing it appeared that curve (A) in *Fig.* 13 would become standard for 3.75 ips, although there is no assurance about this. In view of the uncertainties about playback equalization at 3.75 ips, the following discussion will be conducted in terms of the 7.5 ips speed, for which there seems to be little dispute, if any, about playback equalization.

When playing most 7.5-ips commercial recorded tapes, relatively flat response will be obtained if the tape machine provides NAB equalization. But to this day a substantial number of home machines provide different equali-

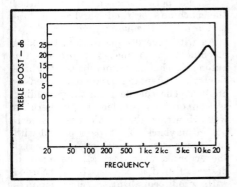

Fig. 11. Typical treble boost employed in recording on a tape machine at 7.5 ips with NAB playback equalization.

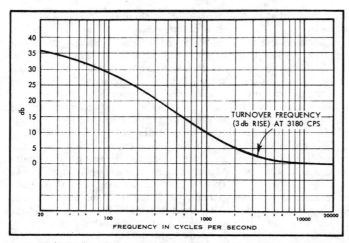

Fig. 12. NAB (formerly NARTB) playback equalization.

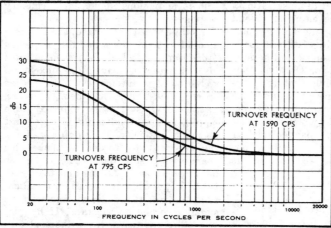

Fig. 13. Playback curves that have been used at 3.75 ips.

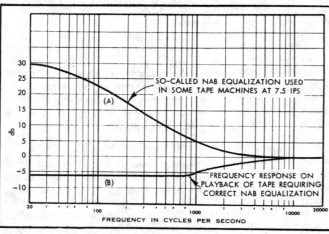

Fig. 14. Effect of improper playback equalization upon frequency response.

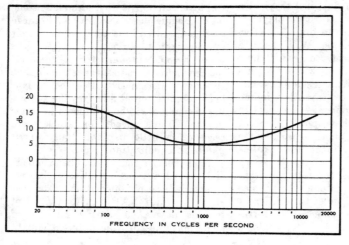

Fig. 15. Example of "half-and-half" equalization employed in some tape recorders for both record and playback.

zation, typically that if curve (A) in *Fig.* 14, which the manufacturers often call NAB equalization even though it is not. If playback equalization is that of curve (A), the result will be too much treble and not enough bass, as shown by curve (B). This frequency imbalance can be more or less compensated through bass boost and treble cut elsewhere in the audio system.

A certain number of home machines employ so-called half-and-half equalization, as in *Fig.* 15, whereby half the required bass boost is supplied in recording and the other half in playback; similarly, treble boost is equally divided between recording and playback. When playing a commercial recorded tape, there will again be a thinness in the bass region, probably to a greater extent than with machines using the playback curve of *Fig.* 14. Moreover, there will be excessive treble, because half-and-half equalization provides treble boost in playback, whereas such boost (except to compensate playback head deficiencies) is not called for under NAB equalization.

All in all, the individual who wishes to play commercial recorded tapes is well-advised to ascertain that the machine he owns or plans to purchase provides accurate NAB equalization. As previously indicated, a number of tape machines or tape amplifiers that advertise NAB equalization fall short of the mark. Some say simply nothing on the subject. If one already has a machine that deviates appreciably from the NAB curve, it is quite simple for a qualified technician to make the necessary circuit change, often requiring replacement of only one component. However, in doing so, one upsets the record-playback frequency response of the machine in question, meaning that the record equalization must also be altered to achieve flat response when playing recordings made on the machine. Since the record equalization curve is a more complex affair than the playback one, it may be time-consuming and expensive to have record equalization adjusted accurately. What one can do instead is to have a switch installed, permitting one to use either the machine's original playback equalization or NAB equalization.

As with record equalization, the tape machine may or may not provide for automatic change in playback equalization when changing from one speed to another. The comments in the preceding section on this problem apply here as well. However, the problem is less serious in playback. If a tape is played with the wrong equalization, one can correct the error; a new start can be made if desired. But if a tape has been recorded with wrong equalization, circumstances often do not permit the error to be retrieved, as when taping a program off the air.

In a number of tape machines and tape amplifiers, the treble drop above 1000 cps is less than shown in *Figs.* 12 and 13. The reason is to compensate for the falling response of the playback head in the upper range due to gap width. Modern heads, however, with extremely fine gaps on the order of .0001 in., generally do not require such compensation. By and large, the best procedure is to obtain a tape machine or tape amplifier with correct playback equalization and then compensate such deficiencies as may occur by means of the tone controls in the audio system.

Cables

High-frequency losses can be due to the cables between the tape machine and the rest of the audio system. Let us first consider the cable that carries the signal from the rest of the audio system to the tape machine for recording purposes.

In a number of control amplifiers or integrated amplifiers, the incoming signal is fed directly to the tape recorder, as illustrated in *Fig.* 16. If the signal source has a low impedance—for example, when a tuner has a cathode-follower output—a substantial run of cable between the control amplifier or integrated amplifier and the tape recorder will have no consequential effects upon treble response. On the other hand, if the signal source has a high impedance, then under the arrangement of *Fig.* 16 high-frequency response can be seriously affected by more than two or three feet of cable between the amplifier and the tape recorder.

To prevent the capacitance of the cable between the amplifier and the tape recorder's input jack from having a deleterious effect upon frequency response, some amplifiers feed the tape recorder from a cathode follower (or other low-impedance circuit), as in *Fig.* 17. Under these conditions a long cable has no appreciable effect.

The audiofan should ascertain whether the signal presented to the tape recorder comes from a low-impedance source— supplied by either the signal source (tuner, TV, etc.) or the amplifier—or whether it comes from a high-impedance source. In the latter case, he should make special effort to keep the cable to the recorder as short as possible, no more than three feet, and he should use low-capacitance cable.

The same kind of problem exists with respect to the cable leading from the output of the tape machine to the input of the control amplifier or integrated amplifier. Some tape machines have a low-impedance output, so that cable length, within reason, does not matter. But a number of home machines have high-impedance outputs, and here it is necessary to be careful about cable length.

The most serious problem, perhaps, occurs when a cable is run directly from

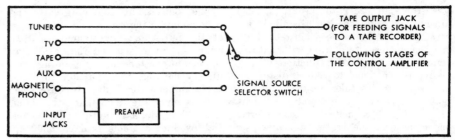

Fig. 16 Method employed in some control amplifiers for feeding incoming signals directly to a tape recorder.

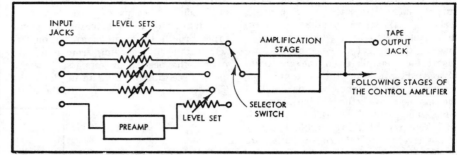

Fig. 17. A second method used in some control amplifiers to feed a tape recorder.

the *tape head* to the control amplifier or integrated amplifier, there to undergo preamplification and equalization. This is the case when the audiofan is interested only in playing tapes, not in recording them, and therefore purchases a tape transport without electronics. In this situation, more than a foot or two of cable can significantly affect treble response.

Kind of Tape Used

Treble response tends to vary somewhat with the brand of tape used and with the kind of tape within the brand. Depending upon the formulation of the magnetic coating on the tape and the thickness of the coating, treble response may vary a few db at the upper end of the audio range.

The thinner tapes tend to have an advantage over standard tapes with respect to treble response. At the middle and low frequencies, the amplitude of the recorded signal increases somewhat with coating thickness. But the higher frequencies, which are recorded closer to the surface of the tape, are less affected by thickness of the coating. Therefore when the tape has a thin coating, as in the case of long-play and double-play tapes, response at low and middle frequencies is reduced in comparison with standard tape. In other words, high-frequency response, relatively speaking, is increased.

The thinner tapes tend to be more limp and therefore conform more easily to the contour of the playback head, assuring close contact between the tape and the head and thereby maximizing treble response. At 15,000 cps at the 3.75 ips speed, differences of as much as 5 db in response have been noted as the result of using thin tapes.

It has been pointed out that an increase in bias current results in a reduction in treble response. However, the extent to which treble response is affected by a slight increase in bias tends to vary somewhat among tapes. In other words, some tapes are less critical than others in terms of setting bias current to the correct operating value.

Location of the Tape Output Jack

In many or most monophonic control amplifiers, the tape-output jack is located ahead of the tone controls so that the setting of the latter has no effect upon the frequency balance of the tape recording. In other amplifiers, however, particularly stereophonic ones, the tape-output jack is located *after* the tone contols, and frequency balance of the tape recordings depends upon how one sets the bass and treble controls (and possibly upon the setting of treble and bass filters as well). Quite possibly, excessive or deficient bass or treble in a recorded tape can be traced to the fact that recording did not take place with the tone controls in flat position. Moreover, the error tends to be compounded in playback. To illustrate, assume one normally turns up the bass control to a substantial degree to compensate for speaker and/or room acoustics. Consequently, a good deal of bass boost gets onto the tape. But if the bass control is left untouched in playback, then the repetition of bass boost can become objectionable.

On the other hand, location of the tone controls (and filters) ahead of the tape-output jack has a decided advantage. It permits one to restore frequency balance prior to recording program material on tape. Thus a substantial number of phonograph records have excessive treble in order to impart a false illusion of high fidelity, so-called. By turning down the treble when recording, one comes closer to a tape with natural balance. In addition, reduction of excessive treble tends to reduce distortion in recording. To take another illustration, AM reception is usually deficient in the high frequencies. Accordingly, one can boost the treble to achieve or approach natural frequency balance when recording a tape.

On the whole, location of the tape-output jack after the tone controls seem to be the more advantageous position. But this is an advantage only so long as the operator remembers to adjust these controls from the viewpoint of recording a tape rather than from the viewpoint of what sounds good at the moment over the loudspeaker. To make this point clear, assume that one wishes to record a tape while listening at low level to an FM program. Pleasant listening may require a substantial amount of bass boost to compensate for the apparent loss of bass at low volume. At the same time, it may be undesirable to have bass boost appear in the signal recorded on the tape.

Microphone and Room Characteristics

When recording through a microphone, one is, of course, greatly dependent upon the range and smoothness of the particular microphone used. In addition, room characteristics are an important factor in determining frequency balance. It is possible to blame a microphone for faulty response when it is as much the room that is to blame.

To illustrate, assume that room acoustics produce a treble peak at 5000 cps when using a microphone that, for the sake of illustration, is perfectly flat through the audio range. This peak is mild enough, let us say, so that it is not disturbing when listening to music taped from a tuner, phonograph, or TV set. But when recording from a microphone, the peak manifests itself twice: once in recording and again in playback. Then, as illustrated in *Fig.* 18, the peak may become sufficiently severe to become objectionable.

In similar fashion, if the room causes a fall-off in treble response because of its sound-absorbing characteristics, the drop may become objectionable only when it occurs both in recording and in playback.

If the microphone has an undesirable peak or droop in response, it is possible that room characteristics may either compensate or not make any difference one way or other. However, should both the microphone and the room tend to peak or droop in the same frequency area, then the results can sometimes become intolerable. In other words, some microphones may sound bad in certain rooms and not in others. Æ

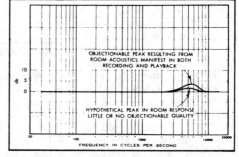

Fig. 18. Effect of a peak in room acoustics upon record - playback response.

DISTORTION

(from page 89)

pear that the equalization patterns ultimately settled upon will vary greatly from those in present use. There is a good chance that the turnover frequency (at which bass boost commences) of 3180 cps commonly used at 7.5 ips will become an official standard. In the case of the 3.75 ips speed, turnover frequencies of either 795 cps or 1590 cps have often been used. At the time of writing there was a proposal before the industry to use a turnover of 1326 cps. This kind of compromise would permit playing older tapes with a fairly minor deviation from flat response; and this kind of deviation could be corrected fairly well by means of audio system tone controls.

Distortion and Bias Current

Figure 8 shows how harmonic and IM distortion vary with changes in bias current. We are concerned with the area to the right of the −7 db point; it is not practical to record in the area to the left—that is, at small values of bias current—because somewhat less signal is then recorded on the tape. Restricting ourselves to the practical area of operation, the following conclusions can be drawn from *Fig.* 8.

1. With increasing bias current, distortion declines steadily to a minimum level and eventually rises again.
2. IM distortion is greater than harmonic distortion.
3. The changes in IM distortion with changes in bias current are sharper than the changes in harmonic distortion.
4. The greater the signal level recorded on the tape the more critical is the bias setting for minimum distortion.
5. An increase in signal level produces the least increase in distortion when bias is set for minimum distortion.

It is apparent from the foregoing that proper bias setting is of great importance in minimizing distortion. On the other hand, as brought out before, it is necessary to take into account that treble losses increase as bias is increased, and that such losses grow more severe as tape speed is reduced.

Distortion and Tape Speed

To reduce treble losses and make it possible to achieve frequency response approximating high fidelity requirements, it is necessary at certain tape speeds to reduce bias below the point corresponding to minimum distortion. To prevent excessive distortion, it therefore becomes necessary to reduce recording level somewhat.

At 15 ips it is generally feasible to set bias current for minimum distortion without impairing frequency response. At 7.5 ips it is usually necessary to settle for a value of bias current somewhat below the minimum distortion point. At 3.75 ips it is necessary to employ bias current well below that which achieves minimum distortion. And the situation grows considerably worse at still lower speeds.

One might ask: Why not compensate the increased treble losses at slow speeds by greater treble boost in recording instead of by a reduction in bias current? The answer is that greater treble boost would tend to overload the tape.

Mechanical Distortion

Distortion may be due to mechanical rather than electronic factors. If the tape does not pass smoothly over the heads, the result may be what is known as modulation distortion. Friction between the tape and the head may cause the tape to undergo a sort of vibratory action, akin to the effect that a bow has upon a violin string. The result is that the frequencies being recorded or played back are modulated at the vibration frequency. Thus the vibration frequency or frequencies are impressed on the audio signal and are manifest as distortion. Friction between the tape and the heads may be due to accumulation of tape oxide or other material on the head, or to a tape that is not sufficiently polished or lubricated.

Wow and flutter produced by the tape transport mechanism may also be considered forms of distortion. Wow is apparent as a quavering in the pitch of a prolonged tone. Flutter, consisting of speed changes that take place hundreds or even thousands of times per second, serves to modulate the audio signal, so that the flutter frequency or frequencies become the distortion products. Accordingly, the sound tends to take on a grainy quality. Æ

Improving the Signal-to-Noise Ratio

HERMAN BURSTEIN

If the performance you are getting from your tape recorder does not come up to the standards you would like, one or more of the suggestions offered may improve your lot. Most of the ideas are fairly simple, but collectively they could make even a poor machine satisfactory.

THE CRITERIA of a tape recorder's quality may be placed on two broad categories: (1) mechanical motion (wow, flutter, timing accuracy); and (2) electrical performance. The latter comprises three basic criteria: frequency response, distortion, and signal-to-noise ratio. Of these three, signal-to-noise ratio is at least as important as the other two and is probably the characteristic most noticeable to the average user. It takes no particular skill and very little time to ascertain whether a given tape machine is excellent, average, or mediocre with respect to keeping noise and hum at a suitably low level with respect to the audio signal.

The problem of obtaining a high signal-to-noise ratio is generally greatest in playback, because the playback head delivers a very tiny audio signal. Hence it is difficult to keep noise and hum produced by the playback amplifier sufficiently below the signal as to be inaudible. In recording, one generally deals with a higher order of audio signal, unless using a microphone of quite low sensitivity, so that maintaining a high signal-to-noise ratio is less of a problem.

Obviously, one should pay careful attention to what a tape machine's specifications have to say about signal-to-noise ratio. Based on a recording level that produces 3 per cent harmonic distortion, a high-quality machine will provide a signal-to-noise ratio of at least 55 db in playback. If the rating is on the basis of 2 per cent harmonic distortion, one can subtract about 3 db, so that a ratio of at least 52 db is called for. If the reference level is 1 per cent harmonic distortion, one can subtract another 3 db.

Since specifications and actual performance can differ, one should, if possible, check the performance of the machine one intends to buy. A quick way to check is to play a commercially recorded tape recorded at a level low enough to result in clear sound. The noise and hum produced by the tape machine should be of about the same order, relative to the audio signal, as encountered in the case of an FM tuner or when playing a phonograph record; at least this is so for a high-quality machine. At the worst, the noise from the tape machine should be only slightly greater relative to the audio signal than in the case of a tuner or phonograph.

Another way to check the playback noise of a tape recorder is to play a blank tape. If the dominant noise is tape hiss—as can be determined by listening to the tape amplifier with the tape still and with the tape in motion—it is safe to say that the tape machine has a very good signal-to-noise ratio in playback.

It is also advisable to check the noise in recording. One should record from a high-level source, as from an FM tuner, and, more important, from a microphone. Of course, the greater the sensitivity of the microphone (the higher its output for a given sound level), the higher will be the ratio of audio signal to noise produced by the record amplifier. One should attempt to use a microphone of average sensitivity, about −55 db/microbar, at fairly close range. In playback, the signal-to-noise ratio should be roughly comparable with that from a tuner or phonograph.

A tape machine that originally had a very good signal-to-noise ratio may fail in this respect with age. Or the ratio may not have been as good as desired to begin with. In either case, there are various factors to be considered and measures that may be taken by the person wishing to improve the signal-to-noise ratio of his tape machine. Some of the measures to be discussed require a slight amount of technical ability, about the same order of ability acquired by those who have assembled an amplifier kit, which many thousands with no knowledge of electronics have done. Hence it does not seem amiss in these pages, intended especially for the layman, to deal with a few simple procedures requiring one to go inside the tape amplifier. Moreover, many readers probably can count among their friends at least one with sufficient knowledge of electronics to assist them.

Tube Selection

Tubes of a given type tend to vary from one to another because of manufacturing tolerances. Two of the respects in which they vary is the amount of noise and hum produced. In these respects, the most important tube is that in the first stage, because its noise and hum are amplified in all successive stages. Hence, to assure minimum noise and hum one should obtain three or four of the tube type employed in the first stage and try these successively. Since tape amplifiers frequently employ the same tube for more than one stage, the extra tubes generally are not wasted.

The differences among tubes of a given type can be fairly profound—as much as 10 db in the case of hum if the heater is operated on a.c. The ear will generally serve as an adequate instrument for rejecting a tube that is poor so far as hum is concerned. Similarly, the ear can enable one to detect tubes that are outright hissy. However, if one is trying to select among tubes that differ by only a few decibels in their hum and noise characteristics, it is probably necessary to measure the output of the tape amplifier with a sensitive vacuum tube voltmeter to be sure which is the best tube. Moreover, when one does find the tube that is best, there is a possible pitfall: be sure that this tube provides as much amplification as the others, because sometimes a tube apparently has less noise and hum only because it provides less amplification. The amplification of a tube can be determined on a suitable tube tester; or, using a test tape with a constant signal, one can measure the output of the tape amplifier when various tubes are used in the first stage.

Tube Substitution

A significant reduction of noise and/or hum often can be achieved by substituting a simialr type of tube for the one originally used in the first stage. Perhaps the best example is the use of an ECC83 in place of a 12AX7, which is frequently the first stage tube. The ECC83, basically a European tube (although some are made in America), is electrically identical to the 12AX7 but on the average has superior hum and noise characteristics. Recently, American manufacturers have brought out the 7025, also intended to be a superior replacement for the 12AX7. Other high-quality replacements include the American 12AY7 and the Telefunken 12AX7.

Sometimes a pentode is employed in the first stage. Here the European EF86 has won a position of very high esteem.

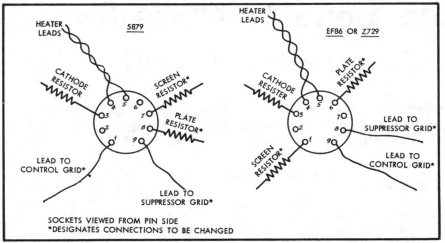

Fig. 1. Rewiring a tube socket to accommodate an EF86 or Z729.

If the tape amplifier presently has a 5879 (American) or Z729 (European) in the first stage, these can be replaced by the EF86. No rewiring of the tube socket is required, if the original tube was a Z729, whereas rewiring is necessary if it was a 5879. *Figure* 1 shows the pin locations of the 5879 and EF86. If the latter is substituted, it is necessary to rewire the resistors, capacitors and other elements of the first stage so that they are connected to the proper tube pins; this task requires probably no more than 10 to 15 minutes.

Reversing the Power Plug

As simple a measure as reversing the power plug in the house outlet can sometimes reduce hum by a significant amount. If the tape machine has a potentiometer for canceling hum, this should be readjusted for each position of the power plug.

Tube Shields

The first stage tube is, or at least should be, shielded against hum pickup. The shield should make firm contact with ground through the socket mounting or through other means. If the shield has worked loose, it may be ineffective; in fact, the loose shield may serve to increase hum above the level produced by an unshielded tube. For maximum protection against hum pickup, shields are available that are made of special material, such as the "Co-Netic" shields made by Perfection Mica Co. If one has access to Mumetal, it is possible to fashion a tube shield out of this material.

Demagnetization of Tubes and Shields

A magnetized tube, particularly in the first stage of the tape amplifier, can be a source of hum. While tube selection can eliminate this difficulty, another technique is to demagnetize the offending tube by means of a bulk eraser, being careful not to bring the tube too close to the eraser lest the powerful pull of the latter dislodge any of the tube elements.

Similarly, a magnetized tube shield can be an unsuspected source of hum. Here tube substitution would probably do no good. However, the shield can be demagnetized by a bulk eraser.

Routing of the Cable from the Playback Head

If one employs only a tape transport, using the control amplifier instead of a tape amplifier to provide the necessary amplification and equalization of the signal produced by the playback head, careful attention must be paid to the routing of the cable between the head and the control amplifier. If the cable passes close to leads carrying a.c., or close to a power transformer, motor rectifier tube, or other component where a.c. is present, it may pick up enough hum to become audible; keep in mind that the signal presented by the cable to the amplifier undergoes tremendous amplification, which is greatest in the vicinity of the hum frequencies, namely 60 and 120 cps.

Sometimes one may run into a crosstalk problem if the cable from the playback head runs close to another cable carrying a high-level signal, for example from a tuner. Should the tuner be on, then one might hear a radio program along with the signal from the tape machine.

Grounding of the Tape Transport

Whether an integral tape amplifier or one's control amplifier is used in conjunction with the tape transport mechanism, there should be a good ground connection between the transport chassis and the amplifier. An increased hum level may result if one depends upon the cable shield for a ground between the transport and the amplifier, as in *Fig. 2*. Instead, a separate, heavy ground connection is generally preferable, as in *Fig. 3*. In this case the playback head should not be grounded to the transport chassis, for this will provide a second path to the amplifier ground point; the dual paths, as shown in *Fig. 4*, constitute a ground loop, which picks up hum.

Lead Dress

Sometimes excessive hum is due to improper routing of the hot lead from the playback head to the grid of the first tube in the playback amplifier; the reference here is not to the routing of the cable between the head and the amplifier but to the path followed by the hot lead of the cable *inside* the amplifier. If the lead comes too close to a.c. heater wiring, other wires containing a.c., a power transformer, or the like, it may pick up enough hum to be audible. Moving the lead by a slight amount, perhaps a small

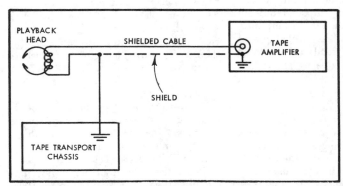
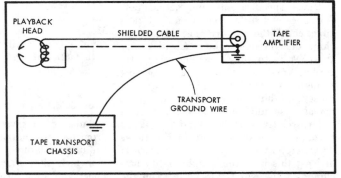

Fig. 2 (left). A method of grounding the tape transport that may produce hum. Fig. 3 (right). Method of grounding the tape transport that generally results in the least hum.

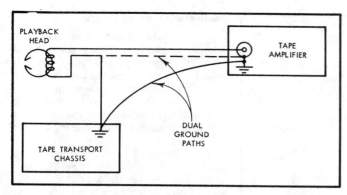

Fig. 4. Formation of a ground loop.

fraction of an inch, can sometimes result in appreciable improvement. Of course this is a trial-and-error procedure, but the results on occasion justify the effort expended. It may be preferable to leave the lead from the playback head in its original position and try moving other leads, such as a.c. heater wiring.

Installing a Hum-Bucking Pot

To cancel hum due to operation of the tube filaments on a.c. instead of d.c., a number of inexpensive tape machines connect a tap at the center of the heater winding on the power transformer to ground, as in *Fig. 5*. While this does greatly reduce hum, seldom does it minimize hum as fully as possible. A superior procedure, illustrated in *Fig. 6*, is to place a potentiometer across the heater winding, with the arm of the pot connected to ground, and adjust the arm for minimum hum. Usually it is not difficult to find room to install a pot. Drilling a hole takes a few moments, and the necessary connections are few and very simple. Note in *Fig. 6* that the center lead of the heater winding is removed from ground.

Construction of a D.C. Heater Supply

The more ambitious home constructor may wish to install a source of d.c. for the heaters of the tubes in his tape amplifier, replacing the original a.c. supply. *Figure 7* shows a suitable d.c. supply, which can be used for the first stage or first two stages of the tape amplifier. It will do no harm to continue operating the remaining stages on a.c.

Re-Orienting the Motor

The transport motor is a major potential source of hum picked up by the playback head. Sometimes hum can be significantly reduced by re-orienting the motor, namely rotating it a number of degrees about its axis. If the motor is suspended by three bolts, as is frequently the case, one can then turn the motor 120 deg. in either direction.

Relocation of the On-Off Switch

It is common practice for the on-off switch to be located on the gain control of the tape amplifier. However, the 60-cps current at the switch creates a minute hum field which may be picked up in significant quantity by the gain control, particularly if the control is situated at an early stage of the amplifier so that there is a great deal of subsequent amplification. Installation of a separate toggle switch, as in *Fig. 8*, for turning the tape machine on and off can sometimes effect a worthwhile reduction in hum.

Adding Filter Capacitance

If hum is of the 120-cps variety—that is, pitched an octave higher than the 60-cps hum encountered when bringing a screwdriver or similar metal object close to the playback head—it may be possible to reduce it by adding filter capacitance, as in *Fig. 9*. Additional capacitance of about 30 to 60 µf will usually prove effective, particularly when added to the filtering stage closest to the rectifier stage. Of course, it is possible that a filter capacitor has opened or greatly decreased in value, so that one is really replacing rather than adding capacitance.

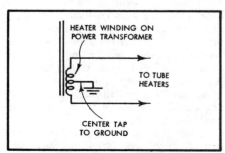

Fig. 5. Method of hum cancellation often used in moderate-priced tape recorders.

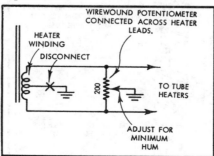

Fig. 6. Installation of a hum-bucking potentiometer.

Shielding the Playback Head

Normally the playback head is partly encased in Mumetal or other special material designed to prevent hum pickup. In addition, the better tape machines surround the heads with a heavy shield with an aperture just wide enough to permit the tape to pass through. In other machines, a piece of Mumetal or other shielding material is sometimes mounted on the pressure pad holder so that when the pad is brought against the playback head the shield guards the face of the head against hum. As illustrated in *Fig.* 10, it may be possible for the handy

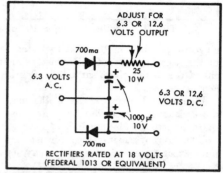

Fig. 7. Construction of a d.c. heater supply.

audiofan to rig up something of this kind himself. In addition to Mumetal, one can use Co-Netic for shielding purposes, or even a piece of silicon steel, fashioned from a transformer lamination. Co-Netic is very easily cut and bent to the desired shape.

Shielding Other Components

If hum is picked up by the playback head from the motor or power transformer, it may be possible to remedy this situation by placing a shield around the offending component. The shield may consist of Mumetal, Co-Netic, silicon steel laminations, or copper.

Defective Playback Head

Most playback heads have two windings which, when properly connected, serve to cancel hum picked up by the head. However, if for any reason the windings are electrically unequal, cancellation will be imperfect, resulting in an increase in hum. Shorted turns in one of the windings could produce such a situation. Replacement of the head is then called for.

Substitution of Resistors

One of the principal sources of noise —high-pitched rushing or hissy sounds as contrasted with hum—is the garden-variety resistor. Imperfect contact of the particles within the ordinary molded carbon resistor results in internal arcing and consequent noise voltages. It is generally possible to obtain a substantial reduction in noise by substituting resistors with low-noise properties in the first

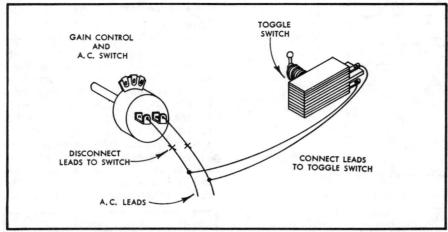

Fig. 8. Installing a separate on-off switch to reduce hum pickup.

Demagnetization of Heads

Noise is often due to heads that have become magnetized. Every time that a recorded tape is played back, noise is added to it by the magnetized heads. In recording, a magnetized head presents a d.c. component that serves to bring up modulation noise on the tape, which is due to unevenness in the coating and/or base.

The tape heads can be quickly and easily demagnetized by exposing them for a few seconds to a head demagnetizer. One should follow a regular course of preventive maintenance in this regard, namely demagnetizing the heads after about every 8 hours of use. Once noise has been added to a tape by a magnetized head, the noise cannot be removed without also removing the audio signal.

Noise can be defined as any *undesired* audible signal. Accordingly, noise may be said to include signal remaining on the tape from a previous recording, left there because of imperfect operation of the erase head. The erase head can be ineffective for mechanical or electrical reasons.

Figure 11 shows how erasure may be incomplete for mechanical reasons. The erase head gap is not positioned vertically in proper relationship to the gap of the record head, so that not all of the recorded track is subject to erasure.

Electrical reasons for imperfect erasure include the following: (1) poor design; (2) shorted turns; (3) a weak oscillator or other defective components so that insufficient current reaches the head; (4) an oscillator frequency that is too high; the higher the frequency, the less efficient the head tends to be.

stage or perhaps in the first two stages of the tape amplifier. Such resistors are called for in the plate, cathode, and grid circuits.

Wire-wound resistors have the best noise characteristics, but they are also quite expensive, as much as $2 or $3 apiece. Moreover, since they consist of a number of turns of wire, they may be-

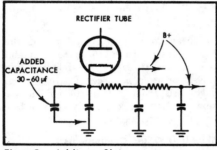

Fig. 9. Adding filter capacitance to reduce hum.

have as an inductance and pick up a slight amount of hum, which is then subject to great amplification in the tape amplifier.

The audiofan seeking to improve the signal-to-noise ratio of his tape recorder will find that some of the deposited-metal-film resistors are virtually as good as the wirewound and cost only about half as much or less. The Davohm 850 series is recommended.

Deposited-carbon resistors have frequently been acclaimed for their low-noise properties, but one must be very careful here. It is true that certain deposited-carbon resistors, particularly of foreign manufacture, produce extremely little noise and are suitable for high-grade applications. Unfortunately, a number of other deposited-carbon resistors are hardly, if at all, better than the conventional molded ones that sell for a few cents apiece.

Accordingly, the audiofan desiring a high degree of assurance of good results will invest in wirewound or deposited metal resistors. If he uses the latter, he should bear in mind that occasionally a poor one will get into the lot, and that failure to get improved results might be due to such an occasional mischance.

If a cost of $1 or more for each resistor seems too high for a not necessarily successful attempt to reduce noise, the audiofan may try another expedient, which is to use conventional resistors of relatively large wattage rating. Whereas ½-watt resistors are usually employed in the plate, cathode, and grid circuits, he may substitute 2-watt resistors and thereby possibly achieve an appreciable noise reduction.

Finding and curing electrical malfunction is largely a task for the electronic technician rather than for the audiofan. However, if the cause is a weak oscillator tube, the audiofan can at least ascertain this and cure the difficulty by substituting another tube. Frequently it is also within his power to adjust the oscillator frequency by turning the screw protruding from the container that holds the oscillator transformer. Turning the screw so that it recedes into the container will lower the frequency and possibly increase the effectiveness of the erase head by a significant amount. However, when the bias frequency changes, this may also result in a change in the amount of bias current that reaches the record head, thereby affecting frequency response and distortion. Unless one possesses the necessary equipment for measuring bias current through the record head, it is best not to tamper with the oscillator adjustment.

If one cannot obtain satisfactory results from the erase head, one may have recourse to a bulk eraser. The drawback, however, is that one cannot erase just

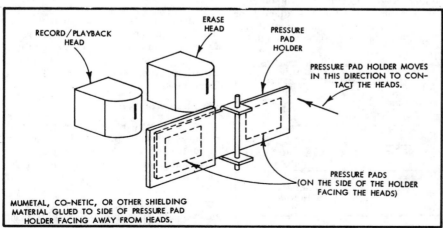

Fig. 10. Example of how a playback head can be shielded against hum pickup.

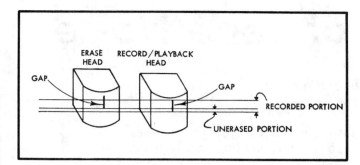

Fig. 11. Incomplete erasure due to improper vertical positioning of the heads.

one track in mono half-track recording or just two tracks in four-track stereo recording; the entire width of the tape is necessarily erased at once.

In connection with imperfect erasure it is necessary to bear in mind that the erase head and oscillator may be functioning properly but that the fault lies in excessive recording level, one that produces a high degree of distortion.

Print-Through

Print-through is a form of noise. The problem may be mitigated or eliminated through one or more of the following approaches: (1) Use of low-print tape, made by at least two manufacturers. (2) Use of a print-through eraser, described in the article on tape accessories.[1] (3) Reduction of the recording level. (4) Storing recorded tape properly in a cool place and away from the magnetic fields produced by motors, transformers, etc.

Tape Hiss

In the case of a very good tape recorder, where the tape amplifier produces a minimum of noise, it may well be that tape hiss is the dominant noise factor distinguished by the ear. (Although hum may seem dominant on an oscilloscope, high-pitched noise may be the only kind apparent to the ear.) Several courses of action are possible. First, subjecting the tape to a bulk eraser may reduce tape hiss by a significant amount. Second, changing to a different brand of tape or to a higher quality within the same brand might result in noticeable improvement. Third, it is sometimes advisable to look beyond the tape machine, namely to other components in the audio system; an appreciable departure from smooth frequency response in the control amplifier, power amplifier, or—most likely—the speaker system, will tend to accentuate tape hiss, particularly if there is peakiness in the range of about 3000 to 5000 cps. The course then is to adjust tone controls, filters, or speaker-level controls as best as one can to remove the objectionable peakiness. The audiofan might even want to consider replacing his speakers, or perhaps moving them to a different location in the room if a change in acoustic environment reduces the treble peak.

Departure from NAB (formerly NARTB) equalization may account for inordinate tape hiss. At speeds of 7.5 and 15 ips, a treble droop of 10 db between 1000 and 15,000 cps is called for in playback. In some tape amplifiers, however, substantially less treble droop or quite possibly an appreciable amount of treble boost is encountered, resulting in emphasis of tape hiss. Treble boost in playback is found in tape recorders employing so-called half-and-half equalization, where the boost required to compensate for severe treble recording losses is provided partly in recording and partly in playback, instead of entirely in recording as stipulated by NAB standards.

Tape Squeal

If tape has dried out and lost its lubricant, it may produce an unpleasant squeal as it passes the heads. This squeal can be recorded on the tape, so that even if measures have been taken in playback to avoid squeal, nevertheless the unpleasant sound will be repeated.

Tape squeal can be avoided by the following measures: (1) purchasing tape of high quality; (2) lubricating the heads, guides, pressure pads, and so on, with substances described in the article on accessories; (3) lubricating the tape with materials described in articles on accessories; (4) replacing worn pressure pads.

Accuracy of the Record-Level Indicator

Thus far in discussing improvement of the signal-to-noise ratio we have dealt with measures to reduce noise. But this is only one side of the coin, and it is also necessary to consider the problem of getting as much audio signal on the tape as possible without running into excessive distortion.

In this connection, accuracy of the record-level indicator is of vital importance. Should the indicator show a high recording level when actually the magnitude of the signal recorded on the tape is small, obviously the result will be a deterioration in the signal-to-noise ratio.

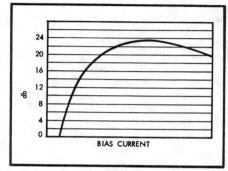

Fig. 13. Variation of output with bias current at 1000 cps at 7.5 ips.

Generally, the technically untrained or unequipped individual cannot correct the calibration of the record-level indicator. Usually this is the province of the technician.

However, what the home recordist can do is to make a more or less rough check whether the record-level indicator is operating properly. Thus he can record a tape at what is presumably maximum recording level, listen to its quality in playback, then try successively higher recording levels. If he can make clean recordings with the gain control advanced well beyond the point where the record-level indicator tells him he should have stopped, this *suggests* the possibility that the indicator is miscalibrated. It is necessary to bear in mind that the point at which some persons find distortion to be offensive may be considerably different than the point at which others find distortion intolerable.

Another procedure is to compare the playback level of a commercially re-

(Continued on page 103)

[1] November, 1959.

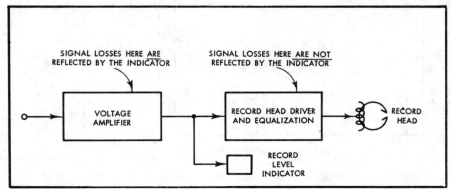

Fig. 12. How the record-level indicator may fail to reflect signal losses.

Characteristics of Tape Noise

WILLIAM B. SNOW

Tape noise is a fundamental limitation in all recording processes. Here are some criteria for judging a tape recorder with respect to noise.

NOISE IS A FUNDAMENTAL limitation in all recording processes. Unless a low noise level is achieved, true high-fidelity sound reproduction is impossible because low passages will be heard against a background of interfering and unwanted sounds. Low noise level with consequent wide dynamic range is a characteristic of modern magnetic tape recording

Signal-to-Noise Ratio

Signal-to-noise ratio in magnetic tape recorders is ordinarily expressed as the ratio of rms single-frequency signal at the level yielding 3 per cent harmonic distortion, to total noise measured over the complete reproducing frequency range. The 3 per cent point represents the maximum permissible recording level for signal peak amplitudes, and is usually measured at 250 cps.

When the signal-to-noise ratio is expressed as a single number in this manner for magnetic tape recorders, it essentially represents a signal-to-hum ratio. Hum reduction is particularly difficult with tape recorders because the magnetic reproducing head must be mounted near motors and power transformers which produce magnetic fields from which the head must be shielded. In addition, the playback equalization for magnetic recording necessitates maximum gain at low frequencies. It was felt that a somewhat detailed examination of the noise from a tape recorder would be of interest. A Movicorder tape recorder was employed operating at a speed of 7.5 inches per second.

Tape Noise Frequency Analysis

First, a portion of tape containing a 250-cps tone recorded at maximum level was reproduced to give a reference output reading. Then, erased tape was reproduced without alteration of the playback amplifier gain while noise output was measured through the electrical filters of two types of frequency analyzer.

Figure 1 shows the results of the noise measurements made with a narrow band (25 cps) and an octave band analyzer. The usual signal-to-noise ratio described above is shown by the line at "Over-all" to be 52 db. With the octave filters, noise was checked for three conditions: tape erased in the machine (Curve A), bulk-erased tape (Curve B) and tape stopped showing only playback amplifier noise (Curve C).

It can be seen that the over-all level is mostly accounted for by the noise in the two lowest octave bands. Above 300 cps the levels are much lower. At low frequencies two sets of "spikes" are shown, measured with the 25-cps band analyzer which could separate the individual hum components. The noise in the two lowest octave bands is contributed almost entirely by the hum components, 60 and 120 cps, and is essentially unchanged when the tape is stopped. Above 200 cps, however, the noise comes principally from the tape, and residual electrical noise (Curve C) is negligible in comparison to it. The amplifier has the capability of playing much quieter tapes in the future as they are developed. Small difference between noise for bulk-erased and machine-erased tape indicates good balance in the erase oscillator.

Comparison With Room Noise

It is important to the success of magnetic recording that the signal-to-noise ratio at higher frequencies is much greater than the usual single number discussed above. The octave-band levels are roughly constant and are about 75 db below the standard 3 per cent distortion level. *Figure* 2 has been prepared to explain the significance of this. Rather than ratios, this figure shows actual sound levels as measured in a room with a sound level meter and analyzer. They have been plotted in the special form of "masking level"; that is, the level which noise from the reproducing system must attain if it is to be detected in the presence of the room noise. If it falls below this value, it will be "masked" by the

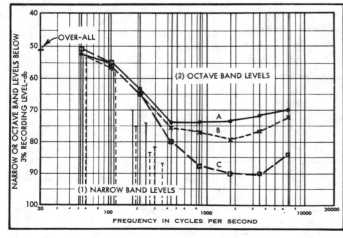 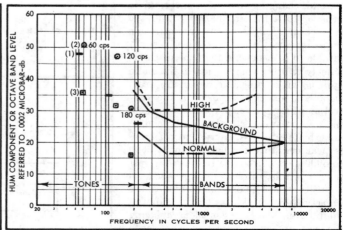

Fig. 1. (left) Results of noise measurements on a typical recorder. Fig. 2. (right) Recorder noise compared to room background noise for typical quiet room. Above 200 cps shown as octave bands; below 200 cps shown as single tones.

room noise and will not be heard.

A typical quiet, listening room background noise condition (1) is shown by the solid lines. This is given in terms of *tones* in the hum region, and *octave band noise* above 200 cps. Of course, the noise level in rooms varies, but the shape remains much as shown.

Figure 2 also shows two reproduced noise conditions during playback taken from Curve A and the "spikes" of *Fig.* 1. The one labelled High (2) is for a volume control setting giving maximum octave band program readings of 95 db in an ordinary living room—a level used mainly to impress friends with the height of the Fi. The one called Normal (3) represents a reproducing level 15 db lower, which is more representative of a usual domestic listening intensity. What stands out is the fact that the shape of the noise curve from the reproduction matches the shape of the background noise with which it competes. At the High setting noise is audible at both high and low frequencies to about the same degree during completely quiet intervals of program. At the Normal setting the reproduced noise would not be audible in this "typical" room, although it might be in very quiet rooms. But, even in a completely quiet room, the *shape* of the reproduced noise curve would remain satisfactory, because the actual minimum sensitivity of hearing nearly parallels the room noise curve.

Weighted Signal-to-Noise Ratio

The effects described above are recognized in communication and noise measurement, where frequency "weighting" of the response of the measuring instrument is used to reduce the contribution by low frequencies. Usually an unweighted and a weighted value are given, and this might be a useful concept for tape reproducers. In the case of the results just presented, for example, and using the A weighting scale of a standard sound level meter, the signal-to-noise ratios measured were

Unweighted, C-Scale   52 db (as given above)
Weighted, A-Scale   74 db

These two numbers show that the most intense noise components are at low frequencies where more can be tolerated, while the high-frequency contribution is a great deal lower. If only the unweighted number is given, a recorder with poor tape or poor construction could measure nearly as well, yet give prohibitively annoying high-frequency noise. Thus by adding one additional number, obtained with a relatively simple addition to standard measuring equipment, a great deal of meaning could be added to signal-to-noise ratio specifications. In the critical listening region good tape recorders give, not the 50 db ordinarily thought of, but 70 db and more.   Æ

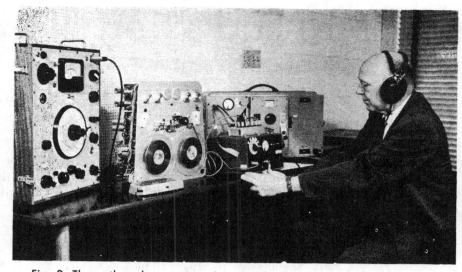

Fig. 3. The author shown measuring signal-to-noise ratio of a Movicorder.

IMPROVING *(from page 101)*

corded tape that sounds clean with the level of a tape the audiofan has recorded himself. If one's own tape appears to have a distinctly lower level in playback, this points to miscalibration of the record-level indicator.

If one is recording at too low a level, however, the fault is not necessarily in the indicator. It may be, as illustrated in *Fig.* 12, that the recording signal undergoes losses at a point in the tape amplifier following the record-level indicator. Thus a weak tube in the stage that drives the record head, or possibly a defective head, may produce such losses. Hence if it appears necessary to adjust the record-level calibration in the upward direction, thus permitting a higher input level, it is advisable to have a check made on the recording stages following the take-off point for the indicator.

Use of High Output Tape

As indicated in the article on Kinds of Tape,[2] one can obtain an increase of nearly 8 db in recorded signal level by using high output tape, although this tends to involve a moderate loss in high-frequency response and may involve greater print-through. However, if one is struggling with a tape machine that produces an undue amount of noise and hum, use of high output tape can produce a very worthwhile improvement in over-all performance.

Bias Current

The amount of bias current fed to the record head is vitally linked to the signal-to-noise ratio that can be attained. The greater the bias current (up to a point), the lower the distortion, so that it becomes possible to put more signal on the tape for a given degree of distortion, resulting in a higher signal-to-noise ratio. But there is a drawback in that increased bias current causes a loss in high-frequency response. Hence, particularly at speeds of 7.5 ips and less, it is not feasible to increase bias current as much as one would like to do from the viewpoint of maximizing signal-to-noise ratio.

On the other hand, if bias current is reduced appreciably below its normal value, this will directly reduce the signal level recorded on the tape, apart from considerations of distortion. *Figure* 13 shows how recorded level varies with bias current at a frequency of 1000 cps at 7.5 ips. Furthermore, distortion will increase with reduced bias, so that it is necessary to reduce the signal level further to avoid an increase in distortion. Hence it is important to avoid letting bias current fall much below its normal value, although sometimes this is an expedient employed to maintain frequency response at the extreme end of the treble range.

Some tape recorders provide a ready means of adjusting bias current. In others, adjustment involves replacement of components, which is a matter for the service technician. In either case, adjustment requires certain equipment, which, together with the procedure, will be discussed in a later article.

Not only is the magnitude of the bias current important to maintaining signal-to-noise ratio, but so is the purity of the waveform. Ideally it should be a sine wave. Departure from a symmetrical waveform results in noise. Checking the waveform requires an oscilloscope. Some tape recorders incorporate an adjustment for balancing the bias current oscillator to minimize asymmetry and noise. Usually in such a machine there are separate record and playback heads, so that one can make the adjustment while listening to playback of a tape which is put through the recording process but without putting an audio signal on tape.   Æ

[2] December, 1959.

Tape Indexing Nomograph

JERRY LERNER

The index counter is a useful device for conveniently locating a specific section of tape. Unfortunately it does not indicate either elapsed time or time remaining on the reel—unless used with the following nomograph.

For anyone making serious use of a tape recorder, an index counter is unquestionably a great convenience. It enables the operator to locate desired sections of tape easily and rapidly, without the bother of either attaching a marker to the tape itself or going through endless repetitions of the stop-listen-rewind cycle. The value of this device is best measured by noting that all of the professional and semi-professional tape recorders come equipped with an index counter, and even the less expensive machines usually have one.

While serving perfectly well for location of the various selections on a length of tape, on most machines the reading of the index counter is *not* proportional to the amount of tape used. This means it does not directly measure either elapsed time or time remaining on the reel. In general, the index counter reading is proportional to the number of revolutions of the feed reel. Since the effective diameter of this reel changes with time as the tape runs out, the effective circumference of the reel also changes, so that one revolution of the supply reel does not necessarily correspond to a fixed length of tape. In fact, the amount of tape used per revolution varies by a factor of about three-to-one between the beginning and the end of the reel for the conventional 5- and 7-inch sizes. Corespondingly, towards the end of a reel of tape the index counter is going three times as fast as it was at the beginning, even though the tape speed is constant.

Offhand it might seem that the engineers in charge of designing tape machines hooked up the index counter at the wrong place. There is good reason, however, for connecting the counter to the reel drive rather than to the tape drive. In record or playback position there is a precise mechanism for pulling the tape past the heads at constant speed. In fast forward or fast rewind this mechanism is generally disconnected from the tape. Instead, power is applied directly to the reels. If a footage counter is to be useful under all circumstances it would have to be connected to the tape at all times and operate without slipping even when the tape is running at high speed. This would surely be very hard on the tape.

The nomograph on the opposite page enables the operator to convert the index counter readings into "elapsed time" or "time remaining" with little effort. It is based on the assumption that the conventional 5- or 7-inch reels are used, though it is appropriate for any reel in which the ratio of full diameter to empty diameter is about three-to-one. For reels with other ratios a formula is given below. Assuming that the counter is reset to zero at the beginning, the operator need only know the index counter reading for the full reel and the total duration of the reel.

A straight line connecting the index counter reading for the full reel (left scale) and passing through the existing index counter reading (center scale) intersects the right hand scale in a point that gives the percentage of the total time used up and the percentage of the total time still remaining on the reel. For example, on some tape recorders an 1800-foot reel of tape (1½ hours at 3¾ ips) totals 1200 on the index counter. If the index counter is reset to zero at the start and reads 600 after a selection has been recorded, then 63 per cent (about 57 minutes) has been used up and there is a little over one-half hour of tape remaining. An operator who expects that only about half of the tape has been consumed is in for a surprise. In fact, if 1200 is the full tape reading, then 450 on the index counter corresponds to the midpoint of the tape.

The nomograph can be useful in many situations. It's most obvious function is to determine directly such things as the duration of a selection, the time required to reach a certain passage, the amount of time remaining on a roll of tape, etc. The nomograph also has other, less obvious applications.

Suppose you borrow a tape from a friend, and want to locate a particular selection. On his machine the index reading is 650–930, with a reading of 1800 for the entire reel. Where does it begin on your machine? From the nomograph you determine that at 650 (with a total of 1800) the percentage of tape used is 48. Draw a line from 48 per cent to the point on the left scale corresponding to the full tape index counter reading for your tape recorder. Since the running time is the same on both machines, the place where this line crosses the center scale is the index counter reading for the beginning of the selection on your machine.

Another common situation is the following. You have a two-track tape recorder and wish to play the section 550–720 of Side 1 on a reel whose total reading is 800 on the index counter. Naturally, the reel is wound so that it is all ready to play Side 2. You could rewind the entire reel, then interchange the full and empty reels, re-thread the tape, reset the index counter, and then run it up to 550. A faster method is to determine from the chart that 550 corresponds to 80 per cent of the full reel. The complement of 80 per cent is 20 per cent, which corresponds to an index reading of 110. Reset the index counter to zero, run the tape until the counter reads 110, then interchange the partly loaded reels and you are at the correct place.

For those who do a lot of recording it might be a good idea to prepare a chart from the nomograph (or the formula) giving the direct correspondence between index counter reading and actual playing time for your particular machine and the tape length and speed you use most frequently. Pasted on the inside of your tape recorder cover, this table provides pertinent playing-time information at a glance.

In using the nomograph, the tape length, tape speed and tape thickness are quite immaterial. The only pertinent factors are the full reel and partly-full reel index counter readings, the total tape time, and the 3:1 ratio of full to empty reel diameters. For other reels or for pre-recorded tape the ratios of diameters may be different. The appropriate formula is then:

$$t/T = n\left[\frac{2-n(1-r)}{1+r}\right]$$

where t = elapsed time,
T = total time,
n = ratio of present index counter reading to total index counter reading,
and r = ratio of inner diameter to outer diameter (both n and r are always less than one).

There are a few tape recorders in which the index counter is connected to the take-up reel rather than to the feed reel. The nomograph can be used for these machines with a slight modification. The index counter reading is subtracted from the full tape index reading and used for the scale reading on the center line. Also, the calibrations of the "Tape Used" and "Tape Left" scales are interchanged.

It should be noted that values for the playing-time obtained from the nomograph are at best approximate. The index counter readings are generally not accurately reproducible, primarily because the tape is not always wound with uniform tightness. Other factors, such as slippage of the counter drive, fluctuations in tape thickness, etc., also influence the index counter readings.   Æ

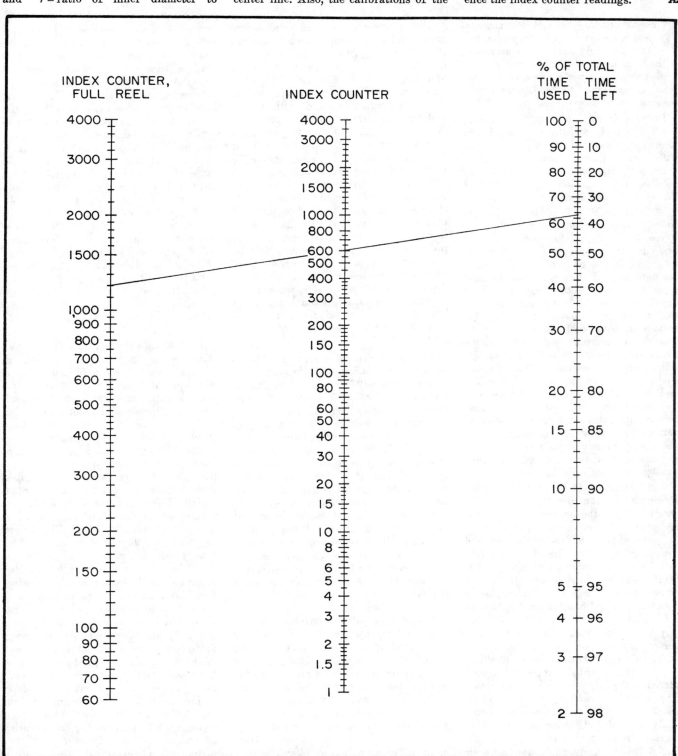

Nomograph for determining elapsed and remaining time on reel

Stereo Misconceptions

JAMES C. CUNNINGHAM

"Move the concert hall into the living room" has been the cry of many stereo "experts." In contradiction to that thesis, a recording engineer asks that we move the living room into the concert hall.

STEREOPHONIC SOUND is the simultaneous occurrence of many phenomena. To study only one phenomenon without an awareness of its inseparableness from all other phenomena is erroneous and leads to fundamental misconceptions of stereo.

Perhaps if some of the popular misconceptions of stereo are reviewed, the interdependence of these phenomena will emerge. Perhaps also some controversy will arise—at least it is hoped for—hence the adoption of a theme of heresy for this article.

In 1933 the Bell Telephone Labs performed their famous experiments with stereo. In the almost 30 years since these experiments, nothing on such a dazzling scale has been attempted—so for many the data arrived at then is still the final word on stereo. In the light of present day knowledge this data holds up well. For present day *usage*, however, this data is of little value, because now the listener we are concerned with is in the living room rather than the auditorium which was the location of the Bell Lab studies. The techniques for recording and reproducing stereo in these two locations is so vastly different that they should probably be given different names. This fundamental difference is the basis for many of the misconceptions we are concerned with and is thus worthy of some investigation.

Auditorium stereo is merely a matter of simulating the directionality of the instruments of the orchestra along with their quality—the acoustics already exist so none should be introduced in the recording system. The ideal system would have n number of channels equal to the number of instruments in the orchestra—n number of microphones with ideal directional characteristics to pick up only one instrument, n number of perfect amplifiers, a perfect recorder, and n number of loudspeakers—each with radiation characteristics identical to the instrument it reproduced and placed in the appropriate place on stage. The recording should be made in an anechoic chamber with musicians especially trained for the ordeal of having to stay on pitch in these surroundings. The Bell Labs study was concerned with the problem of approaching this ideal with only three channels.

The difference between auditorium stereo and living room stereo is obvious if we imagine crowding our n number of loudspeakers in the living room and playing back what was the ideal tape for the auditorium. It would, of course, sound absurd without the auditorium acoustics. Stereo is too often described as "like having the orchestra in your own room". Actually we want the reverse—the living room should be in the auditorium and somehow have its walls and ceiling made acoustically transparent. Such an ideal is impossible to achieve because a wall of reproducers necessary to produce a plane wave would reflect the sound from the opposite wall of reproducers and thus be transparent in only one direction. Any hope of reproducing, in the living room, the exact environment of a concert hall is patently quixotic because the room has characteristics of its own which must unavoidably be a part of the reproduction.

That two channel stereo is able to achieve the illusion of a "living room in a concert hall" to the extent that it does, is remarkable. The credit for this achievement goes largely to the recording engineers who, in spite of a lagging theory, have constantly improved the art. This is not to say that all stereo recorded today is good stereo—unfortunately some of it is not even stereo, but two channels of monophonic sound. This grave misconception is often excused as "commercial sound" to exploit the novelty of stereo. Actually it is probably either indolence or ignorance on the part of the engineer and/or producer. Separation is the battle cry for this kind of stereo, and the more the better.

The reason why separation is the enemy of good stereo is not generally understood if one is to judge by the technical literature of the past few years. If we return for a moment to the idea of the living room in the concert hall as the ultimate goal of reproduction certain facts emerge which are useful. First, the listener is aware that the various sound sources emanate from many different points on stage; second, although he is unaware of it, the direct sound from each source arrives at his two ears first, followed by multiple reflections of this same sound from the walls and ceiling. The important fact here is that the direct and each reflected sound arrive at his ears from *different directions* enabling his binaural hearing mechanism to analyze the sound in the normal fashion. Both of these conditions can be approximated in well-recorded and reproduced stereo. It is obvious that they cannot in mono because with only one channel all directionality is lost. It is not so obvious why two channel mono—the type with almost complete separation—does not fulfill these conditions either. True, there is direction—but only two point sources—the two loudspeakers, not a spread of images between the speakers. There is no possibility of the second condition being fulfilled, hence no real spatial effect, no matter how much the echo chamber is used.

To illustrate the truth of these statements, some diagrams will be helpful. *Figure* 1 shows the well known Haas effect or the relationship between time and intensity existing when two loudspeakers reproduce the same sound but with both time and intensity varied according to the curve to keep the sound image midway between the speakers. There is more, however, to this curve than meets the eye. It does not mean, as some have said, that we can trade intensity differences for time differences in stereo and therefore explain stereo in terms of intensity alone. This brings to mind another canard called "Intensity Stereo" which will be treated later, but for the moment let us look again at *Fig.* 1. As the sound emanating from loudspeaker B is delayed with respect to loudspeaker A, the listener at C hears the sound image D move gradually toward speaker A. This occurs during the first three milliseconds and might be called the primary effect of the Haas effect. After three milliseconds the image stays at A but as the delay is increased, the sound begins to take on "body" or "fullness." This is the secondary effect. When the delay reaches 50 milliseconds the listener hears two distinct sounds or an echo. But why the "fullness"?—Because we are by syn-

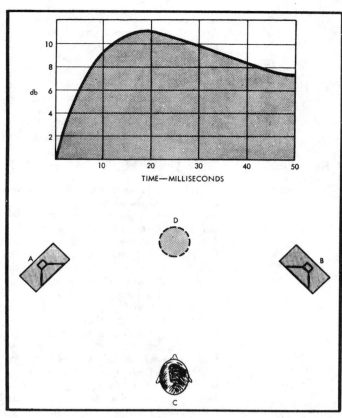

Fig. 1. Relationship between time and intensity when two loudspeakers reproduce the same sound (Haas effect).

Fig. 2. Wide microphone placement.

thetic means giving the listener a taste of what he gets in the concert hall—i. e., a direct sound followed by a reflection of this sound from *another direction* and with the longer delay time associated with the concert hall. Of course, this is not stereo, but this condition can occur in a stereo system if wide microphone spacing is used such as is shown in *Fig* 2. This is a kind of super-real condition because the reflected sound is normally modified in quality by the process of reflection. The more real condition also occurs in stereo as shown in *Fig.* 3 and is the tertiary element of the Haas effect. Here the time difference of two direct waves (solid lines) is small and only influences the directionality of the reproduced sound. Thus the direct sound is shown emerging from the left and the reflected sound (dotted lines) from the right. Under ideal conditions both the direct and reflected sounds will have virtual images effectively "filling in" the space between the loudspeakers. This is still far from the ideal of making the walls and ceiling sound transparent but is a great improvement over monaural or two-channel mono sound.

Two-channel mono sound usually results in a studio recording session when each instrument is close-miked in the normal monaural fashion then the channel switches are thrown to put half the instruments on one channel and half on the other, and adding the proper amount of reverberation via the echo chamber. This is equivalent to grouping half the instruments around one microphone and half around the other and putting a wall between them as shown in *Fig.* 4. The Haas effect cannot operate here, in fact no reflections to speak of exist in either channel and this is why reverberation must be introduced.

While more latitude is possible with this type of recording since it is usually popular music and is not associated with a set acoustical environment as serious music is, the excuse of the engineer that he has no time during a recording session to achieve a good stereo effect is a false one; false because it is false economy to turn out mundane recordings. Actually many of the best selling stereo records have been a compromise between the extremes of good and bad stereo. One example that comes to mind uses the percussion instruments in a superb stereo pickup—a wise choice since the stereo effect is transmitted largely by the transients—and the melodic instruments are close-miked on the two channels. The result is somewhat strange because the percussion pickup wafts the listener into the studio then he is suddenly jolted back into his own living room again when the close-miked melodic instruments enter. Many such acoustic distortions are possible in the realm of two-channel stereo and the enterprising producer in search of a gimmick to sell records might do well to find them.

One thing that makes the job of the producer and engineer difficult is the great variety of stereo systems existing in this country. A stereo system consists of the apparatus and the listening room together. No worthwhile research has been done on this kind of system because no two systems in the world are alike. It is safe to say, however, that the listening room nearly always has a detrimental effect on the apparatus used as far as the stereo effect goes. One only has to move the apparatus outdoors for proof of this statement. There the stereo illusion is not reduced by standing waves, wall reflections, etc., so spatiality and direction are very precise. It is interesting to note that this quality of stereo can be had indoors by increasing the number of stereo channels to at least four. There are, however, a number of other possible

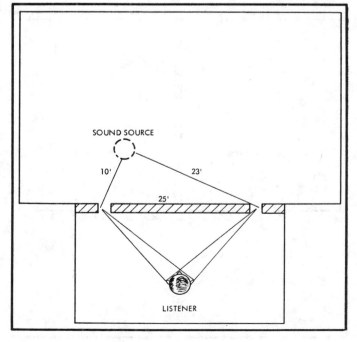

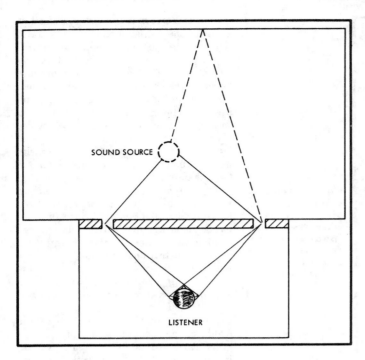

Fig. 3. Better microphone placement.

solutions to this problem of listening room acoustics and it is hoped that the future will bring research in this field. The recent addition of spring reverberators to many commercial stereo systems is certainly a step in the wrong direction if the reverberation is simply added to both channels. One manufacturer, at least, has had the good sense to put this delayed reverberation on a third channel so the speaker can be placed, say in the rear of the room to stimulate the sound from the rear of the concert hall. This principle has been called ambiophonic sound by Philips of Holland. They use a tape delay system to enlarge electro-acoustically the size of a small auditorium. This system works on the principle of submerging the unwanted effects of small room acoustics rather than removing them. Perhaps this has merit, if an economical version can be worked out, in approaching the ideal of a living room in a concert hall.

It was shown earlier that the Haas effect involves three elements to create the stereo illusion and it is related to and inseparable from intensity. Often it is stated that a time difference is merely a phase difference at a certain frequency, so that stereo is phase and intensity differences. A further reduction to absurdity is the so called "Intensity Stereo" which would imply that stereo is nothing but intensity differences. If this were true we would be able to play a mono recording of a piano over two loudspeakers, lower the level of one speaker and have a stereo recording. This does not mean that "Intensity Stereo" is an invalid method of recording —but it does mean the explanation (and the name) given it is an oversimplification which leads to a total misconception of stereo.

"Intensity Stereo" uses 2 directional microphones placed as close together as possible with a matrix network on their outputs which enhances the intensity differences in the two channels. Thus the Haas effect will not operate in the manner shown in *Fig.* 2. Hence the listener will have to depend entirely on intensity differences between channels to get directional perceptions. The tertiary effect of the Haas effect will, however, operate with this system thus providing different paths for the direct and reflected sound and giving the spatial effect of stereo. If any evaluation is to be made as to the relative merits of "Intensity Stereo" as against the spaced microphone system, such an investigation must take into account the fact that music itself is variable with regard to the stereophonic effect. For example, music with rich transients would fare best with a spaced-microphone technique because of the primary and secondary elements of the Haas effect. On the other hand sustained music with few percussive effects would probably do better with an "Intensity Stereo" technique.

The Future Prospects

If two-channel stereo recording and reproduction is to progress, an awareness by all who are responsible for this art is necessary; an awareness of the entire process on both sides of the loudspeakers . . . of exactly what clues *from* these two loudspeakers are used by the brain to form an impression of an orchestra in its own acoustic setting. An understanding of binaural hearing is not enough—it is necessary to know what clues to present, via the two loudspeakers, to give the binaural hearing mechanism the impression that it is *in* the acoustic setting. On the other side of the loudspeakers, any number of microphones, in any setup, and using any tricks at hand are permissable *if* they help to form such an impression.

This article has tried to point out some of the pitfalls in this process and encourage further experiment in what is still a young and challenging art.

Æ

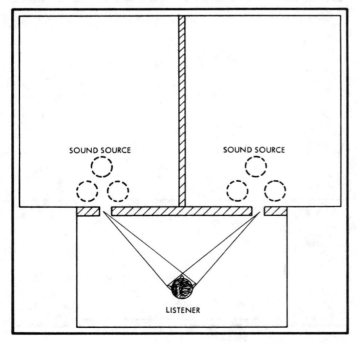

Fig. 4. Two-channel mono.

To Phase Or Not To Phase?

E. A. SNAPE III

To phase, or not to phase? That is the question faced by any audiofan in the process of setting up a stereo system. Whether it is better to ignore an out-of-phase system, or to undergo the agony of attempting to determine proper phasing by listening tests is a problem of stereophony that requires discussion and clarification.

ANYONE, EVEN SLIGHTLY CONVERSANT with the techniques of stereo sound reproduction has heard lip-service paid to the importance of proper system phasing[1]. The idea is to have all speaker cones moving in the same direction at the same time when the system is reproducing a monophonic signal. If, for example, the speakers are out of phase; the left speaker cone may be moving away from you, at the same instant that the right speaker cone is moving toward you. The effect of out-of-phase operation of a stereo system has been called variously: "inconsequential", a "large hole in the middle", and a "complete collapse of the stereo curtain of sound."

In the author's eight-year acquaintance with the vagaries of stereophonic reproduction, it has been found that the effects of improper speaker phasing do in truth vary from inconsequential, to a general destruction of the stereophonic curtain of sound. Factors governing the severity, and even the very noticeability of these effects, are:

(1) *The nature of the recording.* A widely separated two-channel stereo recording of the "ping-pong" variety loses very little by out-of-phase reproduction. It is quite impossible to detect a 180-degree phase reversal with many of the ultra-widely separated recordings being released currently. By contrast, a stereo playback of a recording made with phantom center-channel techniques will be much degraded if playback channels are not properly phased. Similarly, it is very important that proper phasing be maintained in recordings where only two or three microphones are used for the stereo pickup. In many popular music recordings, however, as many as fifteen or twenty microphones have been used. The phase relationships of the sounds captured by these recording methods are generally so confused that it matters little whether playback equipment is properly phased or not.

(2) *The size and acoustics of the listening room, and positions of the loudspeakers and the listeners within the room.* In certain rooms, when the listener's position is away from the stereo center axis (see *Fig.* 1), the out-of-phase mode sometimes sounds better than the in-phase mode.

(3) *The degree of aural acuity and perception of individual listener.* E. T. Canby has suggested, and the author's personal experience seems to confirm, that listeners sensitivity to phasing varies considerably from person-to-person, and within any given individual over a period of time. At certain times, attempts to determine proper phasing of two stereo channels for a given recording can be absolutely frustrating, especially if the listener is fatigued. After a night's rest, and with all other conditions held the same; the same listener has no trouble in determining proper phasing.

The novice stereofan soon learns that determination of proper phasing by listening tests is not always as easy as equipment instruction manuals and popular articles imply. It is likely that more than one of us has been driven to distraction by attempting to follow the inadequate approach and sketchy instructions offered in these manuals and articles.

The following excerpt from a stereo amplifier instruction manual is better than most, but still somewhat misleading to the novice. "To check for proper phasing, play a monophonic recording so that one signal comes from both channels. Then move back and forth between the speakers. If the phasing is not correct, it can be rectified by interchanging the leads between one of the amplifier channels and its associated speaker. There should be no need to change the phasing once the system is set up properly." Fortunately, these instructions advise the use of a monophonic sound source. A few instruction books blithely ignore this fundamental necessity, and one or two make no mention of phasing whatsoever.

Let us suppose that we are novice audiofans in the process of checking our newly installed stereo system. The implication of the foregoing instructions is that by merely walking back and forth between our stereo speakers we will readily be able to discern whether they are working in phase. In actual practice, the chances of being able to do this seem to be poorer than fifty-fifty. Being novices, we are not sure whether the sounds we hear emanating from our speakers are in phase, or not. We decide that we had better reverse the leads to one speaker just for comparison sake.

We know that it is unwise to operate a power amplifier without its output load, so we carefully shut off the a.c. power before crawling behind our loudspeaker to reverse leads.

Two minutes later, if we are lucky; we are once more parading between our speakers and listening for some dramatic change in the quality of the sound. The only trouble is that we have already forgotten how things sounded before, and there is no obviously discernible difference in the sound. Now is the time for the crucial decision. To phase, or not

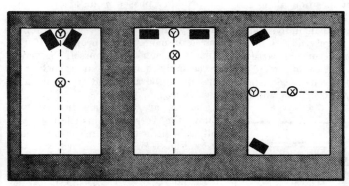

Fig. 1. Three typical stereo setups. Dotted line indicates stereo center axis. "X" indicates approximate position of listener. "Y" indicates apparent position of phantom monophonic source when system is properly phased and balanced.

---

[1] Some authorities, notably Paul Klipsch, refer to "polarity" rather than "phase" in this context. Mr. Klipsch's terminology is technically correct, but for the purpose of this article we will adhere to the more popular usage.

to phase? We can go on reversing leads in this fashion "ad nauseum," or we can decide that it is not worth the trouble for a difference in sound so subtle as to be almost non-apparent.

Fortunately, there are one or two approaches to the problem that are more sane than this.

Color-coded leads. If all components of a stereo system from source to speakers are matched, phasing becomes easier A very simple expedient is to use color-coded connecting leads between the amplifier output terminals and speaker terminals. The color coding merely insures that both speakers are connected to the amplifier terminals in an identical fashion. If the common terminal of amplifier channel A connects to terminal 1 of our left speaker, we must similarly connect the common terminal of amplifier channel B to terminal 1 of our right speaker. It may reasonably be assumed that if the speakers are attached in an identical manner, the system is properly phased.[2] This is all well and good for completely matched stereo channels. What about non-matched stereo systems using components which differ from one channel to another?

The phase-reversal switch. With non-matched stereo systems, a listening test is the most practical way to determine proper phasing. A phasing switch will make such a listening test easier and more practical to perform; and regardless of whether a system is matched, the switch is a useful adjunct. Most recently marketed stereo amplifiers incorporate a phasing switch as one of the front panel controls. While this is convenient, it is not necessarily the best place in the system for such a switch. An otherwise good stereo amplifier should not be rejected merely because it has no phasing switch. Such phase reversal is usually accomplished between the output transformer and the speaker, but at least one recently introduced stereo preamplifier incorporates a phase-reversal control that operates by selecting between cathode or plate output of one of the electron-tube stages in one stereo channel. This method of phase-reversal opens up a whole new realm of control possibilities for stereo, including phantom center-channel output from conventional two-channel amplifiers null balancing of system levels, and positive determination of program source phasing. Discussion of this method of phase-reversal is beyond the scope of this article.

[2] If you are unsure which is terminal 1 on a particular speaker take a flashlight battery and attach leads to the positive and negative terminals of speaker. Touch leads to terminals of speaker. Note which way the speaker cone moves. Mark positive terminal. Repeat the procedure with the other speaker, marking the positive terminal when speaker cone moves in same direction as previous speaker. Consider the positive terminal as terminal 1.

We can make a phase-reversal switch of our own that will be easier to use than the built-in variety. An inexpensive double-pole double-throw toggle switch wired according to *Fig.* 2 is all that is needed to reverse the phase of one stereo channel quickly and at will. The advantage of this home-made variety is that it may be made with leads long enough to reach the stereo center-axis where listening tests must be performed. In most setups the stereo amplifier is located at some distance from this critical listening zone, and if the phasing switch is mounted on the amplifier it will be necessary to enlist the aid of an assistant to throw the switch while you take up a fixed listening position along the stereo center-axis to make A–B listening comparisions.

Balancing the system. Before the system can be phased by ear, it must be properly balanced for levels. Although there are numerous meters and null devices being offered as aids in setting stereo system balance, the best method still involves a listening test using monophonic program fed to both channels. In fact, nearly all meters and null-balance devices must be initially calibrated by ear. The monophonic source can be fed to both channels simultaneously or individually. It is important to sit along the stereo center axis when balancing the system by ear. It will help to have an assistant switch back and forth between channels and adjust levels, while you maintain a fixed position along the axis. If you prefer to feed both channels simultaneously rather than individually, your assistant will merely adjust levels at your direction until you are satisfied that the apparent source of sound is roughly centered between speakers, so that neither speaker is overpowering the other. It is suggested, that a final touch-up balancing check be made from your favorite listening chair, after the system has been properly phased.

Proper use of the phase-reversal Switch. Continue to feed a monophonic signal to both channels. It may help to employ a substantial amount of bass boost, being sure to boost both channels equally. A recording of a single male voice or solo instrument seems easier to phase than some other program sources. In some cases, it may even help to play the recording at a slower speed. Place yourself along the stereo center-axis and at about the middle of its length. Switch the phase-reversing switch back and forth while listening carefully to the sound. If you cannot reach the switch from your listening position, have your assistant switch according to your instructions. The difference between in-phase and out-of-phase operation should be readily discernible. The in-phase mode will cause the sound to come from a definite spot about half-way between the speakers. In the out-of-phase mode the sound will lack this apparent single, centrally-located source; and will float vaguely about the room. It will seem to come from the two speakers that it is really coming from, rather than a single phantom source between the two speakers. Many people find it helpful to close their eyes while listening for this single, phantom-source that indicates proper phasing. If you have trouble determining the properly phased mode of operation; move in or out along the center axis and try again. Also, try a different source of monophonic program. If this fails, and you find yourself becoming increasingly confused; it is better to give-up temporarily and try again at a later time.

Chances are, however, that if you follow the foregoing instructions you will have little trouble in phasing your system. Once properly phased, you will be able to sit back and enjoy your stereo system, fully confident that it is not suffering the sometimes subtle, sometimes acute degradations of improper phasing. Be sure to mark the phasing switch, or remember its normally-phased setting. The only time you need to change this setting will be to accommodate an odd-ball, out-of-phase stereo program source. If you are ever in doubt about the phasing of a particular program source, be sure to return to monophonic operation when making a listening check of that source.

Æ

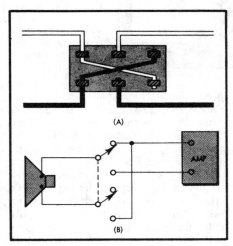

Fig. 2. Wiring and schematic diagrams for double-pole double-throw toggle switch used to reverse polarity in one channel of a stereo system. One pair of leads is attached to the amplifier output terminals, the other pair to the loudspeaker terminals.

Those Crazy Mixed-up Currents

ALMUS PRUITT

In a jocular vein, the author presents a valid clarification of the two concepts—current flow and electron flow—which we utilize in analyzing electrical and electronic circuits.

WHAT WOULD YOU MAKE of a statement to the effect that electrons must flow past a point at the rate of 6.24 million million million per second in order to constitute 1 ampere of current flowing in the opposite direction? (Statements similar to this appear in current textbooks.) No, it's not the astronomical number 6.24×10^{18} to which I call attention. (Numbers like that are like the national debt, anyhow—too big for comprehension.) What bothers me is the phrase, *in the opposite direction*. The textbooks tell us that current is electric charge in motion; electrons carry charge; ergo it appears to me that the above statement means we have current flowing in two opposite directions at once. But electrons are the only charge carriers mentioned; therefore, to my simple mind it appears the only direction of current should be that in which the electrons flow.

Many electronic textbooks nowadays, while acknowledging that negative charges, such as those carried by electrons, actually do travel from the negative to the positive terminals of a load, (A) in *Fig. 1*, nevertheless bow to convention with remarks about *current* being considered as flowing in the direction which positive charges would take, i.e., from positive to negative in the load, as in (B) of *Fig. 1*. In this they commit two grave errors, in my opinion: one, they are apt to give the impression that current and electron flow are two distinct things, whereas they are the same; and, two, they require positive charge carriers, i.e., positive ions, to travel through any kind of load, whereas this is impossible if the load is a solid. A positive ion is an atom which has lost one or more of its electrons. In a solid the position of an atom is fixed, aside from minor vibratory or displacement movements; therefore positive ions cannot flow in solids. As for liquids and gases, even there positive ions do not ordinarily constitute the predominant charge carriers, i.e., there are as many, or more, negative carriers as positive.

Free Grid, writing in *Wireless World*, March, 1954, expressed the contretemps

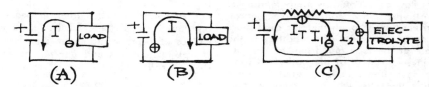

Fig. 1. (A) Electron flow. The charge carriers are electrons or negative ions, as shown by the ⊖ sign. (B) Conventional current flow. The charge carriers are (hypothetically) positive ions, as shown by the ⊕ sign. (C) Here both types of current flow in the electrolyte, but only electrons can flow in the resistor.

One anomaly arising from the current vs. electron concept—how does current (represented by the runners) travel from left to right on a stream of electrons (automobiles) which is travelling from right to left?

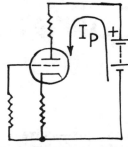

Fig. 2. D.c. circuit of a vacuum-tube amplifier stage, showing conventional current flow as adopted by many textbooks. Questions: How can positive charge carriers flow through metallic conductors from the + terminal of the battery to the plate of the tube and from the cathode to the − terminal of the battery? Granting such flow possible, what happens inside the tube, where it is known that practically all current flows from cathode to anode in the form of electrons?

delightfully in the following excerpt from his "Alice in Solidconductorland":

"But nobody is blind now," protested Alice. "Of course not, child," said the Duchess chidingly, "but out of respect for my great-grandfather's memory we still like to pretend he was right even though it gets us into all sorts of difficulties, so when soldiers are moving from *B* to *A* we always pretend that troops are moving from *A* to *B* even though troops don't really move; at any rate not in Solidconductorland in which we live."

(For soldiers read electrons. For troops read positive ions.)

Benjamin Franklin seems to have started the confusion some two hundred years ago when he advanced his "single-fluid" theory of electricity, in which he proposed that a vitreously electrified body be regarded as positively electrified and a resinously electrified body as negatively electrified. He proposed to assign the algebraic signs + and − to the two kinds of electrification. Although Franklin stated that the decision as to which body was positive and which negative was tentative (actually it couldn't be more than a mere guess at that time), the world ignored such uncertainty. (Uncertainty was regarded as a sign of weakness then, and still is, by nearly all but scientists.)

What Franklin was driving at, although he didn't know how to put it so simply then, is that if a glass rod is rubbed with silk, say, the glass will acquire excess charge (Franklin's fluid) at the expense of the silk. Only it turned out that the excess charge is on the silk, not the glass. We know now that some of the electrons of the surface atoms of the glass are captured (rubbed off, as it were) by the silk.

Even though Franklin made a poor guess, he deserves credit for being on the right track. He perceived that what one substance lost the other gained—a big step forward then. He had a 50-50 chance of being right about the excess charge; it was his and our misfortune that he called "heads" and it came up "tails."

Franklin's theory soon gained wide acceptance, along with the view that any discharge must be from positive to negative, which seemed logical at the time, given the assumption of excess charge at the positive electrode and deficit charge at the negative electrode. It remained for the development of vacuum tube theory, back in the early part of the present century, to raise the first strong doubts; but by that time the positive charge direction of current flow was so strongly entrenched among textbooks, and authorities found it so painful to uproot, that it took a new generation, disgusted with trying to follow a current from the positive plate to the negative cathode of a vacuum tube, to rebel against the nonsense.

Fate of this Convention

Today it appears to me that the convention is well on the way to oblivion, in spite of its retention (often with an apology) in many textbooks, and in spite of views such as L. B. Arguimbeau's, who states in his *"Vacuum Tube Circuits"* (Wiley, 1948): "United States Navy training courses made a commendable effort to do away with the convention, but the result was not a happy one."

Even if there have been unhappy results (mostly as a result of trying to mix the teachings of the older and the

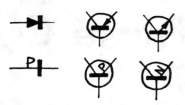

Fig. 4. Standard semiconductor symbols, upper row, may be changed as in the lower row, in order to avoid anomalies.

up-to-date textbooks), one might argue in rebuttal that the results of adhering to convention are apt to be far unhappier. See *Fig. 2*. Furthermore, unhappy or not, the Navy seems to be sticking by its guns with regard to scuttling the convention. I have a copy of *"Physics for Electronics Technicians,"* Navy Training Courses, NavPers 10095, published 1951, which strongly supports electron flow direction, stating, p. 252: "It has been the experience of the Navy that training is simplified when electron flow is used whenever currents are studied." Furthermore, John F. Rider Publisher has recently brought out a series of textbooks based on U.S. Navy training programs which consistently follow the direction of electron flow. Dr. White, of the University of California and the Continental Classroom TV series on physics, does likewise, as does RCA's *"Transistor Fundamentals and Applications,"* published 1958. The Encyclopedia Brittanica in its article on Electricity gives some cogent arguments against what it calls the "false convention."

But even if the convention is on the way out, it appears we are stuck, perhaps for ages to come, with some of its unfortunate consequences.

For example, when the time came to establish the polarity of electron charge, it had to be negative, because Franklin had called the charge on the glass rod positive. So it came to pass that we have a positive charge resulting from a *lack* of something, to wit, one or more electrons. It would appear much better usage to term the excess charge positive and the deficit charge negative. Thus the charge of an electron would be called positive.

Alas, it's too late now. It's like the standard typewriter keyboard. It's very easy to design a much better one, but with millions of the standard models in use, it simply isn't feasible to make the change. Even a child, it seems, would not put often-used letters, such as *a, e, r, s,* and *t,* under the left hand, with letter *a* under the left little finger, and seldom-used letters, such as *j, k, p,* and *y,* under the right hand, with letter *j,* of all letters, given the place of honor under the right index finger, central row. This suggests that the original designer may have been a left-handed, arthritic hunt-and-pecker with missing index fingers.

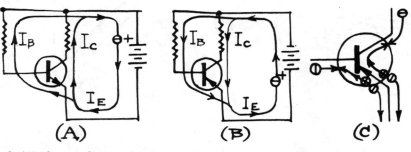

Fig. 3. (A) Electron flow in the d.c. circuit of an n-p-n transistor. Note the opposing arrowheads beside the emitter. (B) Electron flow in the d.c. circuit of a p-n-p transistor; again note the opposing arrowheads beside the emitter. (C) Electron flow outside a p-n-p transistor; hole flow inside. See text.

It is further unfortunate that the positive charge convention of current flow persisted at the time when semiconductor symbols were being formulated, hence all the wrong-way arrowheads, as in *Fig.* 3. Personally, I make the changes as in the lower row of *Fig.* 4 when I draw diagrams for myself. The P's and N's may be interpreted either as denoting the p-type or the n-type side of a p-n junction, or as denoting positive or negative polarity with respect to the opposite element of the junction. With transistors, either interpretation would generally be appropriate; a p-type emitter is normally positive to the base and an n-type emitter is normally negative to the base. Since the arrowhead denotes the emitter, it is this element which should be marked when the arrowhead is replaced with P or N. With diodes, we find some are p-n junction types, either germanium or silicon; some are lower current point contact types, either n-type germanium in contact with a tungsten or phosphor-bronze cat whisker or p-type silicon in contact with a tungsten cat whisker; some are semiconductor-metal junctions such as copper oxide in contact with copper, the oxide acting as a p-type semiconductor, or selenium in contact with Wood's metal, the selenium apparently acting as p-type. So, with diodes, the use of P to denote the element which should be positive to the other element for forward (easy) current flow may be thought of as also denoting the p-type semiconductor, since even the germanium point contact type has an effective p-type element at the junction of the cat whisker and the n-type germanium. But it would not be advisable to use N with diodes unless it were clearly understood that N referred only to polarity, since some diodes have no n-type element.

This seems an appropriate place to add a warning of a booby trap[1] in the form of diode terminals marked with a + sign, or colored red. (This is probably the trap into which the manual referred to in the caption of (B) in *Fig.* 5 fell.) If you conclude, not illogically, that this + terminal should be positive to the other terminal of the diode for forward current flow, you will connect the diode as you would an ammeter, (A) in *Fig.* 5, in the upper branch, with the + side toward the input terminal where the highest positive polarity should be found when the diode is conducting. But your logic will not be that of the manufacturers. Apparently they were thinking of the diode as a substitute for a battery, or a power source. Consequently the + here denotes the cathode, the element where a positive d.c. polarity is found when connected to the positive side of a

[1] S. D. Prensky, "Much fuss about plus." *Radio-Electronics*, April, 1956, p. 102.

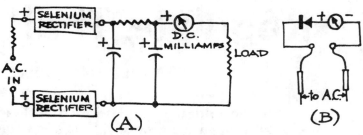

Fig. 5. (Left) (A) Here we have two selenium rectifiers, each with its left terminal stamped +, connected as shown. One rectifier is hooked up backwards and must be shorted out if the circuit is to function properly. (Note the polarities of the meter and the electrolytic capacitors.) Which rectifier should be shorted out? (B) Here the erroneous meter polarity shown is exactly as found on a similar diagram in a manual put out by our armed forces. See text.

filter and load, just as with the usual connection of the cathode of a vacuum tube rectifier. This makes a certain amount of sense, until you put the rectifier in the lower branch of (A) in *Fig.* 5, where it works quite happily in spite of the fact that, when the diode is conducting, the '+' terminal is the most negative point in the circuit.

If you deal much with such diodes (generally selenium types) my advice is to ignore the + or red mark, and think of cathode and anode, as in a vacuum tube diode. This should keep matters straight. Also, mark the anode P, as in *Fig.* 4, if you wish. The main thing is to understand electron flow. With diodes stamped with the conventional arrowhead and bar, the arrowhead denotes the anode, the bar the cathode.

If in doubt about a diode or a transistor, you can easily straighten things out by using an ohmmeter. (The R×10 or R×100 range is generally safest, as neither the current nor voltage applied will ordinarily exceed a safe level.) First determine the polarity of your ohmmeter. (Some VOM's reverse test lead polarity when switched to the ohmmeter function.) Then measure the resistance across the diode or base-emitter junction of the transistor. Reverse the test leads and measure again. One reading should be much lower than the other. The polarity which gave this lower reading is the one of forward conduction. The positive terminal for this lower reading is therefore the anode of the diode (mark it P) or the emitter of a p-n-p transistor (mark it P) or the base of an n-p-n transistor (mark the emitter N).

Returning to *Fig.* 1, it is noteworthy that at (C) the two currents shown flowing in the electrolyte, I_1 and I_2, are not of opposite sign, even though they flow in opposite directions. If one is considered positive, then so is the other. The reason is that both the charges and the directions are of opposite sign; therefore the current in both cases is of the same sign. This must be so, because the two currents add; each increases the total current, I_T. Also, if we take the current of (B) as positive, then the current of (A) is also positive, even though it flows

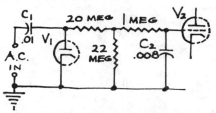

Fig. 6. (right) Can you trace currents and explain why the grid of V_2 goes negative when a.c. input is applied?

in the opposite direction. A little math may help clarify this. The current, I, may be expressed as $I = \rho v A$, where ρ is the charge per unit volume of conductor, v is the average longitudinal, or drift, velocity of the charge carriers, and A is the cross sectional area of the conductor. We take v as positive when its drift direction is clockwise, as in (B). This is to agree with standard electric field theory, which deals with a hypothetical positive test charge. In all three diagrams of *Fig.* 1, the field is such that a positive charge within the conductor is urged clockwise around the circuit and a negative charge is urged counter-clockwise. Therefore when the charge is negative, as with electrons, or negative ions, v is negative; when the charge is positive, as with positive ions, v is positive. With the charge negative, $I = \rho v A$ gives I positive, because ρ and v are both negative. (The product of two negatives is positive.) With the charge positive, I is again positive, because ρ and v are both positive. A of course is always positive.

I fancy by now some would-be heckler is yearning to confound me with the concept of "hole" conduction in semiconductors. Are holes not positive charge carriers? he might ask. If so, is it not logical to use the conventional arrowhead on the emitter of p-n-p transistors? Holes are the majority carriers, except in the base. The emitter injects holes into the base which are swept into the collector and onward toward the collector terminal, which is negative to the emitter. Therefore the arrowhead points correctly toward negative polarity, the current being composed predominantly of positive charges.

But, I argue, what happens outside the transistor? It is hooked up by

(Continued on page 116)

Regulate That Voltage!

WILLIAM G. DILLEY

For the critical or sophisticated application a well-regulated power supply is a necessity. Here's how you can design your own

CONVENTIONAL POWER SUPPLIES utilizing electronic rectifiers with appropriate filter sections provide d.c. power with an a.c. ripple content depending upon the type of filtering employed. Such power supplies may be entirely adequate for some purposes, but exhibit certain characteristics which limit their effectiveness in critical or sophisticated applications. The output from such a supply depends, of course, upon the input, and fluctuations in the line source will be inevitably reflected in a poorly regulated output. From the other end, changes in load current cause resulting changes in output voltage since the output impedance of the power supply is usually quite high. It can be seen immediately that attempts to lower a.c. ripple content by adding more d.c. resistance to the power supply circuit is at odds with voltage regulation requirements for reduced power supply impedance. The ability to employ heavy a.c. filtering while still maintaining a constant supply voltage is, of course, just one of the many advantages of a voltage regulated supply.

The design of voltage regulated supplies for all possible applications can become quite a complex and expensive task, but where current requirements are low, the voltage regulator tube of the cold cathode, glow discharge type, offers a simple and economical approach to the problem.

Principles of Operation

The basic voltage regulator circuit is illustrated in *Fig. 3*. The regulator tube, in parallel with the load, acts as a variable resistor, to maintain a constant tube voltage drop (independent of the current) within its design limits.

In order for the tube to function as indicated, it is necessary to ionize the gas within the tube by increasing the voltage until the gas molecules become ionized. This voltage is usually referred to as the "starting", "firing", or "striking" voltage. Once the tube is ionized, ionization is maintained by the flow of electrons from cathode to plate and the voltage drops across the tube because of reduced tube resistance. This voltage during ionization is called the regulation voltage. Ionization must be sustained to

Fig. 1 Regulated power supply built by author. Unit provides heavily filtered and regulated voltage to bias oscillator and recording amplifier of tape recorder.

effect regulation and this condition is listed as the minimum current requirement of the tube. The maximum current is dictated by excessive heat dissipation or cathode damage because of positive ion bombardment, and is controlled by the selection of the proper value of the

Fig. 2. Typical voltage regulator tubes.

current limiting resistor in series with the VR tube.

The maximum value of the series resistance can be determined from the following equation:

$$R_{MAX} = \frac{V_{MIN} - V_{TMAX}}{I_{TMIN} + I_{LMAX}} \times 1000 \quad (1)$$

where:

I_{LMAX} = maximum value of load current in milliamperes

V_{MIN} = minimum value of d.c. supply voltage (filtered and unregulated)

V_{TMAX} = maximum value of anode voltage drop (see Characteristic Range Values)

I_{TMIN} = minimum value of d.c. cathode current in milliamperes (see Ratings)

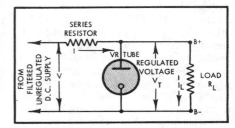

Fig. 3 Basic voltage regulator circuit.

The minimum value of this series resistance can be determined from the following equation:

$$R_{MIN} = \frac{V_{MAX} - V_{TMIN}}{I_{TMAX} - I_{LMIN}} \times 1000 \quad (2)$$

where:

V_{MAX} = maximum value of d.c. supply voltage (filtered and unregulated)

V_{TMIN} = minimum value of anode voltage drop (see Characteristic Range Values)

I_{TMAX} = maximum value of d.c. cathode current in milliamperes (see Ratings)

I_{LMIN} = minimum value of load current in milliamperes

From the two foregoing equations it would appear that, when provided with the tube characteristic data, all information essential to proper design is available. However, these equations provide for current limits *during operation only* and we must further insure that adequate voltage exists for starting under all conditions. The value of starting voltage will vary with supply voltage changes and, since VR tubes are light-sensitive, their ionization characteristics are dependent upon the ambient light conditions. Therefore to provide for adequate starting voltage the following equation is provided:

$$R'_{MAX} = \frac{V_{MIN} - V_{BOMAX}}{I_{LMAX}} \times 1000 \qquad (3)$$

where:

V_{MIN} = minimum value of d.c. anode supply voltage (filtered and unregulated)

V_{BOMAX} = maximum value of anode breakdown voltage (see Charateristic Range Values)

I_L = value of load current in milliamperes

The designer should then select the *lowest* value of series resistance calculated from equations (1) and (3) as the required maximum value of series resistance.

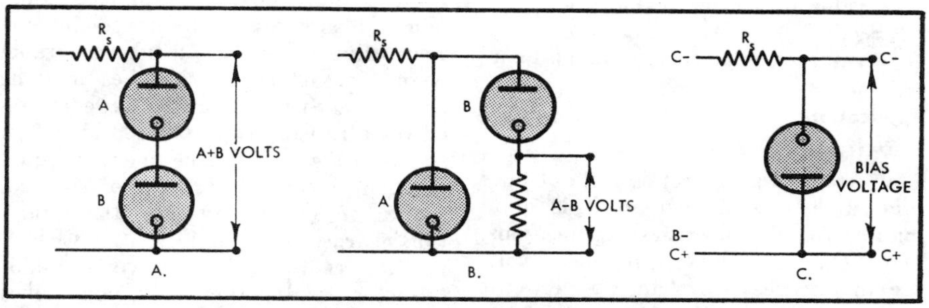

Fig. 5. Various configurations for achieving fixed or bias voltages.

Design Example

Let us assume that we wish to determine the circuit parameters for a regulator which will deliver approximately 150 volts to a load in which the current varies between 25 and 30 ma. The regulator circuit is to operate from a supply voltage which varies between 450 and 525 volts and the regulator tube selected is an OA2. Using the tube characteristics as listed in *Table 1,* we can calculate the minimum series resistance required using equation (1) as follows:

$$\frac{525 - 140}{30 + 25} \times 1000 = 7000 \text{ ohms } (min)$$

Using equation (2) the maximum value will be:

$$\frac{450 - 168}{5 + 30} \times 1000 = 8050 \text{ ohms } (max)$$

The choice of any resistor value between 7000 and 8050 ohms will insure operation of the OA2 within the published minimum and maximum ratings, taking into consideration maximum variations in line voltage, load current, and anode voltage drop. However, to insure proper starting under poor ambient light conditions and low supply voltage, equation (3) must be utilized[1]:

Maximum value for series resistance:

$$R'_{MAX} = \frac{450 - 185}{30} \times 1000 = 8030 \text{ ohms}$$

Selecting the lowest calculated value for the maximum value of the resistor, the final selection to insure both starting and operating limits will be between 7000 and 8030 ohms.

The current (in milliamperes) through the tube may be calculated from the following equation:

$$I_T = \frac{V - V_T}{R} \times 1000 - I_L$$

where:

V = value of d.c. supply voltage (filtered, and unregulated)

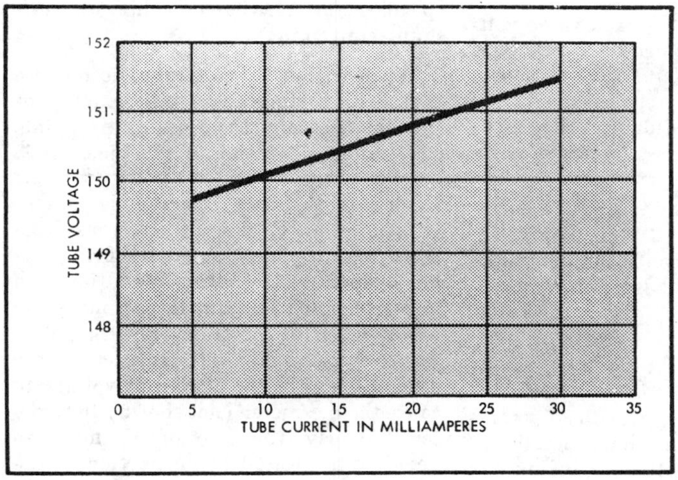

Fig. 4. Measured characteristics of sample OA2.

[1] Where the tube manufacturer lists a higher value of starting voltage for conditions of total darkness, this figure should be used. For example, the OC2 requires 145 volts for starting in total darkness.

TABLE 1

Type	Description		Max. Dimensions Inches: Length	Max. Dimensions Inches: Diam.	Max. Starting Current Ma.	DC Operating Current Ma.: Max.	DC Operating Current Ma.: Min.	Ambient Temperature Range °C	Operating Conditions: Approx. DC Starting Volts	Min. DC Anode-Supply Volts	Approx. DC Operating Volts	Regulation: Current Range Ma.	Regulation: Volts	Type
VOLTAGE-REGULATOR TYPES														**VOLTAGE-REGULATOR TYPES**
OA2	Intended for use in applications where it is necessary to maintain a constant dc output voltage across a load, independent of load current and moderate line-voltage variations.	Miniature button 7-pin base.	$2\frac{5}{8}$	$\frac{3}{4}$	75	30	5	−55 to +90	156	185	151	5 to 30	2	**OA2**
OA3		Octal 6-pin base.	$4\frac{1}{8}$	$1\frac{9}{16}$	100	40	5	−55 to +90	100	105	75	5 to 40	5	**OA3**
OB2		Miniature button 7-pin base.	$2\frac{5}{8}$	$\frac{3}{4}$	75	30	5	−55 to +90	115	133	108	5 to 30	1	**OB2**
OC2		Miniature button 7-pin base.	2.63	$\frac{3}{4}$	75	30	5	−55 to +90	105	115	75	5 to 30	3	**OC2**
OC3		Octal 6-pin base.	$4\frac{1}{8}$	$1\frac{9}{16}$	100	40	5	−55 to +90	115	133	108	5 to 40	2	**OC3**
OD3		Octal 6-pin base.	$4\frac{1}{8}$	$1\frac{9}{16}$	100	40	5	−55 to +90	160	185	153	5 to 40	4	**OD3**

V_T = value of anode voltage drop
R = value of series resistance
I_L = value of load current in milliamperes

Applications

While the VR tube is not an absolutely stable voltage reference and does vary slightly with current, it is quite suitable for all but critical applications. The OA2, for example, provides a voltage that varies less than 1.5 per cent throughout the full range of current operation (see *Fig. 4*). Other typical VR tubes vary as listed in *Table 1*. Available tubes include both octal and miniature types. Examples of circuits suitable for VR tube application are local oscillators in receivers, oscillators in frequency meters, bias oscillators in tape recorders, tuned oscillators in a V-F-O, audio preamplifiers, tape recording amplifiers, screen supplies for audio amplifiers, bridge circuits in vacuum-tube voltmeters, and so on.

Practical Considerations

In actual design practice, a quick look at the tube characteristics will reveal the major limitations associated with circuits of this type:

1. low current carrying capacity
2. fixed voltage ratings

Circuit designs to isolate heavy current requirements not specifically requiring regulation can be effected as in *Fig. 6*. This example allows regulation of voltages to the first two stages and the screens of an audio amplifier and bypasses the heavier current requirements within the power amplifier section. Where it is necessary to exceed the current rating of a given VR tube, a parallel connection of two such tubes may be employed. Such operation, however, requires a resistor (approx. 100 ohms) in series with each VR tube to assure equal division of current. The use of such resistors, of course, may affect the regulation characteristic slightly.

With regard to the fixed voltage limitation, a combination of various tubes connected in series will provide the sum of the voltage of the various tubes employed (see *Fig. 5a*). The value of the series resistance is calculated in the manner described except that $V_{T_{MAX}}$ and $V_{T_{MIN}}$ are the sum of the anode voltage

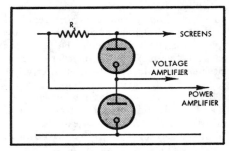

Fig. 6. Voltage regulator circuit for an audio amplifier.

drops for the tubes employed in series connection.

Additional changes in fixed value voltages also may be obtained by a difference rather than a sum connection as shown in *Fig. 5b*.

Combinations of regulated voltages, of course, are obtained by series connections tapped where required. For example three OC2 tubes in series would provide 75, 150, and 225 volts.

Regulated bias voltages may be obtained as shown in *Fig. 5c*.

A word of caution relative to the design of regulated circuits supplying tubes as the load. Filament type tubes do not conduct current until the heaters have warmed up and, therefore, when calculating minimum load current, a zero minimum current should be assumed. This assumption dictates a maximum load current not exceeding the current rating of the tube, since the tube *is* the full load until such time as the load tubes conduct. Under these conditions, a practical method of assuring proper series resistance consists of disconnecting the load and adjusting the series resistor for a tube current of slightly less than the maximum rating of the tube employed.

Higher load currents may be used if the power supply rectifier is changed to a separate cathode type in order to insure that the load conducts whenever voltage is impressed across the VR tube.

Wherever a requirement exists to shunt capacitance across the VR tube, such as to reduce tube noise or reflect a low impedance, a possibility of tube oscillation exists. To prevent such oscillation, the tube manufacturer usually specifies the maximum value of such shunt capacitance.

When wiring the VR tube, do not use pins 3 or 6 since these pins extend inside the tube and their use as tie points could affect operation of the tube. Æ

CURRENTS (*from page 113*)

means of metallic conductors, usually copper wire. We know that only electrons, carrying negative charges, can flow in pure metal. What happens to our direction of current flow? Is it one way outside the transistor and the opposite inside?

Yes, answers the heckler, turning against me my own explanation of how two currents may flow in opposite directions and still be of the same sign, i.e., add. The current inside the transistor is composed predominantly of positive charges. The current outside is composed of negative charges. Therefore there is no anomaly. We simply have the condition of (C) in *Fig. 3*.

To which I answer, the heckler has a point, given the reality of holes, even though it would be awkward to denote change of both direction and type of current every time we entered and left a transistor. But how about n-p-n transistors? Here the principal carriers are electrons, but we still find the arrowhead pointing the wrong way, as at (A) in *Fig. 3*. Then I deliver the crusher. As for holes, these are only a convenient fiction.[2] The actual conduction is via electrons.[3] William Shockley, writing in the *Proceedings of the I.R.E.*, Nov. 1952, p. 1297, warns that "confusion may arise if the model of the hole is taken too literally" and cites an example of such confusion.

Remarks like Shockley's lead me to suspect that the hole concept, useful as it is, is but another outgrowth of wrong thinking about current flow; that is, if Franklin had only gotten us off on the right foot, so that electron charges were designated positive and electrons were customarily accepted as the predominant charge carrier, we would have another, and less confusing, concept to replace that of the hole.

If you want to compare the merits of electron flow vs. conventional flow and at the same time test your understanding of currents in general, *Fig. 6* offers a good opportunity. This represents the a.c. input circuit of a V.T.V.M., on the most sensitive range. As a suggestion, start with the half cycle when V_1 is conducting (upper a.c. input terminal positive to ground), trace currents, then consider what happens on the opposite half-cycle, bearing in mind that an inflow of electrons (carrying negative charge) to one plate of a capacitor is accompanied by an outflow of electrons from the opposite plate, so that the first plate has a surplus of electrons, resulting in a negative potential, and the second plate has a deficit of electrons, resulting in a positive potential, with respect to the other plate. For the a.c. input voltage to cause the d.c. meter (not shown) to register properly, the grid of V_2 must go negative to ground in proportion to the input amplitude. Therefore you should end up at V_2 grid with such a negative d.c. voltage, or more precisely, a filtered composite, mostly d.c., having minor components of the input.

The manual from which (B) of *Fig. 5* is derived states that C_1 of a circuit similar to that of *Fig. 6* "serves to block any d.c. voltage present in the circuit being measured." This seems to imply that if only a.c. were being measured, C_1 could be shorted out. Do you agree? Æ

---

[2] Dewitt and Rossoff: "Transistor Electronics," McGraw-Hill, 1957, p. 42.
[3] K. R. Spangenberg, "Fundamentals of Electron Devices," McGraw-Hill, 1957, pp. 7, 95.

Graphical Solution to the Tracking Problem

This method of determining optimum offset angle, optimum overhang, or the minimum distortion point will appeal to those who do not have the patience or equipment to make the more conventional measurements normally employed.

W. B. BERNARD

WITHIN THE LAST FEW YEARS there have been published a mathematical analysis[1] and a graphical analysis[2] of the stylus tracking problem. The mathematical analysis of the problem is excellent but if you do not have the equations available when you wish to set up the new turntable and arm you still have the problem. The graphical solution referred to will show the errors but will not indicate the optimum overhang or offset angle for a given situation.

By making a change in the graphical solution it is possible to find easily the optimum overhang of the stylus and the optimum offset angle. If we consider the

[1] Dr. John D. Seagrave, "Minimizing pickup tracking error." *Audiocraft*, December, 1956.
[2] F. J. Hennessy, "Tracking errors in record production." *Wireless World*, November, 1958.

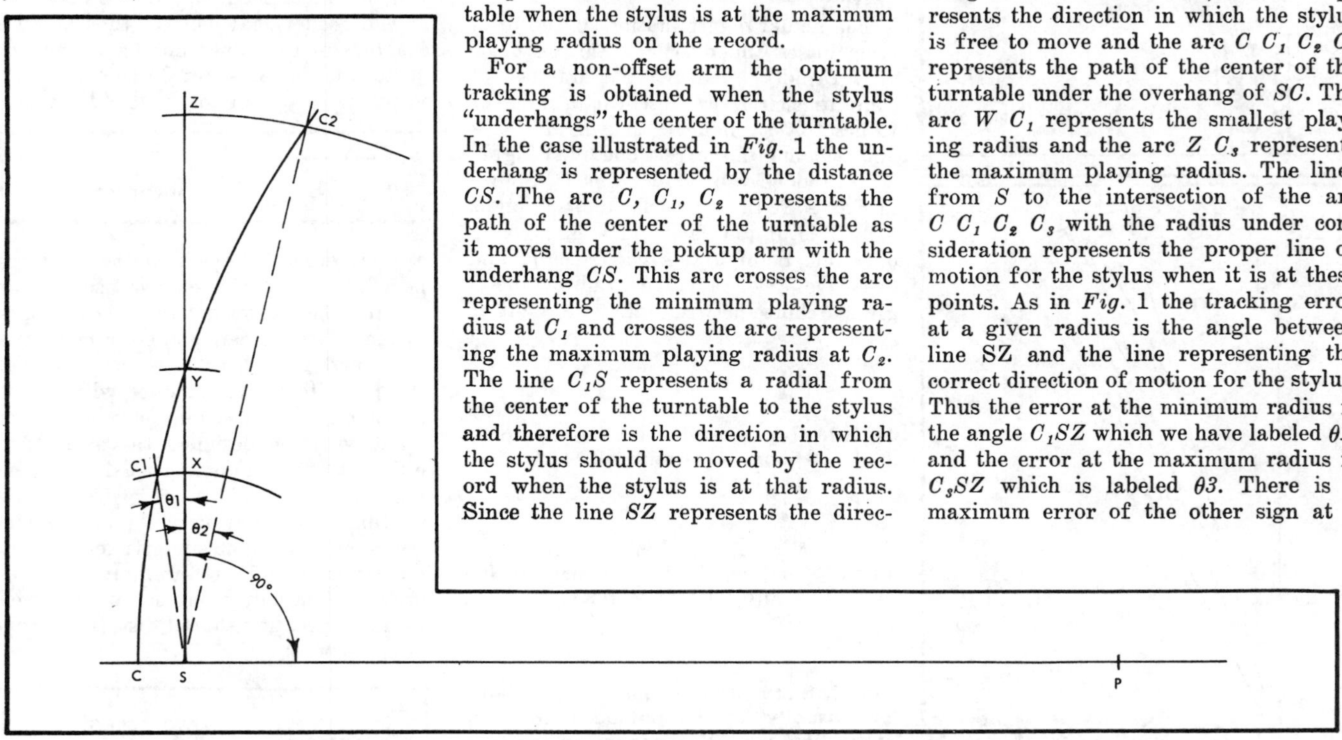

record to be moving under the pickup arm the graphical analysis becomes much simpler and will yield the desired information needed for optimum adjustment of the arm. Let us start with a non-offset arm to simplify the analysis in the beginning. Referring to *Fig.* 1, the point P is the point about which the arm pivots, or in this case, the point about which the rest of the record playing assembly pivots underneath the arm. Point S represents the stylus which moves laterally in the direction of the line SYZ when subjected to the action of the undulations of the record groove. A circle of radius SX, of which the arc $C_1 X$ is a part, represents all possible positions of the center of the turntable when the stylus is at the minimum playing radius and a circle of radius SZ, of which the arc $Z C_2$ is a part, represents all possible positions of the center of the turntable when the stylus is at the maximum playing radius on the record.

For a non-offset arm the optimum tracking is obtained when the stylus "underhangs" the center of the turntable. In the case illustrated in *Fig.* 1 the underhang is represented by the distance CS. The arc C, C_1, C_2 represents the path of the center of the turntable as it moves under the pickup arm with the underhang CS. This arc crosses the arc representing the minimum playing radius at C_1 and crosses the arc representing the maximum playing radius at C_2. The line $C_1 S$ represents a radial from the center of the turntable to the stylus and therefore is the direction in which the stylus should be moved by the record when the stylus is at that radius. Since the line SZ represents the direc-

tion in which the stylus is free to move, the angle $C_1 SX$ represents the tracking error at the minimum playing radius. Similarly the angle $C_2 SZ$ represents the tracking error when the stylus is at the maximum playing radius. For any other playing radius the error can be found by measuring the angle from the line SZ to a line connecting point S with the intersection of an arc at the radius under consideration and the arc C, C_1, C_2. Thus we see that at a playing radius equal to SY there is no tracking error.

By using an offset in the end of the pickup arm it is possible to have a zero tracking error at two radii and to greatly reduce the error at other radii in comparison to errors obtainable with a non-offset arm. *Figure* 2 is a graphical plot of the determination of errors for an arm with an offset angle of ϕ and an overhang of SC. In this case, line SZ represents the direction in which the stylus is free to move and the arc $C\ C_1\ C_2\ C_3$ represents the path of the center of the turntable under the overhang of SC. The arc $W\ C_1$ represents the smallest playing radius and the arc $Z\ C_3$ represents the maximum playing radius. The lines from S to the intersection of the arc $C\ C_1\ C_2\ C_3$ with the radius under consideration represents the proper line of motion for the stylus when it is at these points. As in *Fig.* 1 the tracking error at a given radius is the angle between line SZ and the line representing the correct direction of motion for the stylus. Thus the error at the minimum radius is the angle $C_1 SZ$ which we have labeled $\theta 1$, and the error at the maximum radius is $C_3 SZ$ which is labeled $\theta 3$. There is a maximum error of the other sign at a

Fig. 1. Graphical diagram as described by the author as it appears for a straight arm.

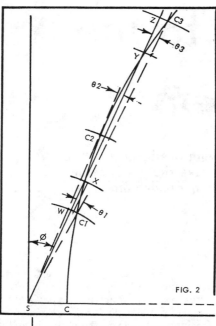

radius of SC_2 which is the angle C_2SZ which is labeled $\theta2$. At radial distances equal to SX and SY there is no tracking error.

In *Fig.* 2 it can be seen that $\theta1$, the error at the minimum radius, is somewhat greater than $\theta3$, the error at the maximum radius. Since the maximum permissible tracking error for a given distortion level is roughly proportional to the radius, the case illustrated is less than optimum. This brings us to the problem of optimizing the offset angle and the overhang distance. We may now apply some dimensions to the problem. The arm under discussion has a pivot to stylus distance of 9½ inches and the desired minimum and maximum playing radii are 2 inches and 5¾ inches respectively. A number of concentric circles are drawn with P as the center. Each circle represents a different overhang distance. For simplicity only two of these are shown in *Fig.* 3. These represent overhang distances of ½ inch and ⅝ inch. Line SX is drawn to show that ⅝-inch overhang is too great since the error at minimum radius is the same as at the maximum radius. It may be seen from inspection of the plot that the error at the minimum radius would be greatly reduced if the minimum radius were increased just slightly but since we wish to maintain low distortion in to a radius of 2 inches let us try an overhang of ½ inch. When we have drawn an offset line SZ and measured the resulting errors we find that $\theta1$ and $\theta2$ are about 1 deg. and $\theta3$ is about 3 deg. These are about one half the permissible standard of 1 deg. per inch radius suggested by Seagrave.

The actual determination of the optimum offset and overhang can be made very rapidly by drawing a number of concentric circles representing a range of values for overhang as shown in *Fig.* 4 and selecting the correct one by swinging a straight edge around point P. Shown in *Fig.* 4 are three tries at an offset angle. Line 1 is at too small an angle, giving only one point of correct tracking and excessive error at other radii. Line 3 gives a choice between only one point of correct tracking or two points of correct tracking with excessive error at small radii. Line 2 gives a reasonably good solution to the problem.

This quick selection by eye may not be as rigorously accurate as the mathematical method, however, from the practical standpoint the difficulties in knowing the actual direction of motion of the stylus and the setting of the cartridge in the arm to an accuracy greater than that which the desirable offset can be determined by the method described tend to make more accurate methods academic. The fact that the determination can be carried out with only a few drawing instruments makes it most useful.

Other information may be obtained from the graphical analysis. It can be seen that the errors in tracking are in-

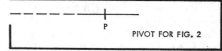

dicated by the divergence of parts of an arc from a straight line, therefore the error can be reduced by shortening the arc or flattening it. It can be shortened by increasing the minimum radius or by decreasing the maximum radius. If the arm is to be used only on 12-inch records it might be profitable to increase the minimum radius to 2½ inches with the result that the tracking error over the remaining part of the record could be even further reduced. If the length of the arm is adjustable the curvature of the arc may be reduced by increasing the length of the arm to the maximum that can be accommodated.

The reader may believe that because his pickup arm does not have any adjustments that it either doesn't need any adjustment or cannot have adjustment

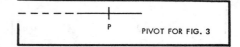

applied even if needed. Neither of these beliefs is necessarily so. Because there is a variation between various cartridges in the relative locations of the mounting holes and the stylus it is possible that the pivot to stylus distance will not be the same as the manufacturer had in mind when he designed the arm. If a different offset angle is needed it may be possible to "steal" a few degrees by slotting the mounting holes in the cartridge or by similar subterfuges. In any case it takes just a few minutes to investigate the situation and it may well result in greatly reduced distortion from your records. Æ

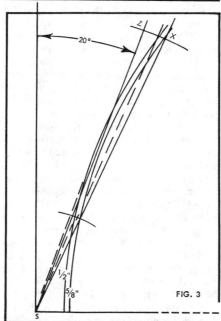

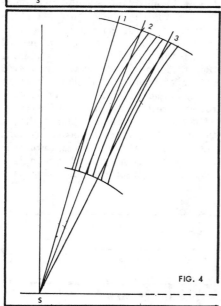

Fig. 2. (Top) Graphical diagram for a phono arm with offset.

Fig. 3. (Middle) Preliminary steps in obtaining optimum offset angle by the graphical method.

Fig. 4. (Bottom) Optimum tracking angle can usually be determined readily by inspection with this graphical construction.

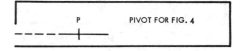

Determination of Tracking Angle in Pickup-Arm Design

NIEL MALAN

A new approach to the tracking-angle vs. overhang problem with methods outlined for determining optimum relations.

IN DESIGNING a pickup arm, the home constructor is faced with the difficulty of deciding the correct angle of offset of the head to the arm, and the correct amount of overhang of the stylus point over the center of the record. In *Fig.* 1, a pickup arm of length L is pivoted at O and swings in the arc indicated. At the particular position shown, the stylus is at a distance r from the center C of the record; as the arm swings in, the needle will eventually be on the line between O and C and will then overlap the center of the record by the distance d. At any spot on the record, the offset angle is b, the angle between line drawn at right angles to the radius r, and the line L between the pickup center O and the stylus.

In the triangle formed by the lines $L, r,$ and D,

$$D^2 = L^2 + r^2 - 2Lr \cos a \quad (1)$$

But from above,

$$D + d = L$$

and also

$$b + a = 90 \text{ deg.}$$

By juggling round with Eq. (1), we can reduce it to

$$\cos a = \frac{r^2 + 2Ld - d^2}{2Lr}$$

All that is needed now for the actual design is a knowledge of the intended length of the arm L, a set of cosine tables, and a slide rule.

Unfortunately, quite a bit of slide-rule work is involved, but this is worth the trouble. The procedure is to assume a series of values for d, the overhang, between say ¼ and ½ inch. For a typical 12-in. arm, values outside this range need not be considered. Then calculate the required values of r^2, $2Ld$, d^2, and $2Lr$ for values of r ranging from 2.5 to 6 in. Very few LP records are recorded to less than 2.5 in. radius; half-inch steps for r are adequate. Also, for values of overhang up to 0.375 in., the value of d^2 is so small that it may be neglected. At the most critical point on this overhang, this causes an error in calculation of only 0.1 deg., about as bad as the usual 0.5-db error in response curves.

For a 12-in. arm and an overhang of 7/16 in., this gives a series of values of angle a, for half-inch steps of r from 2.5 to 6 in., of 73.8, 74.1, 74.3, 74, 73.4, 72.8, 72.1, and 71.2 deg. and by subtracting these values from the right angle mentioned earlier on, we find that the values of b, the offset angle, varies as follows:

Radius r	2.5	3	3.5	4	4.5	5	5.5	6	inches
Angle b	16.2	15.9	15.7	16	16.6	17.2	17.9	18.8	degrees

It will be noted that the angle required decreases, and then increases again. For smaller values the amount by which the angle decreases becomes less, eventually the angle increases from the word "go."

Plotting the Curves

Once the values of angle required for different values of overhang have been

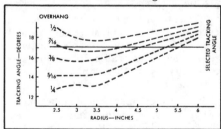

Fig. 2. Curves of tracking angle vs. radius for a 12 in. arm with overhangs ranging from ¼ to ½ in.

calculated, they should be plotted on graph paper against the radius, as in *Fig.* 2. From the collection of curves obtained, it can be seen that: (1) for low values of overhang, the offset angle increases constantly; (2) for high values of overhang the offset decreases and then increases again. Since the average design is limited to one fixed angle of offset, it will be seen that if the overhang is too small, the tracking error, which is the difference between the selected and the required angle will be first negative and then positive. For higher values of overhang, the error will be positive, then negative, then positive again.

The shape of the curve also indicates where the problem of compromise becomes awkward. Using medium values of overhang, the error angle can be made as small as ¼ deg. over the range from 2.5 to 4 inches radius, but rises rapidly after this until it reaches 3.5 deg. at 6 inches radius. This would make a good arm for 7-in. records, but using a 12-in. arm for these would be rather extravagant. Similarly, using a large value of overhang, the average error can be reduced to plus or minus 1.5 deg., but another consideration enters.

Is tracking error more important at small radii than at the outer edge of the record? Presuming that it is—and experience seems to indicate that this is the case—then the problem is keep the variation in error small at the inside of the record, while not allowing it to rise too high on the outside. Here the curves drawn will be useful—by sliding a plastic straightedge up and down the graph paper, one can select a value of average angle, which will then be the tracking angle, which seems to suit the above requirements best.

Those who are not yet tired of calculation can try to work out the ratio of tracking error to radius and use this to find the "best" curve. However, even by inspection one can find that using a 12-in. arm, with an offset angle of 16 deg. and an overhang of 7/16 in., the error varies over plus or minus 0.5 deg. from 5 to 9 inches record diameter, and then rises to 2.5 deg. at 12 in. diameter, all quite acceptable values.

In passing, three points should be mentioned. First, the offset angle is not the angle of the pick-up to the arm, but the angle between the pick-up and a line joining the stylus and the arm pivot. Second, the overhang is measured from the center of disc or turntable spindle. Third (and important), the overhang is quite critical and should be measured accurately.

Æ

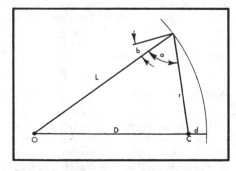

Fig. 1. Basic geometry of the phono arm.

The Silicon Diode in Audio Equipment

Silicon diodes are rapidly becoming popular with designers of audio equipment. The author discusses their use as power rectifiers and as bias regulators.

L. B. DALZELL

AUDIO POWER SUPPLIES using vacuum tube rectifiers have been around for a long time. The rectifier tubes have certainly worked out well, and have been improved over the years. While they are quite reliable today, the rectifier is still the most frequently replaced tube in audio equipment.

Copper oxide and, later, selenium rectifiers have seen some use as substitutes for the tube, but the voltage and current requirements of audio equipment forced the use of large units, as sizable heat dissipating fins are required to cool the rectifying junctions. Copper oxide rectifiers are frequently used in meter circuits, while selenium is common in bias supplies today.

As an offspring of the transistor, a whole new family of semi-conductor rectifiers has developed in the last few years. Germanium diodes are small, but more suited to higher powers at low voltage levels. The comparatively recent silicon diode rectifier seems to have eliminated the problems of size and heat—at least at audio power levels.

It seems fantastic that four tiny wafers of pure silicon with hyper-accurately controlled impurities can do the same job as a conventional full-wave vacuum-tube rectifier. They can even do the job better, for they waste little power, take little space, and in a properly designed circuit they will probably last for years.

Manufacturing methods have been rapidly improved, and silicon rectifiers, which but three years ago were too expensive for serious consideration in audio equipment, are coming into a price range that allows direct competition with the vacuum tube, and there are real advantages in their use. There are problems too, and in this article an attempt is made to cover the design of silicon rectifier power supplies for audio amplifiers. Many of the same ideas apply to the design of supplies for pre-amplifiers and tuners, but in class AB or class B amplifier circuits the power supply requirements are much more severe. The current drain is higher, and usually there is a substantial difference between zero-signal and full-output current—especially when the amplifier is tested with steady-state sine waves.

To simplify our discussion, let us refer to the vacuum tube full wave rectifier as a tube, and the semi-conductor silicon diode rectifier as a diode.

First, consider the conventional full wave center-tapped power supply of *Fig.* 1. The transformer serves to:

(1) Isolate the line supply voltage.

(2) Provide a high-voltage secondary for conversion to the high-voltage d.c. or B plus.

(3) Provide a 5 volt rectifier tube heater secondary (Fig. 1a).

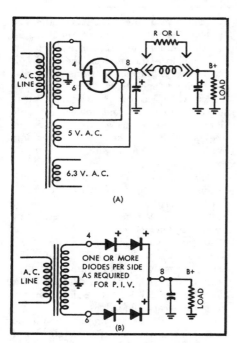

Fig. 1. (A), The familiar full-wave center tapped rectifier circuit using a vacuum tube of the more advanced indirectly heated type. (B) is the silicon-diode equivalent of (A). Pin designations are for the 5AR4 and 5V4 type or replacement rectifier or diode unit.

Fig. 2. Internal construction of an experimental silicon rectifier unit which can replace conventional vacuum-tube rectifiers in certain situations.

(4) Provide a 6.3V secondary to heat the other tubes in the circuit.

The rectifier tube heater may draw up to 3 amperes at 5 volts, which is 15 watts that the transformer has to handle. A powerful amplifier, or a large stereo amplifier, might require two rectifier tubes, thus doubling this portion of the transformer load.

The full-wave center-tapped rectifier arrangement is covered in any number of texts and tube manuals. However, a review of P.I.V.—peak inverse voltage—seems in order, as this will be an important consideration in the use of silicon diodes.

Peak Inverse Voltage

If a variable d.c. supply is arranged with positive connected to the cathode and negative to the anode of a rectifier tube or silicon diode, at low voltage the units will present a very high resistance, and only a minute current will flow. As the voltage of this inverse connection is slowly increased, the current will remain very low until an inverse break-

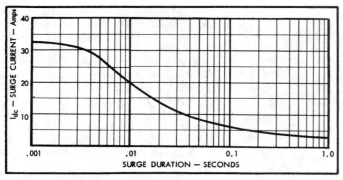

Fig. 3. This curve, which applies to Texas Instrument 1N2069, 1N2070, and 1N2071 silicon diodes, shows the extremely high surge current that these rectifiers can absorb.

down point is reached. The current will then increase drastically, and the tube or diode will usually be destroyed. With an increasing a.c. applied, the units will block on the half cycle when the potential is inverse, until the peak inverse voltage equals or exceeds the reverse break-down point, and again a high current can flow and the units would probably be destroyed.

The peak a.c. voltage is most important, for the usual a.c. meter indicates r.m.s. voltages, and such an indication multiplied by 1.414 indicates peak, but only if the wave is sinusoidal.

Now let us consider the power supplies of (A) and (B) in *Fig. 1*. If the transformer delivers 400-0-400 r.m.s. volts, and the supply is a sine wave, then each tube plate or diode pair would see 400 × 1.414 or 565 peak volts. At the peak instant, one end of the secondary will be positive by 565 volts in relation to the ground center tap; the other end will be negative by 565; and 2 × 565 or 1130 peak volts will be applied in the inverse direction across the half of the tube or one diode pair that is at the most negative point, and this must be blocked—usually with an ample safety margin. With no load connected, the filter capacitors will be charged to 565 volts in both cases.

Here now is one great difference between the tube and the diode. All electrical conductors in audio use have some resistance. The forward resistance of rectifier tubes in general is much higher than that of the silicon diode.

After the capacitors are charged, in *Fig. 1*, there will be only a tiny leakage current flowing until the load is connected. When the load is applied and a current begins to flow, the effective resistance of the transformer plus the internal resistance of the tube or diodes will cause a voltage drop. An increased current causes a higher drop across the supply and any external inductance or resistance in the filter, and the term "regulation" is used to indicate the reduction in B plus with increased current. If the power-supply regulation is poor, there will be a large voltage reduction as the current drain becomes greater.

To study the adaptation of silicon diodes to the conventional full wave center-tapped power supply, the author has undertaken a series of experiments. Directly replaceable silicon diode substitutes for the rectifier tubes in common use are available. A rather complete series is manufactured by International Rectifier Corporation and by Sarkes Tarzian, but in the interest of self-education an experimental unit was constructed.

Experimental Set-up

Single diodes rated at a P.I.V. over 1500 volts are available, but they are expensive and usually carry too low a forward current. However, if diodes are connected in series, the allowable P.I.V. will increase as the number of diodes. They can still carry their rated forward current, and they will be comparatively low in price. About 1200 volts P.I.V. looked like a minimum for experiment, and four Texas Instruments 1N2071 diodes were used. Their characteristics are:

P.I.V.	600 v
R.M.S. Input	420 v
Avg Rectified Forward Current	750 ma
Recurrent Peak Current	6 a
Operating Temp	0 to 100° C
Max Reverse Current	0.2 ma
Max Forward Voltage Drop	0.6 v

These diodes are quite small, the body being about 0.25 inches long and 0.2 inches in diameter. The body is insulated, and there are silver pig tail leads for connection and that act as heat sinks, although no measurable heating has been experienced. Two diodes in series in each leg give the desired 1200 volt P.I.V., and can handle 750 M.A.—about equivalent to three 5U4GB or 5AR4/Z34

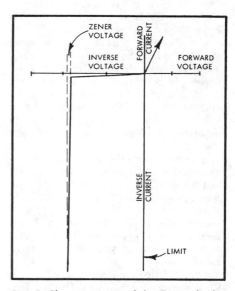

Fig. 5. Characteristics of the Zener diode.

tubes. The four diodes were wired in two series strings, as at (B) in *Fig. 1*, with the anodes connected to pins 4 and 6 and the common cathodes connected to pin 8 of a Vector G2-10 Octal plug-in unit. A 3-inch 6-32 screw acts as the center support, and a modified 2-terminal insulated tie-point acts as the upper support member. An aluminum case covers the whole assembly. The internal construction is illustrated in *Fig. 2*.

The unit was tested in a number of situations; but to illustrate the problems encountered, the experiences with one amplifier are outlined. The amplifier was a modified Mullard, using EL34's and rated at 40 watts. It has a 200-ma 400-0-400 volt power transformer, and a 5U4GB rectifier tube. The input capacitor is a 30-μf, 500 v. electrolytic followed by a 100-ohm choke. A series of rectifier tubes and the diode unit were plugged into the rectifier socket in turn, and the resultant d.c. output was measured at the B-plus primary tap of the output transformer as follows:

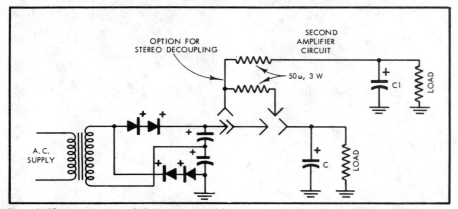

Fig. 4. The conventional full-wave doubler. C is optional for ripple reduction. The number of diodes per leg will depend on the P.I.V. See text.

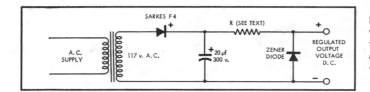

Fig. 6. A regulated voltage standard for shop use Meters may be checked at the output.

Rectifier	B Plus Voltage d.c.
5R4GY	425
5U4GB	450
5AR4/GZ34	475
Diode Unit	505

It is evident that the low forward resistance of the diode unit resulted in a substantially higher d.c. output. When the amplifier was driven to the clipping point the B plus with the 5U4GB dropped from 450 v. to 420 v. With the diode unit the B plus dropped from 505 v. to 490 v. at clipping.

The diode unit gave much superior regulation, but the B plus was now too high. A power resistor connected between pin 8 of the rectifier socket and the first electrolytic capacitor would reduce the voltage to the original 450 v. d.c., but with this improvisation the regulation was about the same as with the 5U4GB. A 350–0–350 volt 200-ma power transformer was then substituted. The resultant B plus was then 455 volts. and the regulation was still 15 volts from zero signal to full drive.

That about winds up the experiment except to say that the diode unit has been in daily operation for well over a year, and the output has not deteriorated a bit.

The experience outlined above brings up a couple of problems. Diodes don't wait for a heater to warm them before they conduct, hence peak voltage is applied to the filter capacitors and the audio tubes until current is drawn by the load, which will then reduce the B plus to its operating level. With the 350–0–350 volt transformer used, the peak is 1.414 × 350 = 495 volts. A purist might want to use a time delay device to hold down the B plus until the audio tubes heat. On the other hand, the diode supply with the lower voltage transformer is easier on the 500-volt electrolytic capacitors than the previous arrangement was.

Surge Current

Rectifier tubes will not handle the surge current that diodes will. *Figure* 3 illustrates the astonishing surge current that the 1N2071 will take in stride. It is not uncommon for someone to "fix" his ailing audio equipment by turning it off for an instant, then right back on. The filter capacitors may be pretty well discharged, the tubes still hot and conductive, and the resultant surge can "fix" a rectifier tube right out of existence. Silicon-diode supplies have repeatedly absorbed such surges without a complaint.

To eliminate surge, or hot switching troubles, tube manufacturers specify the minimum effective plate resistance. This resistance per plate includes the transformers' effective resistance, which may be calculated as follows:

$R_{eff} = R_s + n^2 R_p$

where

R_s is the resistance of the secondary
R_p is the resistance of the primary
n is the ratio:
$$\frac{\text{Secondary voltage}}{\text{Primary voltage}}$$

In *Fig.* 1, secondary refers to one half, i.e. from ground to one end of the winding.

Additional resistance in series with the tube anodes or cathode is usually required to meet the specified condition, but these limits are too frequently ignored in practice. The effective plate resistance of the transformer alone is usually sufficient to limit surges well below the maximum for silicon-diode supplies.

The choke or resistor and second filter capacitor, of (A) in *Fig.* 1a, are completely superfluous with a diode supply —in the author's opinion. Push-pull output configurations may be easily balanced to eliminate the effect of ripple, and a filter capacitor of about 60 μf is all that is required. Chokes and resistors simply make regulation worse and waste power. Of course, this applies only to push-pull output stages.

In fairness to the rectifier tube, it should be pointed out that two parallel rectifier tubes will have one-half the drop of a single tube. Further, the 5AR4/GZ34 rectifier tube is nearly as good as diodes for regulation, and it heats slowly, eliminating the B plus overthrow when it is first turned on.

Voltage Doublers

The voltage doubler supply is gaining popularity, especially when used with diodes. The doubler makes use of a lower voltage transformer without a center tap, which can be less expensive. See *Fig.* 4. It does require two fairly large electrolytic capacitors, but these can be lower voltage units than the full-wave supply requires. The doubler simply charges one capacitor on the first half cycle and the other capacitor on the second half cycle. The capacitors are in series, hence their voltages are additive. Because the source alternates between the two storage capacitors, a 160-volt 500 ma transformer can supply 250 ma at 1.414 × 160 × 2 or 453 volts, less transformer and diode losses.

In practice, the doubler supply using diodes does not have the high peak (zero current drain) voltage of the full wave center-tapped supply. Transformers designed for doubler operation have low effective resistance, hence with good ca-

Fig. 7. Experimental 20-watt amplifier with the outside cage removed.

pacitors doublers regulate better also. (See "A 60-Watt Amplifier with Silicon-Diode Power Supply," AUDIO, March, 1959.) As one storage capacitor is being charged while the other discharges, ripple has a different wave shape, which seems slightly less troublesome than the full-wave center-tapped supply as far as hum is concerned. Finally, the doubler is less sensitive to variation of a.c. source than a full-wave supply.

As a practical example of the doubler power supply, the Triad R93A power transformer, which was designed for this purpose, can be connected to supply 166 v. a.c. to the diode doubler. Peak d.c. voltage at the zero-current condition will be $166 \times 1.414 \times 2 = 468$ v. At a drain of 130 ma, this reduces to a steady 455 v. d.c., while a drain of 250 ma causes the B plus to drop to about 440 v. d.c. This is very good regulation indeed. The diodes in each side of the doubler have to block a P.I.V. of $2 \times E_{sec} \times 1.414$. It is wise to allow a good factor of safety in this case.

The series doubler capacitors should be at least 100 µf each, and greater values give better results. Their voltage ratings should add to a level above the operating B plus.

A ten-volt drop in the a.c. supply reduces the B plus about 40 volts with the doubler supply, while the full wave center-tapped supply voltage drops almost 55 volts.

When silicon diodes are used in power supplies, whether full wave or doubler, there are real improvements. The superior regulation and ability to handle surges have two big advantages. Higher amplifier power can be obtained, and transient peaks have little or no effect on the supply voltage as recovery time is so short. The supplies have to be designed as a unit, as the voltages furnished by the transformer should be lower than when tubes are used. So it will not be possible to simply convert to

Fig. 8. Underside view of the 20-watt amplifier. Note that the output transformer, at left front corner, is mounted on a bracket to reduce above-chassis height and to allow assembly into the cover.

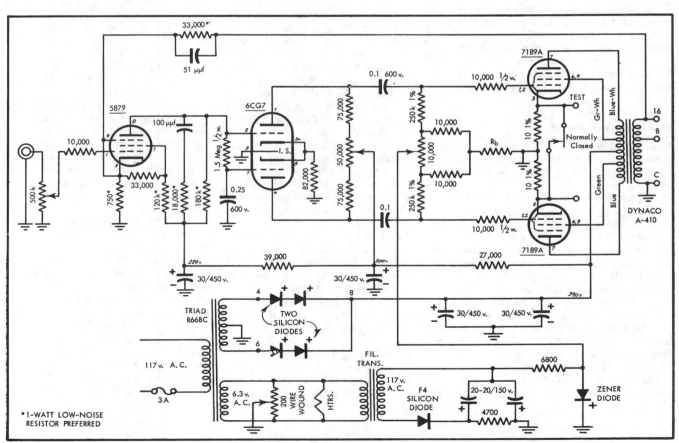

Fig. 9. Schematic of a 20-watt amplifier using a Zener diode bias regulator, and the new 7189A output pentode.

diodes unless wasteful voltage dropping resistors or chokes are added, and the penalty of somewhat poorer regulation is accepted. Of course there are many ways to mount diodes, and only the layout requirements control the method, for heat is no problem at audio power levels.

Diode Regulators

There is a form of silicon diode that can have valuable application in audio circuitry, and that is the Zener diode. These are silicon diodes with a somewhat different inverse voltage characteristic that makes them excellent voltage regulators.

Referring to *Fig.* 5, as an inverse voltage is applied to a Zener diode, there is a point where diode current will increase a great deal with a minute voltage increase. This is the Zener, or regulating point. Of course the maximum current must be limited. Zener diodes operate like gas regulator tubes, but usually at a lower voltage and higher current. They are small units like rectifiers, and may be connected in series for higher voltage regulation.

Figure 6 illustrates a simple and accurate d.c. voltage standard for shop use. An International Rectifier MZ27 Zener Diode was used. Its individual calibration chart, furnished with the diode, indicated regulation at 28 volts with 8 ma through the diode. Resistance R was selected to give the 8 ma with the diode shorted out. A wire-wound resistor was used for safety and stability. At ambient shop temperatures, the unit is simply turned on and meters are checked at the output. International Rectifier Corp. publishes an Engineering Handbook that covers the field quite thoroughly.

New Equipment Design

Now that we have covered many aspects of silicon diode power supplies, let us design a supply for a new amplifier.

The recently announced General Electric Company's 7189A output pentode seems worth a trial. The output tube is similar to the popular EL84, but G.E. has used a new heat-conductive laminated plate (anode) material, and has used all the spare base pins to act as heat sinks for the grids and plate. The tube will stand 440 volts on the plate. The screen will take 400 volts, or even 415 volts with a tapped screen arrangement. The rated voltages will give an output of 20 to 25 watts, which should be about right for most systems.

The popular Mullard circuit, considerably modified through long experiment, is extremely satisfactory, and was chosen. The old Ampex front end, the low-phase-shift 5879, needs about 210 volts for lowest distortion. The long-tailed-pair splitter using a 6CG7 in high-load operation requires 300 volts. The two first stages draw about 4 ma. The 5879 can be, and usually is, a very good tube, but there is an occasional variable that can be very bad, but these "sports" seem less frequent lately.

Depending on the operating mode, class B or AB, the 7189A push-pull pair will draw from 15 to 80 ma at zero signal, and about 125 ma at full drive. Thus a total power supply drain of 150 ma at steady peak drive will suffice for a monophonic amplifier.

To allow a small safety factor, a B plus of 390 volts at zero signal seemed about right. So if there were no power transformer and rectifier diode losses at zero signal, and if we didn't have to draw some current, the secondary voltage for a full wave center-tapped supply would be 390/1.414, or 275 volts on each side of ground for a center-tapped supply. The transformer would obviously have to supply a bit more than 275–0–275 volts, but just how much more was the question.

A 400–0–400 volt power transformer was available for tests. The primary was fed through a 10-ampere Variac to allow exploration of secondary voltage level. Silicon diodes were temporarily connected, a 30–30 µf, 500-volt capacitor was paralleled for 60 µf and was placed across the output. An 80-ma load at 390 v. d.c. would be: 39,000/8 = 4875 ohms, so a string of power resistors was wired to make the test load 5000 ohms, and the experimental supply was fired up. The desired 390 v. d.c. was obtained with 288 volts a.c. on each side of the center tap. A Triad R66BC transformer would fill the bill, as it is rated at 290–0–290 at 270 ma., furnishes 6.3 v. a.c. at 10 amps, and has a 5-volt 3-amp winding which is not necessary with silicon diodes. The transformer would require a rather large rectangular chassis cut-out, but aluminum is easily cut and filed.

The center-tapped full-wave supply was selected, as the amplifier was to be packaged in a California Chassis LTC 469, and space, especially height, was controlling. The output transformer selected was a Dynaco A410, which matches the 7189A's required 8000-ohm load, and will give 25 watts at low distortion.

Now for bias. There are real advantages to be gained in fixed-bias class AB operation, and accurate bias voltage assures consistent operation. An International Rectifier Corp. MZ18 (1N1515) Zener diode could be arranged to regulate the negative bias voltage to a balancing network. A bread-board test indicated that the desired negative 14 volts would hold without variation as the a.c. supply varied from 80 to 130 volts, so here was stability. A Triad F13X transformer has its 6.3 volt winding connected to the 6.3-volt filament supply of the amplifier. The secondary high voltage is rectified with a Sarkes Tarzian F4—another tiny silicon diode unit—and is filtered and fed to the regulating resistor and Zener diode. R_b was selected to give the required −14 volts, and 5000 ohms proved correct with a 19.5-volt regulated source.

The amplifier is shown in *Figs.* 7 and 8 and the circuit is shown in *Fig.* 9. The measured results were:

Power at clipping	22 watts
I.M. distortion at 20 watts	0.6 per cent
I.M. distortion at 5 watts	under 0.1 per cent
B plus at zero signal	390 v. dc
B plus at 20 watts	382 v. dc
Output tube I at zero signal (0.4V at test points)	80 ma
Input signal for 20-watt output	0.6 v. dc

This amplifier has operated daily for a number of months. Periodic measurements have been identical to the initial tests, indicating a very stable design. 7189A tubes operate quite hot, and proper ventilation is required.

This circuit is an adaptation of that described by the author in the March, 1959, issue of AUDIO, and that article covers component and circuit considerations that can be applied to this smaller unit. The 6CG7 may be replaced with a 12AU7/ECC82 with excellent results.

The Triad R93A power transformer can be connected to supply the proper doubler voltage for a stereo pair of the amplifier circuits, and will handle them at full power. In this case the optional stereo-decoupling of *Fig.* 4 has proved satisfactory.

Diode failure is possible, although theoretically it should be extremely rare. Out of sixty diodes in various circuits the author has experienced one failure, and this after 350 hours of operation. The diode was hot, and found to have 100 ohms resistance in both directions. Its mate in one doubler leg continues to operate properly. Other diodes in doubler and center-tapped supplies and as bias rectifiers have given complete satisfaction. Improved manufacturing methods, including diffused junctions, will make diodes more reliable. Theoretically, at least, they should last for years. Æ

Equipment Failure Alarms

ALLAN M. FERRES

A failure alarm will prevent expensive and embarrassing "down time" for equipment which operates without constant supervision.

ELECTRONIC EQUIPMENT can be designed and built with a high degree of reliability. Failures, however, still do occur. Unless the equipment is continually monitored, some time may elapse before the failure is detected and appropriate corrective action started. This undiscovered "down" time of the equipment can be both embarassing and expensive.

Described here are three failure alarms, flexible enough in design so that they can be used with several types of equipment under various conditions. These alarms will activate a buzzer or other warning device when the equipment to which they are connected fails. Corrective action can then be started without undue delay.

The alarm buzzer should be powered by a battery so that a failure of the alarm unit's power supply, or its a.c. line, will not prevent the buzzer from sounding. Under usual conditions, the operating life of the battery is equal to its shelf life. Two of the circuits are arranged so that a tube or other part failure in the unit will cause the alarm to sound. This "fail safe" design adds considerably to the reliability of the warning system.

Figure 1 is the circuit of an alarm used to monitor a 24-hour-a-day recorder. This tape machine is fed from a tuner to provide a continous off-the-air check of a broadcast station's program material. It can be used to monitor any continually operated amplifier or tuner.

When no signal is applied to the input of the unit, the bias provided by $R8$ and $R9$ cuts off $V2b$, and the normally closed contacts of the relay remain

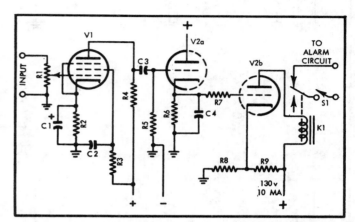

Fig. 1. Alarm for a 24-hour-a-day recorder.

closed. These contacts are wired, through $S1$, to operate the alarm buzzer. The output of the recorder or amplifier to be monitored is connected to the input terminals of the alarm. This output is amplified by $V1$ and then fed to $V2a$, a power detector. The signal voltage causes the cathode of $V2a$ to become more positive and this cathode voltage, applied to the grid of $V2b$, overcomes its cut-off bias. $R7$ is a grid-current limiting resistor. When $V2b$ draws plate current, the relay operates, opening the normally closed contacts and the alarm buzzer circuit.

If the input signal falls below a predetermined level, the relay drops out and the alarm sounds. The time constant of $R6$ and $C4$ provide a ten-second delay before the alarm sounds after the dropping of the input signal. This prevents the normal pauses in a radio broadcast program from sounding a false alarm. If a longer time delay is required, the capacity of $C4$ should be increased. If a balanced line is to be monitored, a suitable bridging transformer, of course, must be added to the input circuit.

With $R1$ set at maximum, the sensitivity of the unit is high enough so that a signal of 12 mv or greater will prevent the alarm from sounding. The relay will drop out if the signal level falls to 4 mv.

The ratio of pull-in and drop-out signal voltages can be controlled by adjusting the relay armature spring.

To set up the failure alarm for operation, connect the input terminals across the output of the amplifier to be monitored, and set the input level control, $R1$, full on. Feed a tone into the amplifier and adjust its level so that the minimum acceptable signal voltage appears at the output of the amplifier. A level 20 db below the normal operating level is usually satisfactory. Slowly reduce the setting of $R1$ until the alarm sounds. The unit is now ready for use. The alarm will stop as soon as normal program level is fed through the amplifier.

If it is necessary to change the output level of the amplifier frequently, adjustment of the alarm unit can be made more quickly if $R6$ is shunted by a 1.0 µf capacitor and a switch connected in series with the positive lead of $C4$. Opening this switch will eliminate the ten second delay while the unit is being adjusted.

Figure 2 is the circuit of a carrier failure or Conelrad alarm. It was designed to be used with a receiver in an

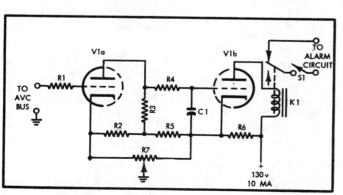

Fig. 2. Carrier failure or Conelrad alarm.

125

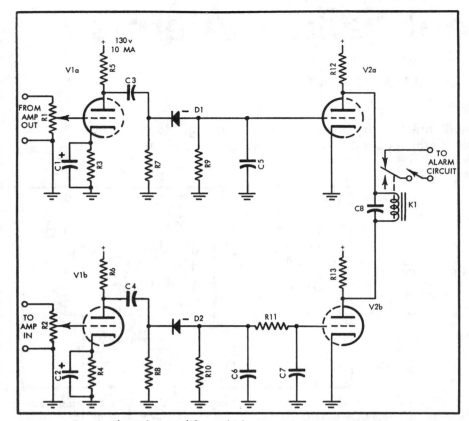

Fig. 3. Alarm for amplifiers which are not in constant operation.

electrically noisy location. The local noise level was so high that the a.v.c. voltage produced by the transmitter carrier was only slightly greater than that produced by noise when the carrier failed. This design is necessary because most carrier alarms are not sensitive enough to detect small changes in a.v.c. voltage. It can also be used with receivers which do not produce enough a.v.c. voltage on weak signals to operate other types of alarms. Its sensitivity is such that only a 0.5 volt change in control voltage is required for reliable operation, and it can handle a wide range of a.v.c. voltages. Its use does not upset the normal operation of the receiver.

The input of the unit is connected to any convenient point on the a.v.c. bus of the receiver. The negative-to-ground voltage, which is developed on the bus by the received carrier, is applied to the grid of $V1a$. A positive-to-ground voltage is applied to the cathode of $V1a$ by $R7$. $R7$ is adjusted so that $V1a$ justs cuts off. For example, if the grid receives a minus 15 volts from the a.v.c. bus when the carrier is tuned in, $R7$ is set to provide a positive 14 volts on the cathode. This places the grid 1 volt negative, with respect to the cathode, cutting off the tube.

With $V1a$ cut off, there is no voltage drop across its plate load resistor, $R3$. Then bias on $V1b$ is only the 0.75 volts developed across $R5$ and the tube draws enough plate current to pull in the relay. The relay contacts are wired to the alarm buzzer so that the alarm is silent with the relay pulled in. When the carrier goes off the air, the voltage on the grid of $V1a$ is less negative and current flows through its plate load resistor. This produces a voltage drop across $R3$ which is negative at the plate end of the resistor. As $R3$ is in series with the grid circuit of $V1b$, it is cut off and the relay drops out, sounding the alarm.

$R4$ and $C1$ are used to provide a ½ second delay in the operation of the relay so that instantaneous carrier breaks, which might be caused by arc-overs at the transmitter, will not operate the alarm. If these short carrier breaks are to be monitored, $R4$ and $C1$ can be omitted.

If the alarm is used with a transformerless receiver, an isolation transformer *must* be used in the power line cord of the set. The receiver's B-minus lead should be connected to the receiver chassis, then this becomes the ground point which is connected to the alarm unit.

To set up the alarm for operation, tune the receiver off the carrier to be monitored (and off any other station), and slowly turn $R7$ from its cathode end to the point where the alarm sounds. When the carrier is tuned in, the alarm will be silent.

The alarm unit shown in *Fig. 3* is used with amplifiers which are not in constant operation. It was designed for use with a remotely controlled tape machine used for making short radio spot announcements. It can also be employed with banks of tape machines used for tape duplication, or with power amplifiers of paging and music systems.

The input to the equipment to be monitored is amplified by $V1b$, rectified by $D2$ and this rectified voltage is used to cut off $V2b$. The output of the equipment is similarly amplified by $V1a$, rectified by $D1$ and cuts off $V2a$. When both an input and output voltage are present, both sections of $V2$ are cut off, and there is no difference of potential between the plates of $V2$, relay $K1$ does not pull in. When there is neither an input or output signal, both sections of $V2$ are drawing maximum plate current and still no voltage appears across the relay. If a signal is fed into the equipment, and, due to some circuit or operating fault, no signal appears at its output, $V2$ is cut off, $V2a$ draws maximum plate current. This places a voltage across the relay and it pulls in, operating the alarm circuit.

When this alarm is used with a tape machine having separate record and playback heads and amplifiers, $R11$ and $C7$ must be added to the circuit if a momentary operation of the alarm is to be avoided at the start and end of each recording operation. Due to the spacing of the record and playback heads, a delay exists between the time the signal is fed into the record amplifier and the time the recorded signal appears at the output of the playback amplifier. At the end of each recording an equal time delay exists between the end of the recording signal and the end of the played-back signal. The length of this delay depends upon the tape speed and the spacing between the heads. $R11$ and $C7$ delay the operation of the relay by the input signal long enough to prevent the alarm from sounding when a recording starts and stops.

The values of $R11$ and $C7$ must be determined by experiment for each type of machine and tape speed. The values listed are suitable for an Ampex 350 operating at 7½ ips where the delay is about 0.25 seconds.

To place this alarm unit in operation,
(*Continued on page 131*)

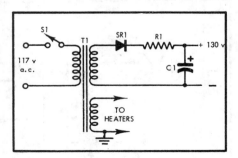

Fig. 4. Power supply adequate for any of the failure alarms.

FM Sweep Alignment Unit—
Austerity Model

ALLEN R. GREENLEAF

To receive FM-stereo in low signal areas it is important that the tuner be properly aligned. Here is a sweep alignment unit which can be built from your junk box.

AFTER SUCCESSFULLY assembling a commercial 3-in. oscilloscope kit, the writer had a hankering for a sweep generator with which to align the intermediate frequency transformers and discriminators of my FM tuner. The "Junk-box Alignment Unit" (AUDIO, December 1960) provides data that can be plotted to provide a frequency response curve, but with a sweep generator, the frequency response curve appears directly on the oscilloscope screen.

It was decided to make the width of sweep 10.2 to 11.2/mc, which is 500 kc on each side of the 10.7-mc center frequency; this should be more than ample to check the response curve of any FM tuner. By restricting the required sweep to this range it is possible to omit much of the switching that is required in commercial sweep generators that are designed to cover a number of bands. It

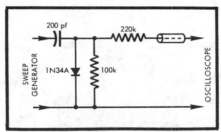

Fig. 2. Schematic diagram of demodulator probe.

is unnecessary also to obtain the sweep frequencies by using both a fixed-frequency oscillator and a tunable oscillator to produce a beat frequency; a single oscillator tunable over a frequency range somewhat greater than 10.2 to 11.2 mc is sufficient.

A considerable amount of literature, both books and magazines, was read, to learn as much as possible about the construction of sweep generators. One type utilizes a small loudspeaker "skeleton" connected to the 6.3-volt line which changes the inductance of the sweep circuit by moving an iron slug in or out of a coil at a rate of 60 cps. Another variation of this moves a plate or plates of a capacitor to change the capacitance of the circuit at the 60-cps rate. Use of either of these methods presented too many mechanical difficulties for a tyro constructor. A sweep generator based on a reactance tube is mechanically simpler, but looks rather complicated electronically. Then information appeared about the new voltage-variable capacitors; this looked like a simple and practical method of sweep-tuning an oscillator. After I obtained such a unit experimentation started on an empirical basis. The original intention was to use the "Junk-box Unit" as a marker generator for the sweep generator; but later I decided to build a duplicate marker generator on the same chassis with the sweep generator.

My junkbox supplied the metal case, chassis, power unit, 12AU7 and 12AV7 tubes, and a number of odd parts from a discarded commercial r.f. generator. Because the junkbox contained only a single-gang capacitor of suitable capacitance range, the marker generator is connected as a Hartley oscillator; otherwise it is essentially the same as the "Junk-box Unit." The only required parts that were not available in the junkbox were the Varicap (PSI, V-20) a new 8- by 10-in. aluminum panel to fit the r.f. generator case, and, as a later addition, a 10.7-mc crystal and its socket.

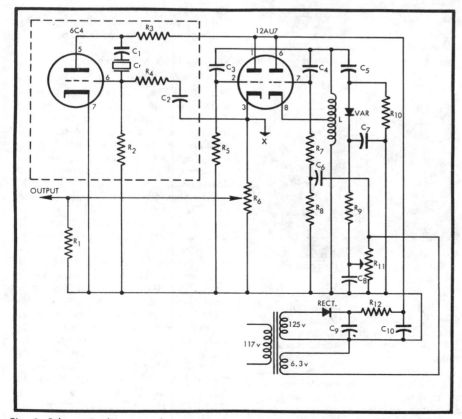

Fig. 1. Schematic diagram of sweep generator and 10.7 Mc fixed marker oscillator.

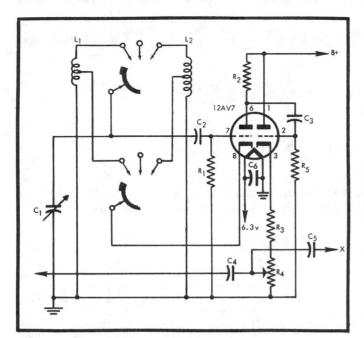

Fig. 3. Schematic diagram of variable marker generator.

Sweep Generator

Figure 1 is a schematic diagram of the sweep generator plus the 10.7-mc crystal oscillator; the dashed lines enclose the portion of the diagram that pertains to the crystal oscillator. One triode section of the 12AU7 tube is used as the sweep oscillator; the other section is connected as a cathode follower for the sweep output. Output voltage is controlled by potentiometer, R_6, and output is through a 50-ohm concentric cable, terminated by the 51-ohm resistor, R_1, at the panel connection, and by another 51-ohm resistor at the output end of the cable.

The oscillating circuit comprises inductor L, the Varicap, and the fixed capacitor C_5. The values of L and C_5 determine the maximum range through which the Varicap can vary the resonance frequency of the circuit. The inductor L was wound with No. 26 enameled wire on a slug-tuned coil form of 3/8-in. diameter, with enough turns (37) to provide a sweep range greater than the desired 10.2 to 11.2 mc. The tap is at 20 turns from the ground end, but that is by no means critical. With the oscillating circuit completed, it was found that the flatness of the output was determined primarily by the value R_{10}. Various values of R_{10} were tried, until a nearly flat output was indicated on the oscilloscope screen; 8200 ohms was found suitable, although that value is not critical. Using that value of resistance, the slug of the coil was adjusted slightly until the pattern on the oscilloscope screen showed the fixed 10.7-mc marker in the center of the pattern.

The desired width of sweep is obtained by adjusting the potentiometer, R_9. The value of R_9 is not critical but

Fig. 4. Front view of sweep generator.

Fig. 5. Rear view of generator.

should be high enough so that it does not impose undue current drain on the 6.3-volt filament transformer; the 900-ohm load shown draws only 7 ma.

Plate voltage on all tubes is about 90 volts.

The sweep of the sweep generator and the sweep of the oscilloscope must be in phase. The writer's oscilloscope provides an internal sinusoidal sweep that is brought into play by setting the "sweep selector" to LINE; then correct phasing is effected by use of the oscilloscope SYNC/PHASE control. If the oscilloscope that is to be used with the sweep generator lacks the internal sinusoidal sweep, a separate phasing device must be used, or one must be incorporated in the sweep generator structure. In the initial setup of the sweep generator it was found that a double marker appeared. Finally a literature reference was found that showed a network similar to the one which comprises R_7, R_8, and C_6. Addition of this network produced a single marker pip, and at the same time produced return-trace blanking.

In the preliminary testing, the sweep generator was connected to the oscilloscope through a demodulator probe, because the frequency response of the oscilloscope extends only to 5.5 mc. *Figure* 2 is the diagram of the modulator probe that was used. If no probe is available, a similar network can be made for the purpose.

Fixed 10.7-Mc Marker

The fixed-frequency 10.7-mc crystal marker oscillator is coupled to the sweep oscillator through R_4 and C_2. If the coupling is too tight, the 10.7-mc pip is so large that it distorts the oscilloscope pattern, and the smallest capacitor available, 2.2 pf, still produced excessive coupling. Insertion of resistor R_4 in series with a C_2 of 10 pf reduced the size of the pip to usable proportions. The value of R_4 was determined by cut-and-try; perhaps a variable resistor for R_4 might offer some advantages.

Variable Marker Generator

This is a Hartley continuous-wave oscillator (*Fig.* 3) that is used to put a marker on the response curve at any desired point over the range of 10.2 to 11.2 mc. It is coupled to the sweep oscillator at the point marked X on both diagrams. In *Fig.* 3, the inductor L_1 covers the 10.2 to 11.2 mc range; inductance L_2 covers a range of 29 to 36 mc, so that the third harmonic covers the FM range of 88–108 megacycles. The center position of the 2-circuit, 3-position rotary switch is left blank, so that it is possible to view the response curve on the oscilloscope screen with only the fixed 10.7-mc center marker in place. With the switch in this position, the sweep width and output, and the controls of the oscilloscope (connected to the sweep generator through the modulator probe) are adjusted to give a symmetrical curve of desired height and width. Then the switch is turned to the 10.2 to 11.2 mc position, and the output of the marker generator is adjusted to produce the smallest marker

(Continued on page 136)

"Ersatz Stereo" Unlimited

C. H. MALMSTEDT

A multichannel monophonic system that gives stereo some impressive competition.

IN A RUSTIC HOME in California there is a monophonic hi-fi installation that, in results achieved, matches the grandeur of the country around it.

"As good as stereo!" the system has been acclaimed. And, by a visiting symphony conductor.

"Magnificent! It is as though I were standing on the podium, the orchestra right here before me!"

Whether or not these accolades are extravagant, the fact remains that "monophonic" applied to this installation is as pleasantly deceptive as the name "Erosion Acres" is for the home and grounds that house this audio system created by its owner, Mr. Harwell Dyer of Carmel Valley.

"How," all ask, "do you get such marvelous sound from an installation that looks so simple?"

The answer is: growth, of more than twenty years' duration; growth born of a constant desire for improvement—of the technical facilities and of an understanding of music, a knowledge of the composition of the sounds that were to be reproduced with the best possible fidelity; things, in fact, that are not come by cheaply, in either time or money.

Considered by the standards of today's hi-fi, Mr. Dyer's beginning was, however, a modest one. A Gilfillan radio-phonograph with a Garrard changer handling only 78-rpm discs was, back in 1938, the first nucleus of the system. Today, the Carmel Valley installation consists of six speakers and twelve electronic units housed in five unit-locations

Fig. 2. The music corner, with part of the large library above and below the turntables. Note the home-built turntable on the left.

in and about the large living room, but so placed that, while everything is readily accessible, little is in evidence to mar the furniture grouping, the decor of the room, and the magnificent view from it.

Interestingly, the original heart of the amplifying system still serves as one of the power amplifiers—proving that the best is always in the long run the cheapest, and that modification intelligently applied can obviate the too-often-assumed necessity for discarding good units merely because of age. Designed and built in 1946 by Dyer and William Hilchey, this 300-watt amplifier utilizing 6L6 tubes feeding two 845's, in push-pull, employed the best components then available. Originally part of a 300-pound rack-mounted composite unit, this amplifier was later converted by James Meagher into a Williamson class-A amplifier of 150 watts full power and 75 watts distortionless output. Following the same desire to preserve and effectively use the worthwhile and the proven, an Altec 515 woofer also was salvaged from the earlier installation (where it had been in an infinite-baffle arrangement stabilized by a half a ton of concrete within a wall) for modification and use in the present system, as was an Altec 604-B coaxial speaker system, along with associated crossover networks.

With these and other units as a starting point, the present unique installation got under way in 1951, many years before the advent of commercially available stereo. As with all lovers of fine music, the goal of the Dyers was not only high fidelity but as well concert hall realism. It soon became apparent that

Fig. 3. Electronic cabinet, formerly a closet.

to get this realism, more was needed than a judicious placement of good speakers. Dyer went back to the first-things-first principle: he decided, first of all, to design and build his own turntable. The result leaves little to be desired, even in these days of many fine commercial turntables.

While few may care to go this far in a do-it-yourself endeavor, this home-made turntable is worth looking at before a view of the entire system.

Constant speed, free of vibration influences was the aim. Parts were picked up here and there. In a wrecking yard was found a 65-pound, 16-inch diameter,

Fig. 1. Not stereo —but magnificent music . . . a magnificent view . . . peace.

Fig. 4. Home-made 16-inch turntable.

two-inches-thick balance wheel once used in a saw-mill—massive enough and heavy enough to resist vibration. To one side of this wheel was bolted a ½-inch-thick disc of plywood. Over the plywood a ½-inch disc of neoprene was then glued. This combination became the turntable. The problem of a motor to turn this table was solved by a Green Flyer motor of the type used in broadcasting station transcription turntables. Set in an "H" saddle constructed at home, the motor was placed on a foam rubber pad within a cabinet under built-in bookshelves. To assure vibration-free drive of the turntable, sections of the drive-shaft were separated by Lord rubber couplings, with a free-wheeling device in one section of the drive-shaft. As a precaution against overheating of the motor during long use, a small rubber-bladed fan of the type used in automobile interiors was added to the motor compartment. To render its operation inaudible, its speed was reduced by the use of series-connected light bulbs, which also conveniently illuminate the enclosure during operational inspections. There was one problem: how to start this heavy turntable spinning without asking the motor to do it. Solution: to

Fig. 5. Midrange horn disperses sound throughout the large room, augmented by a woofer at floor level.

the drive-shaft was fastened a six-inch length of stiff wire protruding straight out; to the end of this wire a small magnet was attached; at one point on the travel-radius of this magnet a microswitch was so placed as to be actuated when the magnet passes it the first time. Result: manually start the turntable slowly—and within one turn or less the microswitch applies the current automatically, and the table smoothly works up to the speed it was set for—33⅓ or 78 rpm.

Both the construction time and the time required to attain full speed (about a minute) are more than amply justified by the performance. "If there is any rumble in evidence, it is inherent in the recording, not in the turntable."

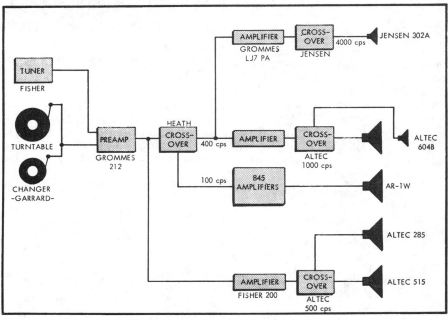

Fig. 6. Block diagram of the four-channel system.

With a GE magnetic cartridge and 16-inch transcription pickup arm on this turntable, it became again a case of one thing leading to another: where a good audio system had inspired the quest for a better turntable, the turntable thus developed now led the way to a demand for an even better audio system: "The best monophonic thing I've ever heard," said one listener.

But it was a desire for a stereophonic kind of realism that brought about the "unorthodox" use of crossover networks that is one of the keys to the success of this system. This, in turn, was brought about by the physical characteristics of the house the Dyers purchased in Carmel Valley. With walls of unsurfaced concrete block, a floor of smooth, waxed cement, and a large picture window in one wall, plus a rather high beamed ceiling, the living room was obviously a "live" one. A member of the infinite-baffle school, Mr. Dyer set to work accordingly—to assure, first of all, adequate but natural reproduction of bass.

More concerned with fidelity than efficiency, he decided to use the big home-made power amplifier to feed an AR-1W speaker in its infinite-baffle enclosure, placed on the floor at one end of the room. With the feed originating at the home-made turntable, the Garrard changer, or a Fisher FM tuner, the AR-1W woofer receives its input through a Grommes 212 preamplifier feeding a Heathkit electronic crossover, the Low output of which was set to decline at 100 cps.

With one bass-response channel thus established, another set of speakers—an Altec 604-B coaxial and a 515 woofer—were mounted on a common infinite baffle half the size of a large closet door. To accommodate this baffle, the door was removed from a closet at the same end of the living room that holds the AR-1W. Bottom half of this 6-feet-deep closet was partitioned off as a housing for the rack of major electronic units and power supplies, with the lower half of the door cut vertically in the center to provide two flap doors that could, without jeopardizing appearance, be left ajar for ventilation. The upper half of the closet thus vacated was lined with absorbent padding and utilized as an enclosure for the two speakers. To the same baffle now was added a Jensen 302-A "bullet" tweeter, thus making this a four-speaker infinite-baffle enclosure about six feet to the left of and about five feet above the AR-1W on the floor.

Input to this speaker system was now arranged through individual channels to which only the Grommes preamplifier and equalizing system are basically common. To feed the 604-B coaxial, the High output of the electronic crossover, set to pass above 400 cps, was fed into an independent power amplifier—a

Fisher 200—and from that through an Altec network with a crossover frequency of 1000 cps between the woofer and the tweeter. To feed the 515 woofer on the same baffle, another Fisher 200 power amplifier was fed direct from the Grommes preamplifier, with the output of this amplifier directed to an Altec network with a crossover frequency of 500 cps, the Low output of this unit feeding the 515 woofer. To assure even better bass response, this old 515 was modified to reduce its cone resonance to 23 cps. This was effected by running a Casco tool around the outside edge of the cone, sawing the spiders half-way through.

To feed the Jensen bullet tweeter, an independent Grommes LJ7 power amplifier is used, its input fed from the High end of the Heathkit electronic crossover and its output feeding the 302-A through a Jensen network with a crossover frequency of 4000 cps.

With the lowest and the highest ranges thus provided for, attention was next directed to the midrange. On hand was an Altec 285 multicellular midrange horn. Properly placed, this would assure not only good response but also adequate dispersion of sound. With an ear to the elimination of gaps, several locations were considered. The final choice: ceiling level of the corner to the far right of the AR-1W floor woofer. With this horn angled at about 30 deg. and pointing corner to corner, dispersion is well-nigh perfect, and virtually free of the influences of parallel reflecting surfaces.

Feed for the midrange horn was taken from the High output of the Altec crossover network the Low end of which was used to feed the 515 woofer, all frequencies above 500 cps therefore being directed to the midrange horn.

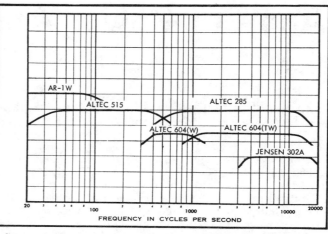

Fig. 7. Frequency response of the 6 speakers.

System response? Here it is, per speaker channel: the AR-1W, flat up to almost 100 cps; the 515, flat up to almost 500 cps; the woofer of the 604-B, flat from about 400 cps to almost 1000; the 285 horn, flat from about 500 cps to about 10,000; the tweeter of the 604-B, flat from about 1000 cps to almost 14,000; the 302-A, flat from about 4000 to almost 20,000 cps.

With all the speakers in operation, the quality of sound is impressive. And little wonder, for, as a glance at the block diagram will show, here we have, in fact, a monophonic system of four channels, with three of these channels utilizing frequency separation before power amplification. Why this expensive departure from the conventional? Well, for one thing, as was said earlier, stereo was not only unavailable but generally unheard-of when the Dyers wanted concert hall realism. For another, Mr. Dyer reasoned: why not utilize the available speakers and equipment in such a manner as to amplify in each channel only the desired range, and thereby at the same time obtain the best possible control and flexibility!

Needless to say, a system of this kind can be even more expensive than stereo. But it shows what can be done to and with a mono system when one demands continuous improvement. It also shows how earlier units can be effectively combined with the latest. Possibly the system will in time give way to stereo. Meanwhile, there is nothing static about this system or the results it produces. Listening to it, however, one quickly forgets the experimental aspects, the techniques that are, and in all art should be, not the end but the means to it.

In that respect this installation has brought about another phenomenal growth—that of a music library which in size and quality can well be the envy and inspiration of many a professional endeavor. And here, heard on this unique system in the cozy livingroom overlooking the mountains, the cottonwoods and the river, the masters of music come truly into their own, with presence, fidelity, definition and perspective about which nothing of the ersatz is discernible.

Æ

FAILURE ALARMS (from p.126)

the input and output connections are made to the equipment to be monitored, and both controls turned to their full off position. With the amplifier operating at its normal level, adjust R2 so that the relay pulls in and operates the alarm. Next adjust R1 so that the alarm stops. If both sections of V2 and R12 R13 are closely matched, these settings will hold for a wide range of operating levels. The minimum input and output signal required for reliable operation is about 50 mv.

The power supply shown in *Fig. 4* is suitable for use with any of the failure alarms described.

PARTS LIST
(*Fig. 1*)

R1	500,000 ohms, audio taper pot
R2	6800 ohms, ½ W
R3	820,000 ohms, ½ W
R4, R6	470,000 ohms, ½ W
R5	2.2 megohms, ½ W
R7	220,000 ohms, ½ W
R8	1000 ohms, 1 W
R9	22,000 ohms, 5 W
C1	20 µf, 25 V elect.
C2	0.5 µf, 400 V
C3	.002 µf, 400 V
C4	30 µf, 150 V elect.
S1	s.p.s.t. toggle switch
K1	s.p.d.t. plate circuit relay, 8000 ohms, 1.6 ma (Sigma 4F-8000-S-SIL)
V1	6AU6
V2	12AT7

PARTS LIST
(*Fig. 2*)

R1, R4	470,000 ohms, ½ W
R2	1800 ohms, 1 W
R3	100,000 ohms, ½ W
R5	100 ohms, ½ W
R6	20,000 ohms, 5 W
R7	50,000 ohms linear pot
C1	1.0 µf, 200 V
S1	s.p.s.t. toggle switch
K1	s.p.d.t. plate circuit relay, 8000 ohms, 1.6 ma (Sigma 4F-8000-S-SIL)
V1	12AT7

PARTS LIST
(*Fig. 3*)

R1, R2	500,000 ohms, audio taper pot
R3, R4	4700 ohms, ½ W
R5, R6	470,000 ohms, ½ W
R7, R8, R9, R10	1.0 megohm, ½ W
R11	470,000 ohms, ½ W (see text)
R12, 13	10,000 ohms, 1.0 W (matched pair)
C1, C2	20 µf, 25 V elect.
C3, C4	.068 µf, 400 V
C5, C6	.1 µf, 400 V
C7	.5 µf, 200 V (see text)
C8	1.0 µf, 200 V
D1, D2	1N34
S1	s.p.s.t. toggle switch
K1	s.p.d.t. plate circuit relay, 8000 ohms, 1.6 ma (Sigma 4F-8000-S-SIL)
V1	12AX7
V2	12AT7

PARTS LIST (*Fig. 4*)

R1	270 ohms, 2 W
C1	20 µf, 250 V elect.
S1	s.p.s.t. toggle switch
T1	Power transformer 125 V secondary at 50 ma; 6.3 V at 2 amp. (Stancor PA 8421 or PS 8416 using half of 250 V secondary)
SR1	Silicon rect. (Sarkes Tarzian M-150)

Æ

Another Power Amplifier!

R. A. GREINER

Here's an amplifier that will loaf through the task of providing music throughout the house—and outdoors too. Also the distortion is extremely low.

THE READERS FIRST REACTION to this article on power amplifiers might very well be, as the title indicates, "Another power amplifier!" Still there are worth-while and interesting things to talk about with respect to vacuum-tube amplifiers. The amplifier described here and some of the design problems are just a bit unusual. In the first place, this is a very high-power amplifier in terms of the power levels which are usually used in high-fidelity systems. The amplifier itself is a relatively straightforward combination of well-tested audio circuits which have been presented in the literature over the years by various authors. This might well be the last of the vacuum-tube amplifiers as far as this writer is concerned. Perhaps a brief explanation of that last statement is in order.

For many years now it has been relatively easy to build essentially distortionless power amplifiers in the 50 or even 100-watt size. The only challenge has been to do it with fewer tubes, at smaller cost, and in the smallest possible space. It would seem that the next logical step would be to completely transistorize the power amplifier and thus make it even smaller and presumably more reliable. It is certainly a worth-while goal—but it is also very difficult. In fact, at the present state of the transistor art it is very expensive to make relatively high-powered amplifiers with wide frequency response and low distortion.

In the case of the amplifier to be considered here, it was felt that the vacuum tube could be used with standard components to give the required design characteristics in the quickest time and with the least amount of effort. There was need for an amplifier which would deliver 100 to 150 watts under continuous duty sinusoidal output over the entire range from 20 cps to 20,000 cps. This requirement plus the inherent challenge presented by designing and constructing a really good high-power amplifier are the primary reasons for the existence of the amplifier design and design discussion which follows.

The final specifications which were set down for the amplifier are summarized as follows: Power output 120 watts continuous duty, frequency range 20 cps to 20,000 cps, 200 watts for short periods (5 minutes at a time), less than 2 per cent total harmonic distortion at 200 watts and less than 0.5 per cent at 100 watts (over the entire frequency range). These goals were attained in the amplifier described here. More detailed performance characteristics are given at the end of this article.

This amplifier is of course much larger than would ever be required in even the largest home audio system contained in a single listening room. There are uses for such an amplifier in addition to those which led to its construction however. For example, it is very useful as a central amplifier for distributing music to an entire house and garden where a large number of individual speaker systems are to be driven. It is also valuable for high-quality commercial sound distribution systems which carry music as well as voice. Even at a 50-watt average

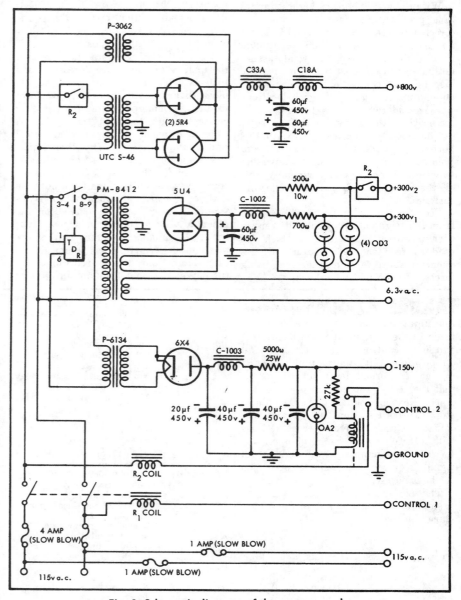

Fig. 1. Schematic diagram of the power supply.

Fig. 2. Top view of the massive power supply.

level this amplifier is loafing along with plenty of reserve for the peaks. The peak output is 400 watts continuous wave or almost 600 watts on the instantaneous peak power basis described below.

Description of the Circuit

Turning first to a description of the circuit, we see in the power supply, *Fig. 1*, a relatively direct, though involved, circuit. There is a bit of complexity due to the need for provision to allow the output-tube filaments and the rectifier-tube filaments to warm up before the main power transformer is turned on. This delay is provided by a time delay relay which delays the turning on of both the 300 volt supply and the negative supply used for biasing the output tubes. Thus the filaments are allowed to warm up for a minimum of thirty seconds after the power switch is turned to the "on" or "standby" positions. Only after the negative supply has reached full voltage and the bias for the output tubes is supplied will the high voltage to the plates and the screens of the output tubes be turned on. The bias current through the plate relay operates a double pole relay which turns on the main high-voltage supply transformer and in addition connects the screen supply for the output tubes to the regulated 300-volt supply which is already in the standby condition. It is convenient to provide two power switches. One closes relay No. 1 which turns the bias supply and the 300-volt supply on. This is then the standby condition. The second switch is in series with the plate relay and thus can be used to override the turning on of the main power if desired. In cases where no long periods of standby operation are required, the second switch may be eliminated and the normal turn-on sequence will be followed each time the amplifier is turned on. The completed power supply is shown in *Fig.* 2 and *Fig.* 3. The chassis is made of aluminum and is a standard 13 × 17 × 3-in. in size. A large amount of filtering is used to keep the total hum in the amplifier to very low levels. In addition, a choke-input filter is required for the main plate supply because the power transformer is is used very near its maximum ratings at the continuous output rating of the amplifier. The main power transformer, UTC S-46, will deliver 300-ma continuously which is enough for 100-watts output. At the 200-watt level, however, the amplifier requires 500 ma. This transformer will supply this current for short periods without difficulty. If continuous operation at 200 watts is required, it would be necessary to use a somewhat larger power transformer.

The driver sections of the amplifier are supplied with 300 volts from a pair of regulators. Two VR-150 voltage regulators are used in series for the driver stages and a second pair similarly connected are used for the screen supply. There is a considerable variation of the screen current in the output tubes as the amplifier is driven from quiescent to full output and hence regulation is almost essential for the screen supply.

The negative supply should, of course, be regulated since the biasing of the output tubes is critical for long-term low-distortion operation. In the present amplifier a single OA2 is used for this purpose with the power supplied by one side of the 300-volt supply transformer through a separate rectifier tube. A separate filament transformer is necessary to avoid any heater-to-cathode voltage difficulties in the 6X4 rectifier.

The power amplifier schematic is shown in *Fig.* 4. Notice that the filament transformer for the output tubes is on this chassis to avoid running the high current required for the tubes through a long connecting cable. The total current required for the output tubes is 6.4 amperes. The filament power for the driver and input stages is taken from a separate transformer winding because it was at first thought necessary to bias the filaments at some d.c. potential with respect to ground. This later turned out to be unnecessary so that they may be connected to the same transformer which supplies the output tubes.

All components are standard. Notice

Fig. 3. Neat, cabled wiring as shown is best practice for the power-supply portion of any amplifier.

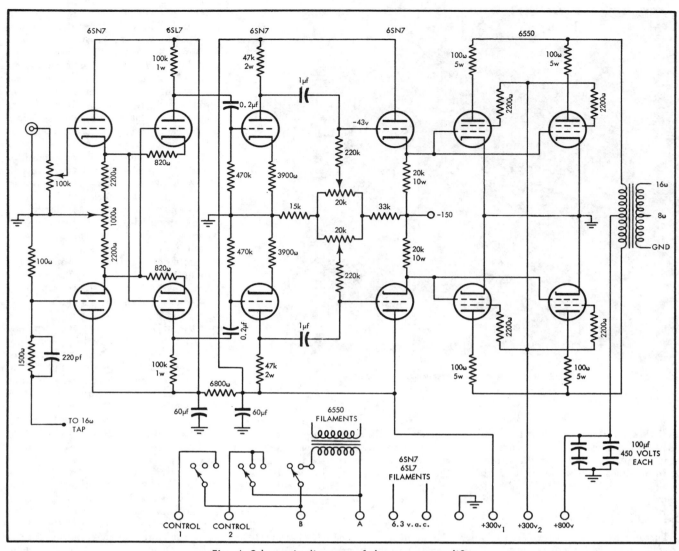

Fig. 4. Schematic diagram of the power amplifier

that there are only four capacitors in the entire amplifier. The high value coupling capacitors are physically quite large unless the "metallized" type is chosen. These may be seen in the photograph of the underside of the amplifier shown in *Fig. 5*. Straight point-to-point wiring is used throughout the amplifier. Good sturdy hook-up wire going from point-to-point and covered with spaghetti will hold its shape and give a neat and permanent job. Several capacitors which were added after the amplifier was completed can be seen on the underside also. These would be more conveniently mounted on the top. They should be protected from the direct radiation from the output tubes, however, since excess heat may be harmful to them. A pair of 20,000-ohm potentiometers are used to adjust the bias on the output tubes to an optimum value. A 1,000-ohm resistor is used for balancing the drive. Adjustment is best achieved by means of distortion measurements, but it is not very critical. Particular care should be taken with the output-transformer leads since they are at high potential and have 1200-volt peak-to-peak signals appearing on them at full output.

The bias voltage at the grids of the output tubes will be about −43 volts. The driver stage is capable of about 140 volts peak-to-peak drive and will thus drive the output stage slightly into the grid conduction region at full output. This will not cause trouble however, since the grids are cathode driven and there is no recovery time involved for bias changes which would normally take place if the output tubes were capacitor coupled to the driver stage. The output clips smoothly and there is no measurable recovery time from even extreme overload. Recovery time from ten times overdrive to one tenth maximum output is less than one cycle at 20,000 cps. Absolutely no signs of high-frequency oscillation or low-frequency instability are present under any signal conditions.

The regulation of the high-voltage supply will have a rather interesting effect on the maximum power output capabilities. The power available for steady-state operation is not as high as it is for sudden short impulses or bursts of sinusoidal signal. Thus one finds that there are different ratings possible for a power amplifier, depending upon the test procedure used. To clarify this problem, consider the following situation. Using the present amplifier as an example, we find that the power supply voltage is 800 volts with no signal, or with a very small signal, but that it drops to 600 volts at the 200-watt level. This drop in voltage does not occur immediately. If the capacitors which are used to filter the plate supply are very large as they are in this amplifier, it will take as long as 1/10 second to drop if full power is suddenly called for. Since the maximum power output possible depends on the plate voltage supply we find that the amplifier will supply considerably more than the steady-state power for a few cycles of applied signal even at low frequencies. The steady-state power in this case is 200-watts rms. This

Fig. 5. Bottom view of the amplifier, in contrast to the power supply, shows the diligent use of point-to-point wiring for all signal carrying connections. Note that all components are rigidly fastened down rather than supported on thin leads.

means that the peak power is 400 watts for a steady sinusoidal signal. However, if only a short burst of sinusoidal signal is applied it is possible to obtain almost $(800/600)^2$ times this power, or about 700 watts. This latter figure might be called the instantaneous peak power.

Now it would be folly to try to rate all amplifiers according to the suggestions discussed above. Manufacturers have difficulty enough with the present sets of advertising claims. It is interesting to note, however, that a rating on the instantaneous peak power available is not completely ridiculous. (At least one manufacturer uses this rating technique.) If we consider normal music for example, we find that the ratio of the average to peak power required for undistorted reproduction is in the neighborhood of 10 to 20. This means simply that the loudest tones during a given interval which is not too long will require a 10 or 20 watt amplifier if the average level is 1 watt. The system is thus required to pass high power peaks for only very brief instants and can spend most of the time recovering from these peaks. All of this consideration would tend to indicate that any amplifier would perform slightly better under music or noise conditions than it does when sinusoidal signals are used to test it.

When we use a sinusoidal signal to test an amplifier which is designed to handle music, we are in fact using a signal which is as much unlike music as possible. The reason for using sinusoids, of course, is that it is easy to do and easy to obtain consistant and relatively interpretable results.

This effect became of interest in the present amplifier design because it is somewhat more exaggerated in this amplifier than in a normal low-power amplifier. If the voltage regulation on the power supply were perfect this effect would not appear and the steady-state peak power would be the same as the instantaneous peak power. This would be an ideal situation which can be achieved in low-power amplifiers rather easily by means of electronic regulation. In the case of a 600-volt 500-ma supply, the added complexity of electronic regulation becomes almost prohibitive.

The filtering shown in this design consists of 4 capacitors of 100 µf each connected in series-parallel to give an effective value of 100 µf at 900 volts maximum voltage. The choke-input filter with a swinging choke at the cathodes of the rectifiers gives as good regulation as one can expect for a simple rectifier-filter combination.

As far as the steady-state output of this amplifier is concerned, it will deliver 100-watts continuously. That is, for hours or even days at a time. It will deliver 200 watts for periods of a few minutes. Periods longer than ten minutes have not been tried. But the output tubes and power transformer warm up considerably after this time.

The distortion for the amplifier for several output levels is shown in *Fig.* 7. The distortion at high outputs is easily measurable and is found to lie at about the level anticipated in the specifications. At low outputs the distortion is below about 0.2 per cent and is impossible to measure with the equipment available to the author. It is in fact very difficult to measure distortion levels below 0.2 per cent because of the residual distortion in even the very best oscillators. Special filters are required for more accurate measurements and these can usually be made only at single frequencies. The noise output of the amplifier, which includes thermal noise and hum, is less than 500 µv at the 16-ohm tap of the output transformer. This is negligible and may be attributed to the very heavy filtering in the power supply and the high voltage on the output tubes.

In conclusion, it is fair to say that the amplifier described here has been found to be a very reliable power source when driven by an oscillator for variable frequency applications as well as a very fine power amplifier for high-fidelity applications where extreme output is required. For the serious amateur it is a very useful addition to his stock of electronic tools.

Æ

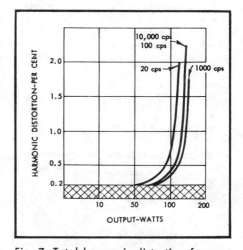

Fig. 7. Total harmonic distortion for several frequencies.

Fig. 6. Top view of the amplifier.

PARTS LIST

In addition to the standard parts which are shown in the schematic diagram, the following list gives the specifications for the transformers and filter chokes which may be replaced by equivalent types:

Output transformer
Dynaco A-450

Power transformers

UTC	S-46	2000-0-2000 volts, 300 ma
Triad	F-21A	6.3 volts, 10 amp
Stancor	PM 8412	400-0-400 volts, 200 ma
		5 volts, 3 amp
		6.3 volts, 5 amp
Stancor	P 6134	6.3 volts, 1.2 amp

Filter Chokes

Stancor	C 1002	15 henry, 75 ma
Stancor	C 1003	16 henry, 50 ma
Triad	C 33A	300 ma, 5/25 henry (swinging)
Triad	C 18A	300 ma, 8 henry

Relays

Potter and Brumfield, Type M LB-5	10,000-volt plate relay
Potter and Brumfield, Type KA11AY	110 volt d.p.d.t.
Potter and Brumfield, Type KL11A	110 volt d.p.d.t.
Amperite 115NO30T	time delay relay

FM Sweep Alignment Unit—
Austerity Model

(from page 128)

pip that can be readily observed. *Figure 4* shows the panel of the unit. The large dial that operates the tuning capacitor reads over 360 deg. but actually turns through its full tuning range in 180 deg. and is calibrated only over the 0 to 50 range of its markings. Both frequency ranges were calibrated by checking against FM broadcasts, as described in the article "Junk-box FM Alignment Unit," and a calibration curve was drawn for each range, rather than marking the frequency values on he dial. The paper vernier pasted on the panel implies the necessiy for reading frequency to a degree of precision that is actually not necessary; the FM tuner is aligned to produce a symmetrical curve of maximum height and width, with as little as possible depression in the center.

Fig. 6. Under-chassis view.

When the unit and the oscilloscope are connected to the a.c. line, it may be found that the variable marker is at the left end of the response curve when the dial indicates 11.2 mc; all that is necessary is to reverse the a.c. plug of either the alignment unit of the oscilloscope (but not both) to make the curve read from left to right. A panel output for the marker generator is provided (lower right on *Fig. 4*) in case it may be desired to use it alone as a CW generator.

Figure 5 is a rear view of the unit, and *Fig. 6* shows the under-chassis wiring. It was considered advisable to shield two of the oscillating tubes, but no other precaution was taken to prevent unwanted interaction of circuits, and no difficulty was encountered. The 12AV7 tube is the one without a shield. The unit must be secured firmly in a metal enclosure to prevent r.f. leakage.

In aligning an FM tuner, it is always advisable to follow the specific directions provided by the manufacturer of the tuner. In general, the grounded side of the sweep generator line is connected to the grounded side of the tuner; the "hot" side of the sweep generator line is connected, through a capacitor of about 100 pf, to the grid of the mixer tube of the tuner. In aligning the writer's tuner, the grounded terminal of the oscilloscope is connected to the grounded side of the tuner, and the "hot" lead from the vertical amplifier is connected, through a resistor of 100,000 to 250,000 ohms to point A on *Fig. 7* to align the i.f. transformers; then to point B to align the discriminator. The i.f. transformers are aligned, secondary first, starting with the transformer nearest the first limiter, and working back toward the tuner input. *Figure 8* shows the sweep generator connected to point A.   Æ

Fig. 8. Aligning an FM tuner.

PARTS LIST (*Fig. 1*)
R_1	51 ohms (optional)
R_2, R_3, R_5	10,000 ohms
R_4	1500 ohms
R_6	1800-ohm pot.
R_7	4300 ohms
R_8	72,000 ohms
R_9	300 ohms
R_{10}	8200 ohms
R_{11}	900-ohm pot.
R_{12}	1800 ohms
C_1, C_6, C_7, C_8	0.005 uf
C_2	10 pf
C_3	5 pf
C_4	30 pf
C_5	30 pf
C_9, C_{10}	20 uf
Cr	10.7 Mc crystal
Var	Varicap, Type V-20 Pacific Semiconductors, Inc.
$Rect.$	Rectifier, silicon or selenium

PARTS LIST (*Fig. 3*)
R_1	8200 ohms
R_2	2000 ohms
R_3	82 ohms
R_4	200-ohm pot.
R_5	22,000 ohms
C_1	23–35 pf max. (variable)
C_2	46 pf
C_3	5 pf
C_4	1000 pf
C_5	15 pf
C_6	2000 pf

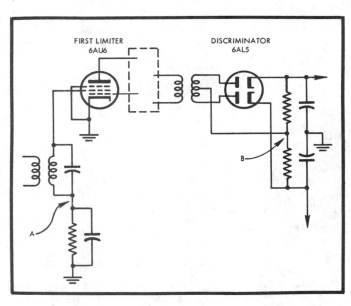

Fig. 7. Locations for aligning i.f. and discriminator.

Computers in Audio Design

R. G. BUSCHER

Through use of computers the audio engineer can materially reduce the amount of time he spends on routine computations and thus increase the amount of time available for handling design problems. Here is a description of the various computer types plus a specific audio design example.

IN RECENT YEARS, names such as PACE, 650, MANIAC, REAC, 704, and ESIAC have appeared more and more throughout our society. These are the designations given to the computers which are used in the areas of accounting, engineering, and research. Through the application of computers, time- and money-consuming procedures are being simplified. Each year more people come into contact with these applications. Utility bills, bank accounts, savings bonds, credit cards, income tax, government checks, and the paper work of many other everyday activities are handled by some sort of computer.

In engineering, the speed of data processing is of extreme importance. By freeing engineers from routine complex calculations so that they may go into new endeavors, these machines are stepping up the rate of progress.

Complete models of complex systems can be computer simulated for engineers to study. In this way the cost of optimizing a design can be reduced.

If a new system were to be built for each design change, the cost of development would be many times what it is now. Each time a design error was made a new system would be required. Computers, however, allow quick and easy changes in design. On computers, "mistakes" are indicated by means of signal lights, horns, or other such means. A flip of a switch will return the problem to its original state. The engineer can then make the necessary changes and start again.

Even the field of audio can benefit through the use of computers. The design or optimization of audio systems can be done on computers more quickly and more accurately than by hand methods.

By introducing circuit equations into computers, the laborious task of amplifier design, for example, can be made easier. Tube characteristics can be placed in the computers in order to determine the amplifier tube operating points. A more complex design in terms of component aging effects is readily accomplished by parameter variation. Changes in resistor, capacitor, and voltage values can be programmed to study the trade-offs between power, bias, and distortion.

Other audio areas such as speaker, cartridge and tone arm, and tone control design can be investigated in similar fashion.

While most of the computers in use today are built to perform a specific function, they all fall into one or the other of two classes: analog or digital.

While each class can do the problems handled by the other, there are basic differences which make necessary a choice of which type to use in a particular case. Such a choice is made on the basis of the problem and its requirements. Such factors as problem accuracy requirements, the number of parameters, and their changes enter into this choice.

The digital computer has its greatest use in numerical analysis work where precise bookkeeping-like routines can be set up. The analog computer has its application in the area of system analysis work where systems must be engineered and optimized.

In order to illustrate the differences between the two classes, each will be discussed in the following sections.

Since audio design work is of interest to the reader of AUDIO, the major emphasis will be on the analog computer.

The Digital Computer

The digital computer is a device that uses discrete steps to represent numbers while it performs mathematical operations. This is similar to the operation of an abacus. On the abacus, beads are used to represent numbers. Addition and subtraction are performed by the shifting of these beads. A similar procedure is used in electronic digital computers. In this case "bits" are used to represent the numbers. A number or quantity is changed to "bit" representation by the use of a code. The "bits" take the form of either the presence or the absence of a signal. The presence is denoted by the number "1" and the absence by the number "0" when setting up the problem. The over-all number or quantity is then represented by some combination of 1's and 0's, according to the code used.

By the use of Boolean algebra and other techniques beyond the scope of this article the various mathematical operations are performed.

The accuracy of these mathematical operations is limited only by the number of bits used to represent the quantity. Six decimal places of accuracy imposes no strain on a typical digital computer.

One of the main features of digital computers is the memory function. By the use of this memory a number can be stored in the machine until it is needed

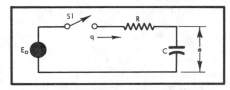

Fig. 3. Shock absorber analog.

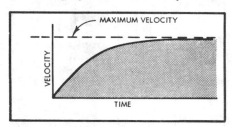

Fig. 2. Velocity of mass when the pin is pulled.

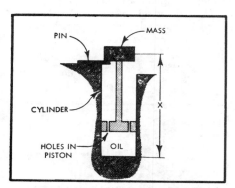

Fig. 1. Shock absorber schematic.

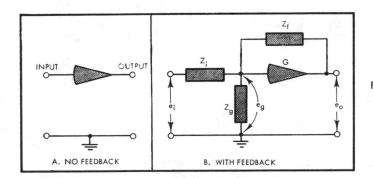

Fig. 4. Operational amplifier.

for computation. Upon being changed it can again be stored until further needed. For example, a bank account balance could be kept in a computer's memory. When the depositor makes a deposit, the teller, by pushing buttons, could call the depositor's balance from the memory. The new deposit could be added and the new balance put back in the memory where it remains readily available for future transactions.

These memories take three common forms: magnetic core, magnetic drum, and magnetic tape. The rate at which information is required in the computation determines which type of memory is used. The types are listed in order of decreasing accessibility. The core is the "fastest" memory. The bits take the form of a magnetized or unmagnetized core to indicate respectively the "1" or "0." The core memory immediately supplies the number it contains as often as desired. The magnetic drum is a metal drum coated with magnetic oxide. The bit is recorded as the presence or absence of magnetization in a particular area on the drum. The recording is done by heads similar to tape recording heads. The drum rotates at a high rate of speed. As the area passes under a fixed writehead a pulse is applied to magnetize or demagnetize the area as required. For readout, another head is used to detect the presence or absence of the signal as the area passes.

The information on the drum is available only once per revolution thus its "access time" is longer.

The tape storage is essentially the same as the drum except that a reel of tape is used. Some sort of searching technique has to be used to find the area in which the information is found, thus this is the slowest of the three methods described.

For such applications as missiles where high speed calculations are necessary, the core memory would be used. In the example of a bank account, the tape memory would probably be adequate.

The Analog Computer

The analog computer operates, as its name implies, by providing an analog of some physical process. One of the best examples of the analog computer is the slide rule. On the slide rule numbers are represented as lengths. The operations of adding or subtracting of lengths accomplish various mathematical operations.

In its engineering form the analog computer normally is used in the simulation of entire systems. While there are mechanical, electromechanical and electrical analog computers, this discussion will be confined to the electrical type.

In setting up an analog simulation the equations which represent the behavior of the actual physical system should be available. Electrical circuits are then made up which obey the same type or class of equations. Voltage and current variables within the electrical circuits

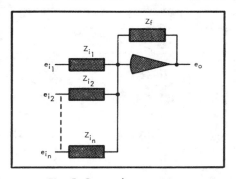

Fig. 5. General summation.

will behave in the same manner as do the variables of the physical system.

Changing resistors, capacitors, and voltages in the electrical circuit will then correspond to changing various physical parameters; mass, spring force, or length, for example. By making these circuit changes the engineer can easily optimize his design. The final circuit quantities can then be related back to the physical parameters.

The types of problem which are appropriate for an analog computer include aircraft simulation (of which the "Link Trainer" is a classic example), atomic reactor control, automobile suspension systems, and power transmission systems.

In order to show the theory of the analog computer it is necessary to have a problem. For the present, the problem to be discussed is that shown in *Fig. 1.* A mass is raised above some reference and held by a pin. The mass is connected to a piston which is immersed in a cylinder of oil. The piston has holes in it to allow the oil to flow through. This piston-cylinder combination is a typical shock absorber. What does the velocity of the mass become when the pin is removed? From experience it can be surmised that the mass will fall with increasing velocity until the oil is going through the piston holes as fast as it can. At this time the mass will have its maximum velocity and will keep this velocity until the piston strikes the bottom of the cylinder. The velocity will take the form shown in *Fig. 2.*

In order to set up an electrical analog of this problem it is necessary to write an equation of the system. Using conventional laws of mechanics this equation is:

$$M\ddot{x} - W + K_d\dot{x} = \Sigma \text{ Forces on mass} = 0 \quad (1)$$

where

M = Mass of the body
W = Weight of the body = Mg
K_d = Damping coefficient due to the oil
$\ddot{x}$ = Acceleration of the mass
$\dot{x}$ = Velocity of the mass
g = Gravity acceleration

solving for $\ddot{x}$ and substituting Mg for W yields: \hfill (2)

$$\ddot{x} = g - \frac{K_d \dot{x}}{M}$$

Now if an electrical circuit can be determined which obeys an equation of the same form the analog is found.

Using "20-20 hindsight" consider the circuit of *Fig. 3.* Writing the charge expression for the voltage drops around this loop when S1 is closed we find equation 3,

$$E_o = R\dot{q} + \frac{1q}{C} \quad (3)$$

where $\dot{q}$ is the current and q is the charge. Solving for $\dot{q}$ yields:

$$\dot{q} = \frac{E_o}{R} - \frac{1q}{RC} \quad (4)$$

This equation is of the same form as equation 2. As $\ddot{x}$, the acceleration, is the time rate of change of $\dot{x}$, the velocity, so is $\dot{q}$, the current, the time rate of change of q, the charge.

The analog relationships of the quantities are given below with constants inserted to conserve units.

$$\ddot{x} = k_x \dot{q} \text{ where } [k_x] = \frac{\text{feet/second}^2}{\text{coulomb/second}} \quad (5a)$$

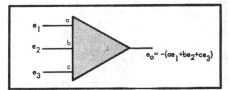

Fig. 6. Adder symbol.

$$\dot{x} = k_x q \quad \text{where} \quad [k_x] = \frac{\text{feet/second}}{\text{coulomb}} \quad (5b)$$

$$g = k_g' \frac{E_o}{R} \text{ where } [k_g] = \frac{\text{feet/second}^2}{\text{coulomb/second}} \quad (5c)$$

$$\frac{K}{M} = k_t \frac{1}{RC} \text{ where } [k_t] = \frac{1/\text{second}}{1/\text{second}} = 1 \quad (5d)$$

The small k constants are known as scale factors and as seen by their units are used to relate the various quantities of the actual and the analog system, k_t, is seen to be nondimensional. The numerical value of these scale factors depend on the expected range of the various variables.

It can be seen that the act of closing S1 in the electrical circuit is the same as the act of pulling the pin in the actual system. Further, it is evident that the actual system can be optimized by changing the electrical quantities until the desired operation is achieved. Then by using the scale factors the values of the actual system parameters that will cause the same response are established.

In analog computer work the basic building block is the operational amplifier. Through the use of RC networks in conjunction with the amplifier various mathematical operations can be formed.

The symbol most commonly used for the operational amplifier is given in (A) of *Fig.* 4. In (B) of that figure this amplifier is shown with an input impedance, Z_i, a feedback impedance, Z_f, and a grid impedance, Z_g. The amplifier gain is G. The input voltage is e_i, the grid voltage e_g, and the output voltage is e_o. An expression for the output voltage e_o will now be found by using Kirchoff's rule that the sum of the current into the grid point is zero. This leads to the expression

$$\frac{e_i - e_g}{Z_i} + \frac{e_o - e_g}{Z_f} - \frac{e_g}{Z_g} = 0 \quad (6)$$

The relation of the grid voltage to the output is given by:

$$e_o = -G e_g \quad (7)$$

$$e_g = \frac{1}{G} e_o \quad (8)$$

By combining equation 6 and equation 8 the output over input relation becomes:

$$\frac{e_o}{e_i} = -\frac{Z_f}{Z_i} - \frac{\frac{Z_f}{Z_i}\left(1 + \frac{Z_i}{Z_f} + \frac{Z_f}{Z_g}\right)}{G} \quad (9)$$

It is seen that if G is made very large, the second term vanishes leaving:

$$\frac{e_o}{e_i} = -\frac{Z_f}{Z_i} \quad (10)$$

This is the basic expression for use in analog computer work. The impedances can take various forms of RC networks giving a large variety of functional relationships between the input and the output voltages. These relationships are useful in complex problems.

The most common forms in use are the adder and the integrator. If more than one input impedance is added to the grid point as shown in *Fig.* 5, the output voltage will be the sum of all the inputs with gains determined by the impedance ratios as shown in equation 11.

$$e_o = -\left(\frac{Z_f e_{i1}}{Z_{i1}} + \frac{Z_f e_{i2}}{Z_{i2}} + \ldots + \frac{Z_f e_{in}}{Z_{in}}\right) (11)$$

This addition applies regardless of the form of the impedance. If the imped-

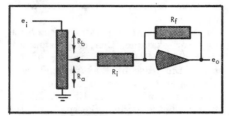

Fig. 7. Potentiometer gain adjustment.

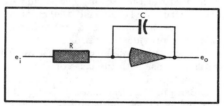

Fig. 8. Integrator operational amplifier.

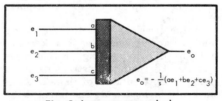

Fig. 9. Integrator symbol.

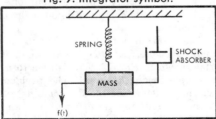

Fig. 10. Spring, mass, damper system.

ances are resistors, the unit is an adder. For simplicity the adder is shown in *Fig.* 6 where gains a, b, and c are the respective ratio of the input and output impedances. (In this figure and from now on all voltages are with respect to ground.)

Potentiometers can be used to obtain gain values. The input voltage, if fed into a potentiometer, will appear at the arm with a gain between 0 and 1 depending on the potentiometer setting. Thus, in *Fig.* 7 the input voltage appears at the amplifier output with the overall gain, $k_1 a$, where "k_1" is the ratio of R_a to $R_a + R_b$, and "a" is the ratio of R_f to R_i.

The integrator is shown in *Fig.* 8. In this case the feedback impedance is a capacitor. The relation from e_i to e_o becomes:

$$\frac{e_o}{e_i} = -\frac{Z_f}{Z_i} = -\frac{1}{Rj\omega C} = -\frac{1}{RC}\left(\frac{1}{S}\right) = -\frac{k}{S} \quad (12)$$

In LaPlace notation the $\frac{1}{S}$ signifies integration. A simple intuitive approach is given to show that this circuit does integrate. From equation 6 the currents into the grid point equal zero. Then if e_i is providing current, $\frac{e_i}{R}$, a current of equal magnitude must be flowing in the capacitor. This is possible only if the voltage across the capacitor is constantly changing. For a constant e_i the output e_o is then constantly increasing. This is integration. If e_i is removed, e_o will remain constant. For simplicity the integrator is given the symbol shown in *Fig.* 9.

The discussion of the integrator immediately leads to an analogy. If e_o is considered in a particular problem to be the position of a body, then e_i is the velocity of the body. The relation, again using LaPlace notation, is expressed by:

$$e_i = -k_1 S e_o \quad (13)$$

For this expression the S signifies differentiation. By the same reasoning the acceleration, which can be denoted by e_3 is:

$$e_3 = -k_2 S e_i = +k_1 k_2 S^2 e_o \quad (14)$$

The minus sign is associated with the minus gain of the amplifier. That is, a positive voltage at the input causes a negative voltage at the output. This reversal of sign is an important point to remember in setting up a system simulation.

In physical systems one relation appears more often than any other. This is the so-called "quadratic" response. The response of a spring, mass, shock absorber system as shown in *Fig.* 10 is of this type. The expression for this system response to an input F(t) is:

$$M \ddot{x} + K_d \dot{x} + K_s x = F(t) \quad (15)$$

This expression states that the input force is balanced by the acceleration, velocity, and displacement forces of the mass.

If LaPlace notation is used the expression becomes:

$$M S^2 x + K_d S x + K_s x = F(S) \quad (16)$$

which rearranges to:

$$\frac{x}{F(S)} = \frac{1/K_s}{\frac{M S^2}{K_s} + \frac{K_d}{K_s} S + 1} \quad (17)$$

From mechanics the general expression for the quadratic response to an input, y, is:

$$\frac{x}{y} = \frac{K}{\frac{S^2}{\omega^2} + \frac{2\delta}{\omega} S + 1} \quad (18)$$

where
K = gain
ω = natural frequency
δ = damping ratio

By comparing (17) and (18) it is seen that:

$$\omega = \sqrt{\frac{K_s}{M}} \quad (19)$$

$$\delta = \frac{K_d}{2} \sqrt{\frac{1}{K_s M}} \quad (20)$$

$$K = \frac{1}{K_s} \quad (21)$$

In order to show how an analog circuit is constructed, equation (16) is employed.

Solving for the $S^2 x$ term equation 16 becomes:

$$-S^2 x = \frac{K_d}{M} Sx + \frac{K_s}{M} x - \frac{F(S)}{M} \quad (22)$$

Putting equation 22 into computer form is done by using scale factors as previously shown in equation 5. When this technique is used, equation 23 is obtained.

$$-k_x S^2 e_x = \frac{K_d}{M} k_x Se_x + \frac{K_s}{M} k_x e_x - \frac{k_f}{M} e_f \quad (23)$$

Constants such as K_d, M, and K_s are not scaled as they will appear as fixed gains. It is also seen that the scale factor k_x is independent except relative to k_f since it cancels in all the e_x terms.

Assume that the voltages on the right hand side (Se_x, e_x, e_f) of equation 23 are available in the simulation. These are then added in an adder amplifier through gains proportional to the coefficients $\left(\frac{K_d}{M}, \frac{K_s}{M}, \frac{k_f}{k_x M}\right)$. This is shown at amplifier 1 of the basic analog circuit shown in Fig. 11. The output of amplifier 1 is $+S^2 e_x$. Amplifiers 2, 3, 4, and 5 are straightforward integrations and sign inversions to find Se_x and finally e_x itself. By using these signals as "feedbacks" to amplifier 1 with the proper gain ratios as indicated, a complete analog of equation 22 is found. Since equation 22 is equivalent to equation 18, this is also an analog of equation 18, the quadratic response.

The main characteristics of the quadratic are the frequency and damping. These quantities are seen to be determined in the analog by the gains into amplifier 1.

Variations in these gains are analogs of variations of frequency and damping

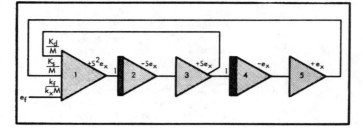

Fig. 11. Complete analog of a quadratic response.

of the system described by equation 22. Optimum values can be quickly determined and translated back to the actual system parameters.

The general use of the quadratic response has been described because it is the most common expression in audio work.

In the tone arm-cartridge combination, for example, this expression appears three times. The spring of the stylus, the mass of the arm and cartridge, and the damping of the pivots form one resonant quadratic system. The spring of the stylus, the mass of the stylus, and the damping of the signal pick-off form the second. The spring of the record vinyl material, the mass of the stylus assembly, and the pick-off damping provide the third.

The speaker system contains a quadratic response with the mass of the cone, the spring of the cone mounting, and the damping of the magnet being the parameters that contribute.

Other examples can be found in microphones, amplifiers, tape transports, and tuners.

In the previous part the theory behind the analog computer was described in terms of simple spring, mass, damper systems. To further illustrate its use a problem representative of those in audio work will be looked at. The intent is not to solve a particular problem but only to show how the computer could be used for audio work.

The problem is that of stabilizing an audio amplifier. Assume the amplifier is that shown in Fig. 12. This is seen to be a simple feedback circuit. The problem will be to determine the RC combination to be used in the feedback of the amplifier. It is assumed that the individual stages have been previously designated and are unchangeable.

Assumed also is that the three stages operated open loop appear as two lags in series; that is, the amplifier stage is flat out to some frequency, ω_a, and then drops off at 6 db/decade. Likewise, the inverter-power stage is flat out to some frequency, ω_p, then it too has a 6 db/decade drop-off. The power stage is assumed to include the output transformer. It is to be expected then $\omega_a > \omega_p$.

As typical numbers it will be assumed that $\omega_a = 2\pi 30$ K rps and that $\omega_p = 2\pi 20$ K rps. The cycles per second are $f_a = 30$ K cps and $f_p = 20$ K cps respectively.

In block diagram form using LaPlace notation, the amplifier is shown in Fig. 13. The feedback RC combination takes the form shown in the feedback block. The gains k_a, k_p, and k_f are the respective stage gains. For convenience the feedback block can be rewritten to be $k_{rc} (S + k_s)$. It is apparent that RC and k_f are implicit in k_{rc} and k_s, thus can be easily determined once k_{rc} and k_s are determined.

The inverter-power stage will be set up first. The input-output relation is:

$$e_o = \frac{k_p}{\frac{S}{\omega_p} + 1} e_n \quad (24)$$

Solving for the highest order of e_o:

$$Se_o = \omega_p k_p e_a - \omega_p e_o \quad (25)$$

By integrating this expression, e_o will be determined. The literal circuit for this is shown in Fig. 14. This circuit could be reduced to a one integrating amplifier circuit but as will be seen later, it is convenient to have Se_o available thus the circuit shown will be used.

From the block diagram of Fig. 13 the amplifier stage is seen to have the same form of equations as the inverter-power stage. This circuit can then be directly drawn as shown in Fig. 15. In this case the reduced circuit is shown since Se_a is not needed in the problem. The reduction is seen to be the elimination of the two inverter amplifiers of Fig. 14.

If the relation between e_f and e_o is examined, the following expression arises:

$$e_f = (k_{rc} S + k_s) e_o \quad (26)$$

multiplied out this becomes:

$$e_f = k_{rc} S e_o + k_s e_o \quad (27)$$

The need for $S e_o$ is seen at this point. The circuit for e_f is given in Fig. 16.

At this point the individual parts can be put together to produce the circuit shown in Fig. 17. This circuit is the complete simulation of the block diagram. With the proper computer facility it would be possible to go much further in simulating the amplifier by putting in more exact expressions for the amplifier. This can be done even to the extent of simulating tube characteristics for a complete design study.

The values of ω_a and ω_p being in the neighborhood of $2\pi 25$ K rps require operation at frequencies and gains well beyond the capability of most computer

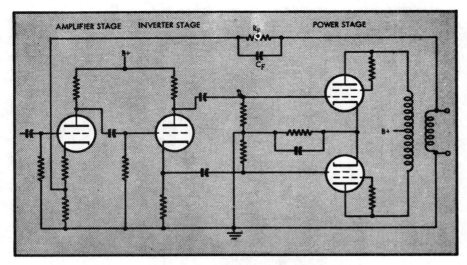

Fig. 12. Typical audio amplifier.

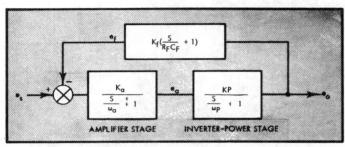

Fig. 13. Block diagram of typical audio amplifier.

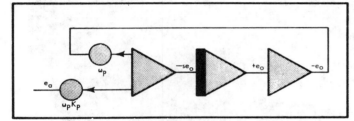

Fig. 14. Literal analog circuit for inverter-power stage.

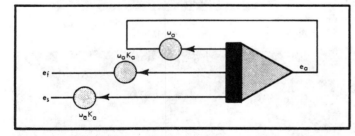

Fig. 15. Reduced analog circuit for amplifier stage.

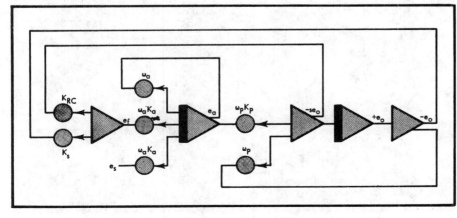

Fig. 17. Analog circuit for the audio amplifier of this problem.

amplifiers. This introduces another scaling technique called time scaling. This means that a problem which occurs in time t_p can be simulated to occur in time t_c on the computer. This can be done to

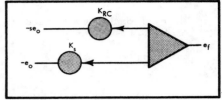

Fig. 16. Analog circuit for feedback.

lengthen or shorten the problem time. A 10 minute problem time can be scaled to take 10 seconds on the computer. Likewise, a 10 microsecond problem time could be scaled to take 10 seconds. Only one time scale per simulation, however, is allowable. This time scaling is done by changing the integrator gains. Going back to equation 12 (Part One) for the basic integrator, the RC product is seen to determine the integrating rate of the integrator. This is normally set at one volt per second per volt of input (R = 1 megohm, C = 1 microfarad). Thus, if one volt at the input represents, for example, one foot per second of velocity, the output will increase at one volt per second. At the end of 10 seconds the output will be 10 volts which represents 10 feet, but if the RC is decreased by a factor of 10, the integrating rate is 10 volts/sec/volt and the 10 volt level would be reached in one second. Since the volt per foot relation has not been changed it is apparent that the seconds of computer to seconds of problem have changed or, more simply, a time scale change has been made. This time scale factor will be denoted by k_t. In the problem in question k_t is applied by dividing the inputs to the integrators. (This problem is to be "slowed down"). It is noticed that each integrator input has an ω_a or ω_p associated with it. This leads to the logical and correct conclusion that ω_a and ω_p are to be reduced by the factor k_t. For this problem let k_t be 1000. Some rearrangement of gain terms was made to enable the setup of this problem on an actual analog computer.

The numbers used in this problem require a number of 10 gain inputs and reductions by 20 of various computer amplifier output quantities in order to keep potentiometer settings below one. This leads to the final simulation circuit shown in *Fig.* 18.

The circuit of *Fig.* 18 was set up on a Heath Educational Electronic Analog Computer Model EC-1 which is available in kit form for about $200. This computer contains nine operational amplifiers, five potentiometers to aid in the gain adjustments, and three voltage sources for fixed or step voltages for

Fig. 18. Final analog circuit used on Heath EC-1 computer.

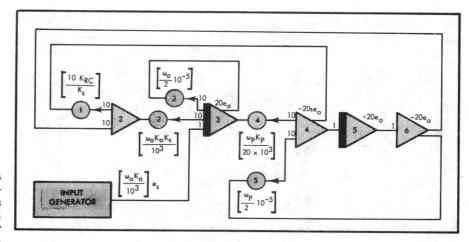

setting up the analog. In addition, a voltmeter is provided to read the amplifier outputs. The control circuitry is such that repetitive solutions can be made; that is, the problem is started, then after a time determined by a dial setting on the computer the problem is stopped, returned to zero and started again. This process, if done at a fast enough rate, allows the complete time solution to be displayed on an oscilloscope. Pictures or sketches of this display provide a record of the problem.

Figure 19 shows the computer and a scope as wired to examine the sample audio problem.

The results of this simulation are shown in *Fig.* 20. A step voltage was applied as an input. The first response is the open loop circuit; that is, no feedback applied. As the feedback gain, which is controlled by potentiometer 2 (denoted by P2) is increased, the amplifier rise time becomes shorter. The damping, however, gets less and the circuit begins to "ring." If P2 were increased too far the amplifier would become unstable. At a value of one for P2 the damping ratio is in the order of 0.1. The value of P1 was then increased. This provides the stabilizing necessary to reduce the ringing and yet maintain the short rise time. The "optimum" response is obtained for P1 equal to 0.11. Comparison of this response to the original open loop response shows that considerable improvement has been made in the amplifier. Using the values found for P1 and P2 it is possible to determine the R and C needed in the real circuit and the problem is finished.

This example has illustrated the procedures used in analyzing a problem and setting up an analog computer simulation.

In this article the subjects of digital and analog computers have been lightly covered. The purpose has been to illustrate somewhat simply how they work and what their applications might be.

As time goes on, computers will perform even more complicated tasks than they do now. As this computer revolution continues, the components produced by audio manufacturers will improve and the high performance specifications of today may become the "so-so" product of yesterday.

Æ

Fig. 19. Heath Educational Analog Computer, model EC-1, showing the hook-up used to study the sample audio problem.

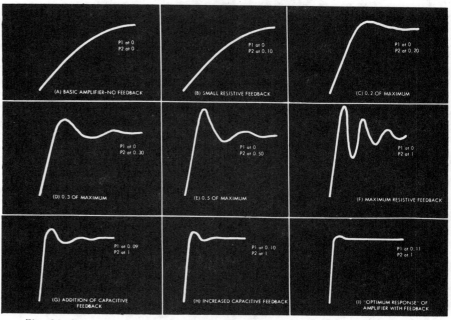

Fig. 20. Step responses of the audio amplifier simulation with variation of the simulated RC feedback.

Speaker Power

PAUL W. KLIPSCH

One virtue of high efficiency in a loudspeaker is reduced distortion at high listening levels.

To produce music realistically, Massa[1] (Chart 72) states that peak sound pressures of the order of 100 dynes per square centimeter are required, corresponding to approximately 115 decibels intensity (dbi)[2]. To produce such a pressure in a 3000 cubic-foot room with a reverberation time of 0.8 seconds requires approximately one acoustic watt of speaker output. (Massa, Charts 70, 72).

A cone excursion of 0.21 in. at 50 cps is found to produce 0.35 rms frequency modulation distortion which produces an intolerable harshness.[3] Kellogg[4] in 1931 proposed limiting the diaphragm excursion to 1/16 in. to limit distortion. This seems prophetic in view of the fact that that frequency modulation distortion was not a part of audio techology and terminology until 1943.[5]

To produce one acoustic watt at 32.7 cps with diaphragm excursion limited to 1/16 in. requires a piston radiating into hemispherical space (infinite flat baffle) of about 60-in. diameter. (Massa, Chart 64.) For the same frequency and 1/16-in. total motion, a 14-in. piston working into a horn throat of 84 square inches is required (Massa, Chart 78).

Typical horn loudspeaker efficiency ranges from 10 to 50 per cent. One loudspeaker of reference quality may be considered to exhibit 16 per cent efficiency so that approximately 6 watts are needed from the amplifier. A 30-watt amplifier of high quality would be operating with an ample margin of surplus power.

Typical direct radiators range from 1 to 8 per cent efficiency. One model of good quality exhibits about 4 per cent efficiency. With this unit, to deliver one acoustic watt requires 25 electrical watts output from the amplifier. Practically it has been found that lesser power levels are desirable. This particular speaker utilizes a 15-inch drive unit with 0.7 kg^2 per watt pull factor.

Direct radiators which are operated in tightly-enclosed-back air chambers have to be weighted to reduce their resonant frequency. A 12-in. cone-type speaker will have about 10-in. effective diaphragm diameter; the air-mass loading it will be only about 10 grams (one side only; other side assumed to be facing the acoustic capacity of a box). (Massa, Chart 55.) The mass of the piston and voice coil may be expected to be another 10 grams or a total of 20 grams. For a box of 1.7 cubic feet or about 2900 cubic inches, and a 10-in. piston with a total load of 20 grams, the resonant frequency will be 100 cps. (Massa, Chart 59.) To reduce the resonance to 33 cps requires increasing the effective mass by a factor of 3^2 or 9 and the efficiency is reduced by the same factor. Typical 12-in. direct radiators with pull factors from 0.1 to 0.4 kg^2 per watt exhibit 1 to 2 per cent efficiency. Hence such a weighted-cone direct radiator should be expected to afford an efficiency of from 0.1 to 0.2 per cent, the lower value applying to one commercial speaker tested. Thus to produce one acoustic watt would require a rather ridiculous 1000 watts of amplifier power. Practically, input power exceeding about 30 watts raises the distortion from fine to gross levels so one must be content with some 30 milliwatts acoustic output. One could multiply the number of speakers and amplifiers but, then, even the apparent advantage of bulk and cost disappear.

The higher the efficiency of a loudspeaker, the lower the distortion. In the absence of weight-loading, the distortion may be expected to be inversely as the square of the efficiency. In the case of weighted diaphragms, the major penalties are the power required to accelerate the extra weight and the resulting looseness of coupling between the electrical power and the air being moved.

Transient response has to do with peak power output available with linearity (freedom from amplitude distortion) and the ability of the speaker to produce sound pressures proportional to applied instantaneous power.

Much effort has been expended to reduce weight of moving parts such as the diaphragm, and so forth—even to the extent of using aluminum ribbon voice coils instead of copper. There is seen to be a premium placed on high efficiency. This significantly applies not only to speakers but to amplifiers.

High efficiency results in reduced distortion in the speaker and less demand on the amplifier. Generation of heat in the speaker is one by-product of inefficiency. One unsympathetic high-fidelity dealer drove a low-efficiency speaker to a cooking temperature and then applied a surge that tore the voice coil loose, all at moderate listening levels. By contrast a horn of medium efficiency was driven to an output of about 20 acoustic watts without damage except to the listeners' ears at the 130 dbi level.

The principles of physics still apply.

Æ

---

[1] Frank Massa, "Acoustic Design Charts," Blakiston, Philadelphia, 1942.

[2] Levels of 110 to 120 decibels intensity are usually encountered in various recording sessions.

[3] Paul W. Klipsch, "Subjective Effects of Frequency Modulation Distortion." *Jour. Audio Engr. Soc.*, Vol. 6, No. 2, April, 1958, p. 143.

[4] E. W. Kellogg, "Means for Radiating Large Amounts of Low Frequency Sound," *Jour. Acous. Soc. Amer.*, Vol. 3, No. 1, July, 1931, pp. 94–110.

[5] G. L. Beers and H. Belar, "Frequency Modulation Distortion in Loudspeaker," *Proc. IRE*, Vol. 31, No. 4, April, 1943, pp. 132–138.

THE SEARCH FOR MUSICAL ECSTASY & THE ARCHAIC AURAL REVIVAL
Harvey "Gizmo" Rosenberg

BKIM1
$24.95

VINTAGE HI-FI VIDEO: THE GOLDEN ERA, 1947-1965
Vintage Hi-Fi Productions

VDVHFP1
$29.95

This is a bizarre book. Then again, it has a bizarre author. Founder of New York Audio Laboratories, gadfly, guru of all aural and musical, prolifician *sans pareil*—these are just a few of the labels that could be pinned on Harvey Rosenberg, if you could catch up with him and keep him still in the first place. This new book is hard to describe—other than to say that it will surely be a classic—but let us say that it is one very sensitive, knowledgeable, fearless, experienced audiophile's wacky yet brilliant ramblings about his life, your life, and about a thousand other things—all infused and intertwined with an iconoclastic respect and reverence for beautiful sound and beautiful sound technology.

If you like tubes, you'll love this book. If you love music, you'll treasure it. If you treasure weird, one-of-a-kind things, you'll sleep with this under your pillow. But it IS hard to describe. Parts include "My Search for the Audio Grail" and "Tweaking Ecstasy." Among the chapter titles are "Celebrating the Tribe of Audioxtasists" and "Speakering Down Untrodden Paths." One subtitle is "I Was A Teenage Mutant Audio Nerd." As Ken Kessler of *Hi-Fi News & Record Review* has written about Rosenberg's "crazed musings," "He sees music as an aphrodisiac, musical ecstasy a measure of one's development, a yardstick of maturity. Rosenberg overturns the standard values, telling us that sheer realism and total accuracy are not the final goals. The goal is musical ecstasy. Nearly all that has been written about hi-fi has been of the 'what' and 'how' variety. *The Search for Musical Ecstasy* tells us why."

So there you have it. This book is not for everybody, but it most certainly would be enjoyed by anybody who cares enough about quality audio to find himself reading these words. It's Julius Futterman Gets Yippie-ized. It's The Zenmeister Meets Tubezilla. Or something like that—we're still not sure. But this book should be yours—it imparts rare knowledge and makes for great fun. 1994, 350pp., 6 × 8-1/2, spiralbound.

Divided into two segments, "Mono: 1947-57" and "Stereo: 1958-65," this nostalgic masterpiece is subtitled, "The Story of Classic American Tube Hi-Fi from Post-War to the Mid-Sixties." Included are more than 65 amps, preamps, tuners, and turntables (plus one speaker, the Stephens 106AX), created by such venerable names as Altec, Fisher, McIntosh, Scott, Dynaco, and Marantz. A worthwhile addition to any video collection! 1994, VHS, NTSC, color, 34:00.

THE AUDIO DESIGNER'S TUBE REGISTER, VOLUME 1
BKMC2
$17.95

This brand-new book, based on fresh data from author Tom Mitchell's lab, features complete electrical and mechanical specifications for the 6C4, 6C10, 6CG7, 6DJ8, 6EU7, 6K11, 12AT7, 12AU7, 12AX7, 12AY7, 12BH7, 12DW7, 5751, and 6922. Each tube type is illustrated by eleven separate graphs: plate characteristic curves; transfer characteristic curves; μ and gm for both Ec and Ib; rp for both Eb and Ib; and three useful graphs not given by any manufacturer—constant current curves, Rdc(Eb), and Rdc(Ib). These graphs can be cross-referenced to seven different data tables for quick analysis. 1993, 144pp., 8-1/2 × 11, spiralbound.

OLD COLONY SOUND LAB PRODUCT SHOWCASE

OUR DISCOUNT POLICY

ORDER VALUE	DISCOUNT
<$50.00	0%
$50.00-$99.99	5%
$100-$199.99	10%
>$200.00	15%

A HISTORY OF MARSHALL VALVE GUITAR AMPLIFIERS, 1962-1992
BKHL2
Michael Doyle
$32.95

Marshall amps have defined the sound of rock for a generation. This book explores the British company responsible for that sweet overdrive sound, tracing the impressive lineage of its tube guitar amps. Doyle is the acknowledged authority on the subject, and here he combines detailed chronologies of the various model and serial numbers, straightforward explanations of their features and construction, and aesthetic evaluations of the results. Includes hundreds of black-and-white photos, plus a 32-page color section, plus another 32-page color appendix that reproduces parts of the Marshall catalogs of the '60s. 1993, 254pp., 9 × 12, softbound.

75 YEARS OF WESTERN ELECTRIC TUBE MANUFACTURING
BKAE5
Bernard D. Magers
$19.95

Subtitled "A Log Book History of Over 750 WE Tubes Including Dates of Manufacture," this enlightening volume was written by the WE Senior Engineer charged with the sad duty of conducting the phaseout of the Kansas City Works in 1988. With a listing and description for each, the book chronicles the history of the 785 tubes Western Electric produced during 1913-1988. And in addition to the tube specs and backgrounds, also included are sections on Nomenclature, References, A Brief History, "D"-Tube Specifications, Other WE Tubes, and A Summary of Manufacturing Notes. Many photos and drawings. 1992, 144pp., 8-1/4 × 11, softbound.

THE TUBE AMP BOOK, FOURTH EDITION
BKGT1
Aspen Pittman
$29.95

This long-awaited fourth edition of the Groove Tubes classic contains hundreds of pages of amp diagrams from more than two dozen manufacturers. In addition to the use of color photos, coverage of mods, maintenance, and history also add interesting and valuable dimensions to the book. If you love tube guitar amps, this book is for you—all two and a half pounds of it! 1993, 768pp., 5-3/4 × 8-1/4, softbound.